AIR-CONDITIONING SYSTEMS DESIGN MANUAL

Prepared by

The ASHRAE 581-RP Project Team
Harold G. Lorsch, Principal Investigator

American Society of Heating, Refrigerating and Air-Conditioning Engineers
1791 Tullie Circle, NE • Atlanta, Georgia 30329

ISBN 1-883413-00-1

Printed in the United States of America

ACKNOWLEDGMENTS

This volume was prepared as a result of Research Project 581-RP of the American Society of Heating, Refrigerating and Air-Conditioning Engineers (ASHRAE). A team of experienced professionals developed the manual to assist entry-level engineers and students in "design studio" courses in the design of air-conditioning systems. The impetus for this project was due to the tireless efforts of Erv Bales and Herbert Becker, members of ASHRAE Technical Committee TC 9.1, who continued their plea for a design-oriented publication that differed from available analysis-oriented textbooks.

The team members acknowledge the assistance of the Project Monitoring Subcommittee of TC 9.1 (Charles J. Procell, Chairman) that supervised the preparation of this document. They are also indebted to other individuals who read parts of the draft of this volume and made suggestions for improvements and clarifications.

If this volume eases the transition of an engineer from schoolbook learning to practical design, it was well worth the effort. It is also hoped that universities will create undergraduate and graduate design-oriented courses to encourage knowledgeable and motivated engineers to enter the HVAC&R industry.

J. Marx Ayres
Carl N. Lawson
Herbert P. Becker
Wilbert F. Stoecker
Fred S. Dubin
Robert T. Tamblyn
Douglas C. Hittle
Robert G. Werden
Gershon Meckler
Arthur H. Wheeler

Harold G. Lorsch, Editor

Contents

Chapter 1 INTRODUCTION

1.1 PURPOSE OF THIS MANUAL

This document was prepared to aid entry-level engineers in the design of air-conditioning systems. It is also suitable as a senior- or graduate-level text in a university course for HVAC design. It bridges the gap between actual design and the theoretical knowledge acquired in the usual college courses and generally available textbooks. All the former concentrate on theoretical calculations and analytical procedures or on the design of components. To the best of our knowledge, no other book stresses design of air-conditioning *systems* rather than analysis. This volume is designed to fill that gap.

We assume the reader is familiar with the theory and analysis of refrigeration equipment and possesses, or has access to, the ASHRAE *Handbook* series and appropriate ASHRAE special publications. Information contained in the *Handbook* series and in special publications is referenced but not repeated here. Exceptions are sometimes made where needed in the interest of continuity of the text.

The most difficult task in any design problem is how to begin. The entry-level professional does not have the experience of similar projects to fall back on and is frequently at a loss as to where to start attacking a design. To aid the reader in this task, a step-by-step sequence of design procedures is outlined for a number of systems. Where applicable, simple rules are given to assist the beginner in the choice of equipment type and size. In addition to specific references given throughout the text, a list of useful general references is presented at the end of Chapter 1.

In Chapter 2, the difference between analysis and design is pointed out. That chapter covers the basic operations that occur during the design phases of a building, and it discusses a number of factors that influence building design, such as the interaction between different building systems and economic considerations. Conditions of human comfort are treated, and their implications to HVAC systems design are discussed in Chapter 3. Performance and application of load calculations are reviewed in Chapter 4. No methodologies are presented since they are thoroughly covered in the *Cooling and Heating Load Calculation Manual* (McQuiston and Spitler 1992) and/or suitable computer programs. Components of HVAC systems are briefly described, and their influence on systems design is discussed in Chapter 5.

Chapters 6 through 8 cover the design of all-air, air-and-water, and all-water systems, respectively. Here, again, the corresponding chapters in the ASHRAE *Systems and Equipment* volume should be read, since a conscious effort was made not to duplicate material from those chapters except in the interest of continuity. Chapter 6 is the largest and most detailed chapter. Its thorough treatment of the air side of air-conditioning systems is equally applicable to other than all-air systems. Those details are, therefore, not repeated in Chapters 7 and 8.

Chapter 9 covers a variety of special systems, and controls are treated in Chapter 10.

The appendixes contain detailed descriptions and design calculations for a number of executed HVAC building projects. They serve to illustrate the procedures discussed in the main body of the manual. The projects in the appendixes were chosen to cover a variety of building applications and HVAC system types. They help to give the entering professional a "feel" for the size of HVAC equipment, and they indicate how a designer tackles a particular design problem.

A word of advice: Do not hesitate to make initial assumptions. No matter how far off the final answer they might be, they enable you to start on the project task and to gradually improve the design until a satisfactory solution has been obtained. Frequently, more experienced colleagues will assist by giving counsel and the benefit of their experience, but do not hesitate to plunge ahead on your own. Good luck!

1.2 HOW BEST TO USE THIS MANUAL

This volume is divided into two parts: (1) a narrative description of design procedures and criteria and (2) a series of appendixes with illustrative examples described in greater detail.

The following suggestions are made to obtain the maximum benefit from this manual.

1. Determine the general building category and read the appropriate chapter in ASHRAE *Systems and Equipment* (1992) and *Applications* (1991) to determine likely systems to consider for your application.

2. Familiarize yourself with the theory and the basic functions of the equipment. The best sources for this information are HVAC textbooks and the ASHRAE *Handbook* series.

3. Read the appropriate chapter in this manual.

4. Read the illustrative problem in the appropriate appendix of this manual.

5. Study state and local building codes, ASHRAE standards, and applicable NFPA documents.

To start, proceed with the steps in the sequence shown. After some practice, your understanding of these systems will increase if you combine steps 2 and 3 and/or steps 3 and 4.

Remember that this manual, in general, does not repeat information contained in the ASHRAE *Handbook* and special publications series. Therefore, you cannot rely on this manual as the only reference to use in your design work. As you gain experience, make notes of what you found, what worked, and what did not work, and keep these notes in a separate notebook or loose-leaf folder. This manual points the way toward building such a data base.

The best reference available is the experience of your peers. While an attempt has been made in this manual to incorporate the experience of design professionals, no written material can replace face-to-face interaction with your colleagues. Use every opportunity to pick their brains, and let them tell what did *not* work. Often, more is learned from failures than from successes.

1.3 USEFUL GENERAL BIBLIOGRAPHY

In addition to the specific references listed throughout this manual, the following publications are useful to HVAC designers. They should be available in every design office. ASHRAE publications are available from the American Society of Heating, Refrigerating and Air-Conditioning Engineers, Inc., 1791 Tullie Circle, NE, Atlanta, GA 30329.

ASHRAE. 1975. *ASHRAE bibliography on available computer programs in the general area of heating, refrigerating, air conditioning and ventilating.*

ASHRAE. 1981. *ASHRAE Standard 55-1981, Thermal environmental conditions for human occupancy.*

ASHRAE. 1981, 1985. *ASHRAE Standard 100 series, Energy conservation in existing buildings.*

ASHRAE. 1987. *ASHRAE Pocket handbook for air conditioning, heating, ventilation, and refrigeration.*

ASHRAE. 1989a. *1989 ASHRAE Handbook—Fundamentals.*

ASHRAE. 1989b. *ASHRAE Standard 62-1989, Ventilation for acceptable indoor air quality.*

ASHRAE. 1989c. *ASHRAE Standard 90.1, Energy efficient design of new buildings except low-rise residential buildings.*

ASHRAE. 1989d. *ASHRAE Guideline 1-1989, Guideline for commissioning of HVAC systems.*

ASHRAE. 1991. *1991 ASHRAE Handbook—Applications.*

ASHRAE. 1992. *1992 ASHRAE Handbook—Systems and equipment.*

ASHRAE weather tapes, available for specific locations in the U.S. and Canada.

BOMA. Experience exchange reports. Washington, D.C.: Building Owners and Managers Association International.

Climatic atlas of the United States. 1968. Washington, D.C.: U.S. Government Printing Office.

McQuiston, F. C. and J. D. Spitler. 1992. *Cooling and heating load calculation manual.* Atlanta: American Society of Heating, Refrigerating and Air-Conditioning Engineers, Inc.

SMACNA. 1988. *Duct system calculator.* Chantilly, VA: Sheet Metal and Air Conditioning Contractors National Association.

SMACNA. 1990. *HVAC Systems duct design manual,* 3rd ed. Chantilly, VA: Sheet Metal and Air Conditioning Contractors National Association.

USAF. 1988. *Engineering weather data,* AFM 88-29. Washington, D.C.: U.S. Government Printing Office.

HVAC design manuals by a number of equipment manufacturers:

NFPA. 1990. NFPA 70-90, National Electrical Code. Quincy, MA: National Fire Protection Association.

NFPA. 1989. NFPA 90A-89, Standard for the installation of air conditioning and ventilating systems. Quincy, MA: National Fire Protection Association.

NFPA. 1988. NFPA 92A-88, Recommended practice for smoke control systems. Quincy, MA: National Fire Protection Association.

NFPA. 1991. NFPA 101-91, Code for safety to life from fire in buildings and structures. Quincy, MA: National Fire Protection Association.

Some of the more frequently used estimating references include:

Means facility cost data. Kingston, MA: R.S. Means Co.

Mechanical estimating guidebook for building construction, 5th ed. New York: McGraw-Hill Book Co.

Rule-of-thumb cost estimating for building mechanical systems. New York: McGraw-Hill Book Co.

The addresses of BOMA, NFPA, SMACNA, and other organizations of interest to HVAC designers are given in the Codes and Standards chapter in the back of each ASHRAE *Handbook.*

Chapter 2 THE DESIGN PROCESS

2.1 DESIGN VERSUS ANALYSIS

Anyone who has taken a course in mathematics or any of the physical sciences is familiar with the process of analysis. In a typical analysis problem, a set of parameters is given that completely describes the problem, and the desired solution is uniquely determined. There is only one *right* (correct) answer (solution to the problem), and all other answers are *wrong*.

Design problems are inherently different. A design problem may or may not be completely defined, some of the parameters may be missing, and there are any number of answers. Some may be better than others, but there is no such thing as a right or wrong answer to a design problem. There are degrees of quality to the solution of design problems, and some of these solutions may be better than others for some people at some time. For a different application or a different client, other solutions may be better. It is important to clearly understand the difference between *analysis* and *design*. You may be used to looking for the correct answer to a problem (analysis), and you are suddenly overwhelmed with problems that have a variety of answers (design) and do not know which one is *correct*. Therefore, learn to use your judgment (or the advice of your experienced colleagues) to weigh the merits of a number of solutions that seem to fit a particular design problem in order to select the best one among them.

Figures 2-1 and 2-2 illustrate the differences between analysis and design. While analysis proceeds in a simple one-way fashion from given data to a final answer with the aid of certain analytical tools, design is an iterative process. Although there are certain "givens" to start with, they are not immutable but are subject to modification during the iterative design process. For example, the owner or the architect may be confronted with the energy implications of excessively large glass walls that had been originally specified and may decide to reduce their size. Also, the mechanical designer may try various system components and control strategies before finding one that best suits the particular conditions. Thus, design consists of a continuous back-and-forth process as the designer selects from the universe of available components and control options to synthesize the optimum solution within the given constraints. This iterative design procedure incorporates repetitive steps of analysis; therefore, analysis is an important part of any design.

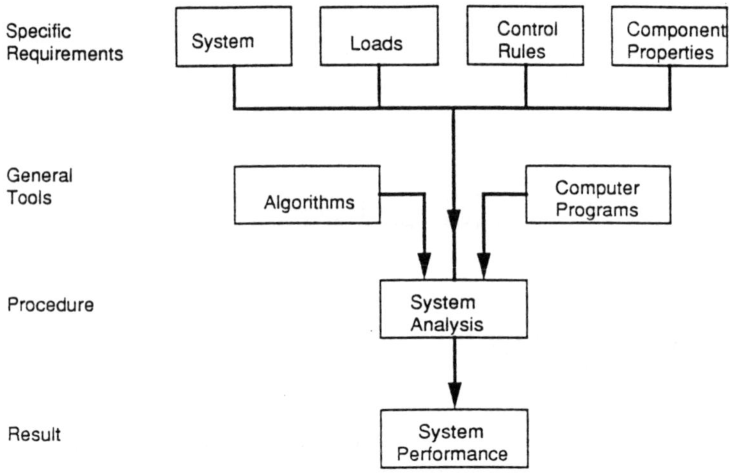

Fig. 2-1 Analysis Flow Chart

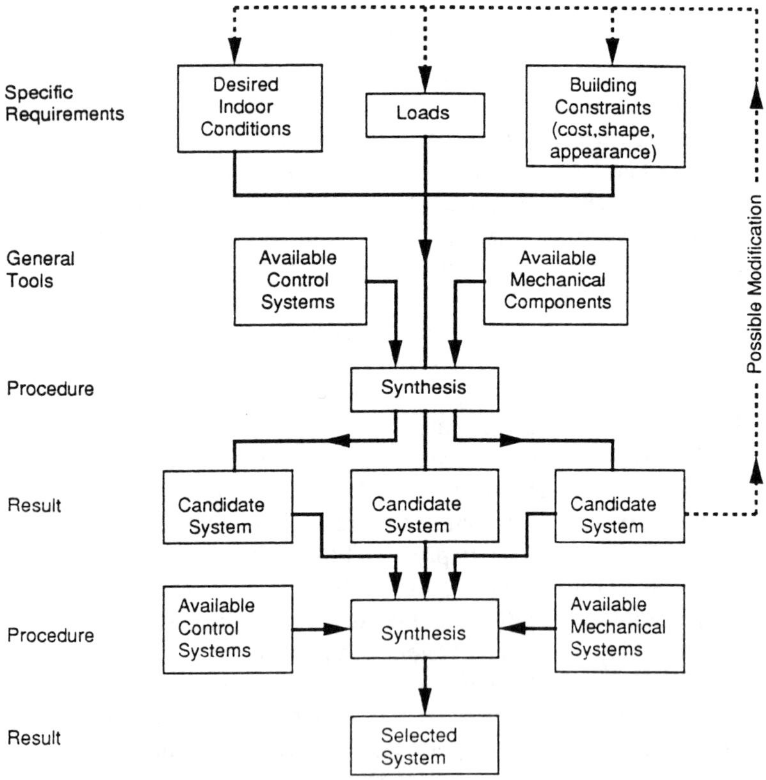

Fig. 2-2 Design Flow Chart

Since the first step in design is to map out the general boundaries within which solutions are to be found, you may not know where and how to start because you have no background. To overcome this obstacle, make initial assumptions and improve on them through subsequent analysis. To assist you in making these initial assumptions, simple rules are given throughout this manual and illustrative examples are contained in the appendixes.

2.2 PHASED DESIGN STAGES

The engineer must understand how real buildings are designed. Construction documents (working drawings and specifications) for a building are developed as a team effort. The architect usually acts as the prime professional, coordinator, and general manager, although many experienced owners and developers deal directly with preselected HVAC consultants. The architect interfaces with the owner, directs the in-house architectural staff, and manages the work of outside or in-house mechanical, electrical, and structural engineers. The negotiated design fees for the consultants' work establish the allowable level of effort. This fiscal constraint, therefore, seriously limits the amount of time that can be allocated to studies of alternative systems.

2.2.1 Program Phase

Before the mechanical engineer can design an HVAC system, a total building program must be created. This is usually prepared by the client or consultants. The architect establishes a program of space needs, develops a construction budget, and prepares conceptual drawings of the building size and shape required for approvals by governmental agencies and interfacing with utility companies. The total building program must include, but need not be limited to, the following:

- The client's objectives and strategies of the initial and future functional use of the building, whether it be a single- or multiple-family dwelling or a commercial, industrial, athletic, or other facility. Functions must be detailed for each discrete area within the building.

- The geographic site location, access means, and applicable building and zoning codes.

- The building area, height, number of stories, internal transportation, roof and wall materials, and type and amount of fenestration.

- The owner's capital costs and operating costs budgets.

- The number, distribution, and occupancy patterns of permanent occupants and visitors.

- Conceptual architectural drawings.

While some of this information may not be available before you, as the mechanical designer, start your work, you should try to obtain it as soon as possible to ensure that you do not consider HVAC systems that are not compatible with the building plan.

While the architect prepares the overall building program, the mechanical engineer has the responsibility of developing an environmental program even though some of this information may be provided by the architect or owner. The building program and use profile provided by the owner or architect and the environmental program developed by the engineer in response to the building functional program should be recorded in a journal for future reference (see ASHRAE 1989a). All changes made to either program during the design process should be recorded so that the journal is always up-to-date.

The data that should be contained in the environmental program include:

- Peak, average peak, and design peak outdoor dry-bulb and wet-bulb temperatures (absolute and coincidental)

- Heating and cooling degree-hours

- Direct and diffuse solar radiation for each month of the year

- Peak and average wind velocity and direction for each month

- Hours of sunlight for each month

- Surface and subsurface site conditions, including type of soil, water table, and rock levels

- Applicable zoning and building, fire, sanitary, and energy codes

- Availability, rate structure, capacity, and characteristics of the utilities, including gas, electricity, coal, oil, or other fuels

The environmental conditions to be maintained for each building space should be defined by:

- Dry-bulb and wet-bulb temperatures during daytime occupied hours, nighttime occupied hours, and unoccupied hours

- Ventilation requirements
- Indoor air quality requirements
- Any special conditions, such as heavy internal equipment loads, noise- and vibration-free areas, humidity limits, and redundancy for life safety and security
- Acceptable range of limits for each of the above

A knowledge of the functional use for each area is essential for selecting appropriate HVAC systems and suitable control systems for each area (zone) because the capabilities of candidate systems must be evaluated and compared to the indoor requirements. For example, some rooms in a building may require humidity control, while others do not, so the HVAC system must be able to provide humidification to areas requiring it without detriment to the enclosure. Some areas may require cooling, while others need only ventilation or heating. For example, a heating system with direct radiation only could not provide humidification, ventilation, or cooling.

2.2.2 Schematic Design

The schematic design stage involves the selection and comparison of appropriate candidate systems. All candidate systems must be able to maintain the environmental conditions required in each area. The ability to provide adequate thermal zoning is also mandatory. For each candidate system considered during this phase, evaluate the relative space requirements for equipment, ducts, and piping; the fuel and/or electrical use and thermal storage requirements; the initial and operating costs; the acoustical requirements; the compatibility with the building plan and the structural system; and the effects on indoor air quality, illumination, noise, and vibration. Also consider energy code impacts in this phase.

Early in this phase, you may be asked to provide information regarding the impact of large glazed areas (*i.e.*, more than 30% of the envelope area), heavy lighting (*i.e.*, more than 2.5 W/ft^2), and other internal loads (*i.e.*, 6 W/ft^2 for computers) on HVAC system requirements, and where to locate large mechanical equipment, considering aesthetics and acoustical controls. Depending on the level of information available, you may be asked to prepare hand (or computer spreadsheet) calculations based on rules developed through experience or based on results from previously designed similar projects. Some design criteria useful for a first analysis are given in Table 2-1. Additional criteria appropriate for this design phase can be found in the *ASHRAE Pocket Handbook for Air Conditioning, Heating, Ventilation, and Refrigeration* (ASHRAE 1989).

If architectural and internal loads are reasonably well defined, peak load and rough energy calculations of alternative HVAC systems may be prepared at this time using outside air temperature bins or hourly data for 12 typical months to present design options to the owner. Although they are preliminary and will change as the building design proceeds, the preliminary loads are definitive enough to compare system performance because all systems will be sized to meet the same loads; as you gain experience, you will be able to ascertain the probable magnitude of the loads for each area in the building with little calculation effort.

Table 2-1 Selected Design Criteria for First-Cut Analyses

General:	450 ±100 ft^2/ton
	1.5 cfm/ft^2 for exterior spaces
	0.6 cfm/ft^2 for interior spaces (minimum)
	400 cfm/ton
Offices:	500 ft^2/ton based on:
	lights—1.5 W/ft^2
	fans—0.75 W/ft^2
	pumps—0.25 W/ft^2
	miscellaneous electrical—1.0 W/ft^2
	occupancy—200 ft^2/person
High-rise apartment houses:	1000 ft^2/ton for north-facing apartments
	500 ft^2/ton for others
Hospitals:	333 ft^2/ton based on 1000 ft^2/bed
Shopping centers:	400 ft^2/ton average
	department stores 2 W/ft^2
	specialty stores 5 W/ft^2
Hotels:	350 ft^2/ton
Restaurants:	150 ft^2/ton
Central plants:	
Urban districts	380 ft^2/ton
College campuses	320 ft^2/ton
Commercial centers	475 ft^2/ton
Residential centers	500 ft^2/ton

Aids to be used during this phase of the project include design manuals, textbooks, equipment literature, and existing installations. Frequently, this eliminates all but a few systems capable of providing the environment required and compatible with the structure.

If the client requests it, if architectural details have been sufficiently developed, and if the mechanical engineer's fee has been set at a level to warrant it, comparisons between construction (capital) costs and operating costs and the performance of different HVAC systems are made in greater detail. Using one system as the base, all other candidate systems are then compared to the base system. Such an analysis would proceed by the following steps.

1. Estimate the probable capital costs of each system using square foot allowances, a rough selection of equipment and sketches of system layouts, and such tools as
   - Cost-estimating manuals
   - Recently completed similar projects (many technical journals contain case studies that provide such information)
   - Local HVAC contractors
   - Professional cost estimators
   - Design office files
   - Experienced design engineers

2. Select the energy source or sources available and their cost per million Btu, both present and anticipated. Determine local utility tariffs, energy charges, demand charges, and off-peak rates, if available.

3. Calculate the number of operating hours and hourly operating costs for each subsystem of each candidate HVAC system. This can be done manually or by spreadsheet using temperature bin data, or by computer using the ASHRAE simplified energy analysis method, or by proprietary programs offered by equipment manufacturers.

4. Using the local utility tariffs, calculate monthly utility costs and add them up for the year.

5. If required by the owner, perform comparative life-cycle (or other) cost analyses, as described in Section 2.8.

The annual efficiency of equipment is not equal to its full-load efficiency.

In the manual calculation of energy requirements include efficiency at part-load conditions, the number of hours at 100%, 90%, 80%, and 60% of full load, etc.

As an annual average, cooling equipment operates at 50 to 85% of capacity; fans and pumps operate at 60 to 100% of capacity.

In order to determine the actual equipment operating profile, hourly weather variations throughout the year are required. The hours of dry-bulb temperature occurrences in 5° increments or bins and coincident monthly average wet-bulb temperatures at many locations can be obtained from the Air Force *Weather Data Manual* (USAF 1988). Bin and degree-hour weather data are also available from ASHRAE (1986b).

Operating costs are a function of the way in which a building is operated—how much is the indoor temperature allowed to drop, and for how many unoccupied hours in the winter? Will the air-cooling system be shut down at night and on weekends? Do special areas, such as computer rooms, require cooling year round? If such information is not available, educated estimates must be used. All systems must be analyzed under the same assumptions.

At the conclusion of the schematic design phase, a recommendation regarding HVAC system selection is made to the owner. That recommendation is usually approved if the engineer's reasoning is sound and reflects the client's objectives. Should the owners select another approach—probably because of weighing factors differently from the design team, perhaps construction dollars over performance features—that is the owner's prerogative. Design decisions should be clearly understood and documented whether they constitute approval of, or deviation from, the designer's proposals.

2.2.3 Preliminary Design

The preliminary design is used to coordinate the HVAC system design with the architectural and structural systems, which are also in the preliminary stage, to resolve potential conflicts. Close integration of the mechanical and electrical systems with structure, plan, and building configuration requires the cooperation of each team member with the others—architect, mechanical engineer, electrical engineer, structural engineer, acoustical consultant, and all other team members from each discipline.

The requirements of state or local building or energy codes (see Section 2.7) must also be considered in this phase because of restrictions on the amount of glass and other envelope materials, lighting, HVAC systems and equipment, and, in some instances, on the annual energy budget. If required, energy budgets are established. In some states or localities (*e.g.*, California, New York City), a simplified prescriptive compliance calculation is required that can be prepared by hand, but annual energy budget compliance calculations must be prepared by computer. It is wise to select a computer program that will minimize the inputs required for both the load and the energy programs. Currently, programs are available that share the inputs for both loads and energy programs. Simple calculations may be performed on personal computers to compare the energy implications of architectural alternatives.

Architectural floor plans and elevations are developed in greater detail; structural, mechanical, and electrical systems are designed in compliance with applicable building codes; and drawings in preliminary form are prepared. Heat loss, heat gain, and ventilation calculations are refined and used to design the air and water distribution systems and to select equipment. System sizes and capacities are selected to match design and part-load conditions. The designer has the choice of sizing pipes and ducts manually or using a variety of computer programs. The main objective, however, is to develop system layouts for space requirements and cost estimates. If required by code, more detailed energy studies are undertaken at this time and the accompanying calculations performed. Construction details and cost estimates are refined based on additional forthcoming information. After preliminary drawings, outline specifications, and cost estimates have been approved by the owner, the project scope is defined, and only minor changes are allowed in the final phase of the design work.

2.2.4 Final Design, Preparation of Construction Documents

In the final design stage, specific equipment, model numbers, and sizes are selected. Duct and piping systems and control strategies are finalized. Final special owner requirements are determined. The budget is refined, and some of the earlier contingency costs can be eliminated. It is possible that trade-offs occur at this stage based on performance and costs. For example, the type of insulation may change and the number of panes of glass (U-values) or shading coefficients (SC) may change if you can show that these changes reduce the cost of the mechanical systems accordingly or provide a more energy-efficient building with, possibly, higher investment costs justified by lower operating costs.

The documents submitted to the owner at the end of this phase include architectural drawings showing space requirements for the mechanical, electrical, and structural systems; specification of materials; and estimates of construction cost. After approval by the owner, the documents (including energy code calculations, if required) are submitted to governmental agencies and, after approval, to contractors to obtain firm bids, with the contract awarded to the lowest responsible bidder or to the one who may be able to complete the construction in the shortest time.

At this time, or at the end of construction, the owner may ask for additional computer simulations to establish future energy costs and operating budgets. These simulations should be performed on the most comprehensive energy program available because the results will be compared to actual utility bills. Studies of this type require extensive and detailed inputs and are very costly. Extra services of this type and/or last-minute owner-required changes constitute additional work for the HVAC design engineer, who must be reimbursed for such efforts.

2.2.5 Postdesign Phase

During construction, engineers:

- Check shop drawings for equipment, piping, and other items submitted by manufacturers and contractors to ascertain that the equipment and subsystems that will be installed conform to plans and specifications

- Make periodic visits to the building under construction to observe and maintain a log on the work being installed by the contractors

- Conduct or witness tests for system performance, such as airflow volume and temperature, equipment efficiency, control sequence strategies, and indoor air quality, if called for in the professional agreement and considered in the fee

- Ascertain proper workmanship and stages of completion to ensure that contractor invoices for work completed each month are correct

2.2.6 Commissioning Phase

Recently, the process of building commissioning has received increased attention because of the ever-increasing complexity of modern building systems. While a simple building can be built by the contractor and turned over to the owner with only a minimum of instructions, this is no longer possible for large buildings and their complex systems and numerous subsystems. Just as the U.S. Navy will not accept a ship without a sea trial, so a building owner no longer will accept a large building on completion without a series of tests that demonstrate the performance of the building systems, preferably over a full season of weather extremes. A new type of engineers and technicians—building commissioning specialists—has come into existence. It is their task to test the proper functioning of all building systems and subsystems. This constitutes the owner's acceptance test.

Frequently, the design engineer will have to be present when the HVAC system is being tested, particularly when a system does not perform in accordance with design specifications. The designer is the person best qualified to troubleshoot such a situation and to determine where the fault lies. Complete design documentation should be on hand when that situation arises. Consult the ASHRAE Commissioning Guideline (1989c) for additional information.

2.2.7 Postoccupancy Services

Ideally, an engineer should be retained for postoccupancy services to check energy use, operating costs, user reactions, system performance, and (with other team members) "total building performance." Unfortunately, this service is not usually done, but it is gaining ever more acceptance. In some cases, the commissioning engineer will support building operations, familiarize operating personnel with the system, and assist them in operating it for a period of one year after building completion. This may include adjustments and modifications to the HVAC system (usually concerning its control subsystem). Whether these tasks are performed as part of the design package or under a separately negotiated contract depends on the individual circumstances, and no general rules apply.

The building owner/operator should be made to realize that modern large buildings are complex systems requiring skilled personnel to operate them. Unless skilled operators are hired to start, newly hired operators must be trained to understand and operate the HVAC system, and the person best qualified to train them is the one who designed the system. Videotapes and other training aids are available to assist the engineer in that task.

2.3 INTERACTION BETWEEN HVAC AND OTHER BUILDING SYSTEMS

HVAC systems cannot be designed separately. Other building systems, singly and in combination, profoundly affect the functional performance, physical size, capacity, appearance, operating efficiency, and maintenance and initial costs of the HVAC system. Thus they are essential elements in the HVAC design process. Conversely, the HVAC system has an impact on all other building systems. This must be taken into account by the HVAC system designer during the design process.

2.3.1 Building Envelope

2.3.1.1 How Building Envelope Affects HVAC System

For each compass orientation, the thermal characteristics of the envelope—opaque wall, fenestration, roof, and even the floor—affect the magnitude and duration of the building heat loss and heat gain due to thermal transmission, infiltration, and solar radiation. Heat gain and heat loss directly influence the required capacity of the energy conversion devices (boilers, compressors, chillers) and the size, complexity, and cost of the distribution systems (ducts, fans, pipes, duct insulation). The need for perimeter heating is a function of climate, wall insulation, window area, type and number of panes of glass, and tightness of the structure to control infiltration.

A heavy envelope of brick or masonry provides thermal mass, which affects the operating cycle of the heating and cooling units. The greater the building mass, the less extensive are the indoor temperature excursions. The placement of the insulation, whether on the exterior or the interior surface of the envelope, affects heating and cooling seasonal and hourly loads, with insulation on the exterior generally being more effective. The envelope construction and the location of the vapor retarder (if required) also determine the transfer of moisture between the building spaces and the outdoors, thus affecting humidification and dehumidification requirements.

A building can be configured to provide solar and wind shading, which influences heating and cooling loads and control. By reducing transmission and radiation transfer at the perimeter, the envelope reduces the climatic influences on the building interior and affects the need for zoning. If zones in a building do not vary much with time of day, solar intensity, and wind velocity and direction, the HVAC system can be less complex and less costly to install.

A secondary effect of the envelope on HVAC design is the amount of daylight that can supplant electric lighting. Less heat is gained with daylighting than with electric lights for the same illumination intensity, so that cooling requirements are reduced.

2.3.1.2 How HVAC System Affects Building Envelope

The HVAC system can affect the appearance of the building envelope. Fresh-air intake louvers in walls, window air conditioners, unitary heat pumps, and through-the-wall units obviously affect the appearance of the building. This factor may limit system selection where this change in appearance is not tolerable to the architect or client.

Cooling towers, usually located on the roof, affect the physical appearance of the building. Other rooftop equipment, such as packaged air conditioners, air-cooled condensers, and air intakes, have similar effects. Attractive architectural enclosures can reduce the aesthetic disadvantage of rooftop equipment. Under certain conditions, rooftop equipment can be placed at ground level and camouflaged by shrubbery. Discharge from cooling towers can be corrosive, leave drift residue on automobiles parked nearby, and may carry bacteria, so location is of prime importance. Cooling towers located below condensers are subject to overflow unless precautions are taken.

The noise of outdoor HVAC equipment can be muffled by acoustical fencing, or equipment of slower speed and quieter operation can be selected. Some localities have laws prohibiting the installation of noisy, unmuffled equipment.

2.3.2 Structure

All HVAC equipment requires structural supports. Therefore, the HVAC designer must inform the structural designer of the location of major pieces of mechanical equipment early in the design phase. Heavy equipment, such as boilers, tanks, chillers, compressors, and large air handlers, should preferably be located on the lowest building level on a concrete subbase, with vibration bases or other isolating bases for rotating equipment. This causes the least amount of vibration to be transmitted into the building structure and constitutes the simplest method of structural support for this equipment. In earthquake-prone areas, additional physical restraint of HVAC equipment may be required. In high-rise buildings, it is usually necessary to install HVAC equipment at intermediate floors as well as on the roof. In such cases, as the mechanical design engineer, you must inform your structural counterpart as early as possible regarding the size and weight of mechanical equipment located at intermediate floors or on the roof.

Rooftop and intermediate-floor locations for HVAC equipment require special structural supports as well as acoustical and vibration control measures. In some cases, this may mitigate against locating equipment at these locations. When buildings are constructed with crawl spaces, without basements, or with lightweight flooring, heavy machinery should be placed next to the building instead of within the building envelope, provided sufficient ground area is available.

2.3.3 Lighting and Other Electrical Loads

Heat from lights may be as much as 85% of the heat equivalent of the power to the lights, depending on the efficacy of the lamp or tube. The amount of heat from lights contributing to the air-conditioning load or reducing space-heating requirements also depends on other factors, such as the thermal mass that is exposed to the light, air distribution patterns and volume, type of luminaire, and whether ceilings are exposed or suspended, forming an air plenum.

It is essential to include the heat from lights as internal sensible heat gain in the design of the HVAC system. Since these heat gains are a major air-conditioning load, closely coordinate with the lighting designer so that you are immediately informed of any changes to the lighting system.

Internal heat gains from electric or gas-fired appliances, motors, food service, and special equipment are normally sensible heat (except for food or steam, which contributes latent heat). Supply air volume is a function of sensible heat only, affecting the size of fans, coils, ducts, and equipment. When supply air or return air troffers are used, the lighting system actually becomes part of the HVAC system, and air supply temperatures to the room as well as supply air and return air distribution are governed by luminaire location.

Design rules for lighting loads are given in Chapter 26, Air-Conditioning Cooling Load, in ASHRAE *Fundamentals* (1989d). For examples of lighting load calculations, see Appendixes B and C.

2.3.4 Fire and Smoke Control Systems

NFPA *Standards* 45, 90A, and 101 (NFPA 1989, 1991b) govern fire protection and smoke control requirements on a national level. State and local fire codes incorporate and often modify the national codes. Where they exist, their requirements supersede the NFPA code. The HVAC system can be used to provide air supply and exhaust requirements in place of using separate systems. When doing so, the air distribution patterns are likely to vary from straight HVAC space-heating and cooling requirements. In large buildings, fire zones are separated by fire walls and doors. Therefore, the air distribution/exhaust systems must also meet the fire zoning pattern. In stairwells, corridors, lobbies, or other areas where positive pressure during emergencies is required, the HVAC system must bring in fresh air to pressurize the spaces. The smoke exhaust system and the pressurization system generally require more air than is required for HVAC environmental control; therefore, the system must be capable of changing airflow volume and maintaining pressure relationships when used in the fire protection mode. Consult *Design of Smoke Management Systems* (Klote and Milke 1992) for detailed information.

Standpipes and sprinkler systems located in unheated spaces may require freeze protection. These systems have alarms so that failures can be promptly remedied.

2.4 HVAC SYSTEM SELECTION CRITERIA

2.4.1 Equipment Location and Space Requirements

From the mechanical engineer's point of view alone, equipment location is governed by:

- Space for equipment components, with room for servicing, removal, and reinstallation
- Distance and pathways from prime movers (chillers, boilers, furnaces) and air-handling equipment to terminal locations
- Access to outdoor air intakes for ventilation (supply and exhaust air, combustion air)
- Headroom and support for ducts and pipes
- Acoustical considerations (noisy equipment should not be located near occupied areas)

Whenever possible, equipment requiring heavy electric service, water, oil, or gas should be located near the point of entry into the building of that utility service. Air-handling units and decentralized air-conditioning systems require outside intake and exhaust air duct connections, and the source of fresh air may dictate the location of the equipment. Exhaust louvers should be located sufficiently remote from intake louvers to prevent reingestion of exhaust air into the building. This complicates heat recovery, unless "runaround" aqueous, glycol, or refrigerant loops are used [see Chapter 8, Applied Heat Pump and Heat Recovery Systems, in ASHRAE *Systems and Equipment* (1992)]. Laboratory and kitchen exhaust hood systems require make-up air and code-specified fire protection [see Chapter 47, Fire and Smoke Control in ASHRAE *Applications* (1991)].

In large buildings and, generally, for central systems, try to locate air-handling units relatively close to the areas they serve in order to reduce distribution duct runs and duct and insulation costs. Piping from boilers and chillers to air-handling units is generally less costly than ductwork from air-handling units to terminal outlets, and it requires less building space or headroom.

Horizontal duct and pipe distribution must be accommodated above the ceiling, under or through structural members, or, in the most innovative buildings, within the raised floor. Vertical shafts for ducts and pipes and some control equipment are usually accommodated within the building core space, a location defined by the architect. Raised floors can carry power, communications, and data cabling and can also be part of the HVAC system. The raised floor plenum, 12 in. to 15 in. high, can be used as an air supply plenum with a ducted return air system either in the plenum or elsewhere. As an alternative, the supply air ducts are run in the raised floor with the raised floor plenum serving as a return air plenum. In either case, supply air can be delivered to the room with specially designed floor outlets through risers connected to the raised floor or to ductwork in the raised floor plenum. Cool air supplied at floor or low levels need not be as cold as ceiling supply or high wall supply air since it cools the occupants before it has picked up other space heat. This decreases the load on the cooling unit.

Table 2-2 lists values for a first approximation of the space requirements for HVAC equipment.

Table 2-2 Approximate Space Requirements for Mechanical and Electrical Equipment (Percentage of gross building area, unless otherwise noted.)

Total mechanical and electrical equipment	4 to 9%
Apartments, schools	5%
Office buildings	7.5% (up to 50% more at lower levels of high-rise buildings)
Research laboratories	10%
The above total may be broken down as follows:	
Plumbing	10 to 20% of lower-level gross building area
Electrical	20 to 30% of lower-level gross building area
Heating and air conditioning	60 to 80% of the above 4 to 9%
Communications (intelligent buildings may require much more)	5 to 10% of the above 4 to 9%
Fan rooms	50 ft^2 per 1,000 cfm supply air, 10 to 20 ft high
Interior shafts (either)	2 to 5% of gross floor area, of which: 70% for HVAC, 15% for electrical, 15% for plumbing
Interior shafts (or)	1 ft^2 per 1,000 cfm (supply + return + exhaust air)
Air-handling openings should provide these maximum air velocities on the net free area:	
Outside air intake louvers	650 fpm because higher velocities may entrain water and ice
Exhaust air louvers	1,500 fpm to prevent excessive pressure on walls and frames
Return air and mixed-air plenums	1,500 to 2,000 fpm but with adequate pressure drop across control dampers and adequate turbulence (0.25 in. water) to mix outside and return air in air-side economizers

2.4.2 Access for Maintenance and Repair

The lack of adequate access to equipment is a major contributor to poor system performance, ranging from breakdown and unsatisfactory comfort control to poor air quality. The following list is a general guide for design.

Building dimensions are not as precise as machinery dimensions; therefore, always add a few inches for good measure when specifying clearances.

- Provide clear space around fans, pumps, chillers, and other equipment for service access and component removal.
- Keep a passageway clear for the removal and replacement of equipment.
- Include access doors and service space between components of air-handling equipment and to intake louvers and screens.
- Specify adjustable slide rail bases for motors to permit belt drive replacement and alteration of the desired driven speed and belt tension.
- Require control components, such as actuators, sensing bulbs, and instruments, to be accessible and protected from damage.
- Mount pressure gages and thermometers on vibration-free supports.
- Insist on allocation of adequate building space for air-conditioning equipment.
- Plan the maintenance procedure for terminal equipment during design—repair in place or replace and shop repair?
- Consider specifying quick disconnects for space units, ducts, piping, and electrical equipment.
- Avoid locating air-conditioning units above ceilings, especially those above occupied spaces. If this cannot be done, place large water-collecting pans beneath these spaces to intercept the inevitable drain pan overflow and leakage.
- Provide access by catwalks or interstitial decks for any equipment located above high ceilings (*e.g.*, auditoriums).
- Design condensate pans and drains to be corrosion resistant, properly pitched, and cleanable.

- Include access for the measurement of air and water flow, temperature, and pressure in major ductwork and piping.

- Provide access for inspection and cleaning of major ducts, especially lined ducts and return air ducts.

- Specify accessible service shut-off drain and vent valves for all water coils and risers.

- Incorporate appropriate chemical treatment equipment for open- and closed-loop water systems or at least allocate space for later installation.

2.4.3 Noise and Vibration Control

Noise and vibration from HVAC equipment can be reduced by proper selection of equipment, vibration control, and shielding. However, residual sound may still be objectionable when equipment is near occupied areas; therefore, both the HVAC system and building design must go together since acoustical control may involve equipment location, floor and wall materials, and room function. The HVAC equipment may influence the functional planning so that areas requiring a noise-free environment are not located next to HVAC equipment. The HVAC engineer should alert other members of the building design team regarding the location of noisy equipment. On large, acoustically sensitive buildings (*e.g.*, theaters, museums), an acoustical consultant should be part of the design team.

The noise and vibration transmitted to the occupied space by system components are important factors in selecting the system. Even after the type of system has been selected, component selection will significantly affect system performance in this regard. Noise can be transmitted to the occupied space from central station equipment along several paths: through air or water flows, along the walls of ducts or pipes, or through the building structure. If central station pumps or fans are used, each of these paths must be analyzed and the transmission of sound and vibration reduced to an acceptable level.

An initial step in noise control is to establish space noise criteria. These criteria should be communicated to the client early in the design process. All noise-generating sources within the air-conditioning system must be identified. Then, the magnitude of those noises must be controlled by

- Equipment selection

- Air and water distribution system design

- Structural containment
- Dissipative absorption along the sound path
- Isolation from the occupied space, where feasible

These control techniques and the information in ASHRAE *Fundamentals* (1989d) and *Applications* (1991) on how to apply them provide a design procedure that can achieve the desired objectives. However, the acoustical design of systems and the buildings they serve is frequently quite complex and is the proper province of specialists known as acousticians. This is especially true for spaces with exacting requirements, such as auditoriums, or where noise-generating components must be located adjacent to occupied areas. Fan and other equipment noise emanating from cooling towers and exhaust or intake ports also affects the ambient noise level of the neighborhood surrounding a building and may need evaluation. Many urban municipalities have codes governing equipment noise.

The major factors governing the sound characteristics of all-air system distribution are shown in Figure 2-3. The occupied space on the floors directly above or below the space housing the air-handling unit will also be in other sound paths. Radiating through the air from the fan to the unit casing walls and from the casing to the room walls and slabs and passing through the structure to the spaces above and below, the sound energy may encounter additional barriers in the form of dropped ceilings. However, most barriers, such as floors and ceilings, while in themselves potentially quite effective, frequently "leak" due to penetrations for pipes, light fixtures, and similar items. Such structural components do not constitute effective sound barriers unless their penetrations are carefully sealed.

All-air systems, particularly pressurized systems, must be examined during all stages of design and installation to ensure quiet systems. The two principal sources of noise in an all-air system are the fans and the distribution duct system itself. Most manufacturers of fan equipment can provide a sound power spectrum for a particular fan operating under a specific set of conditions. With this information, the designer can select an acoustical absorption treatment to reduce this sound source to an acceptable level. The fan noises most difficult to remove are those in the lower octave bands. Therefore, sound attenuation in those bands is an important objective for acoustical treatment at the fans. Sounds in the higher bands will normally be absorbed in the duct distribution system, particularly if the ducts are lined. For quiet operation, fans should be selected for maximum static or total efficiency. When used in variable volume systems, sound characteristics should also be checked for maximum sound pressure levels at the minimum system flow condition when dampers, inlet vanes, or blade pitch fan control are used. In general, large fans at high static conditions produce the highest noise levels.

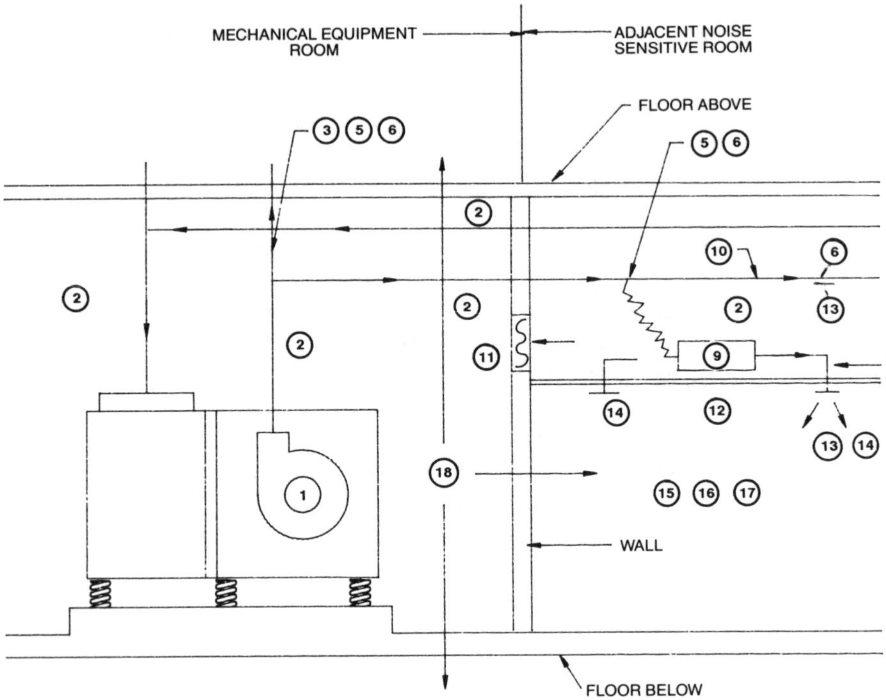

Fig. 2-3 Noise Sources and Propagation Paths in Duct System

Noise and vibration can also be generated within the system by the movement of air or water. These sources can also be controlled by velocity limitations, distribution layout, use of attenuators, and equipment selection. (For piping design, see Section 5.6; for duct design, see Section 5.7.) Several noise sources exist within the air distribution system itself. In general, components with higher pressure losses will produce higher sound levels. Some of the causes of increased sound levels in the occupied space are abrupt transitions, turbulent conditions caused by poor duct fittings or improper duct taps, partially closed dampers used for balancing purposes, improperly located dampers in duct shafts, sound traps improperly selected (more than 2000 fpm face velocity) or installed too close to fans or fittings, sharp bends, crimping of flexible connections, and duct leakage. A pressure-reducing device, damper, or pressure regulator located in a terminal unit may generate objectionable noises as the energy expended in pressure reduction is converted to sound. This is why oversizing of terminal air devices is undesirable. Large-volume terminals (> 2000 cfm) and those with higher pressure differentials produce greater noise levels. Pressure-reducing devices should be installed within the duct system with sufficient downstream ductwork to absorb the sounds generated by the device. Large terminal units with pressure-reducing devices should not be installed in occupied spaces without considering acoustical treatment downstream and in the radiated sound path from the terminal to the room.

Sound travels through ductwork. For example, an air-conditioning system that will serve a series of music practice rooms requires ductwork with sound baffles between rooms, lined ducts, or duct turns to attenuate noise. Sound control then influences duct configuration, size, and system static pressure.

The sound contributed by room terminal equipment cannot be easily reduced. Control of this source starts with system selection and entails careful choosing and sizing to reach, or at least come close to, the noise criteria goals for the conditioned space. The more moving equipment in the terminal, the noisier it will be. Thus, air-cooled unitary terminal equipment is likely to be near the high end of the noise scale. Water-cooled terminals, including water source unitary terminals, can be significantly quieter. Terminals in ascending order of noisiness are air diffusers, variable-air-volume boxes, fan-coil units, high-induction-ratio terminals, and packaged terminal air conditioners. Continuous terminal noise is usually less annoying than intermittent or alternating noise. Thus, on-off control of terminal refrigeration equipment and air-circulating fans may produce annoyance even though the sound level during equipment operation is within acceptable levels.

A greater opportunity for noise control through attenuation (*e.g.*, duct lining) and radiated sound barriers (*e.g.*, solid ceilings) is possible when terminal equipment can be located outside the occupied space (*e.g.*, air-mixing boxes above a ceiling with low-velocity ducts connecting to diffusers). Air ducts passing through adjacent rooms can be transmission channels for cross talk, as can unsealed openings around either ducts or pipes. Occasionally, partitioning will be planned to divide a room terminal or outlet. This creates a virtually uncontrollable path for sound transmission between rooms. Cross talk through such paths can be controlled through building design.

Even inherently noisy terminal equipment can be engineered to meet acceptable noise levels for most applications. Different products vary in their acoustical performance. Often such equipment is not rated, at least on a basis that permits comparison with other equipment using catalog data. When in doubt, consider visiting operating installations or arranging for prototype testing to ensure that the design objectives can be met.

Not only sound produced within the HVAC systems but also vibration from fans, pumps, refrigeration compressors, and other moving equipment must be kept within tolerable levels. As in the case of sound, levels of satisfaction vary depending on the function of the occupied space. Extraordinary precautions must be taken to protect sensitive activities, such as electron microscopes and research animal colonies.

Vibration from imbalanced forces produced by the fan wheel and its drive, unless suitably isolated, will pass undiminished into the structure and be transmitted to the occupied spaces, where, to the consternation of all, less stiff building members (center points of structural spans, window panes, the chandelier above the president's desk) will respond with noticeable excitation. ASHRAE *Applications* (1991) contains excellent resource material for guidance in the isolation of moving equipment.

Every member of the building design team must contribute toward achieving a satisfactory acoustical and vibration-free environment. It is up you as the HVAC design engineer to alert the others to their role in this endeavor.

2.4.4 Integrated versus Stand-Alone Systems

Integrated HVAC systems generally include central systems or central systems with decentralized zone equipment including central chillers, boilers, cooling towers, and air-handling unit(s), while stand-alone systems may be window air conditioners, packaged heat pumps, and unitary or water-cooled decentralized units without a central system. Centralized equipment requires a few large spaces, while decentralized equipment requires smaller unit spaces but more of them.

Central boiler and chiller plants consist of higher grade industrial or large commercial-grade equipment. All major maintenance can be done in one location. The integration of heat recovery from one system to another is facilitated. The larger equipment is usually more efficient. It can be located in one or more locations within the building, with hot water or steam and chilled water distributed from the central equipment room to air-handling units or fan-coils throughout the building. The central or integrated system provides better vibration and noise control since the major equipment is not located in or near occupied areas. Temperature control is provided by mixing boxes, VAV boxes, control valves, or dampers, depending on system design.

Stand-alone installations can provide room or zone control without any auxiliary control equipment, but they may be noisier, present more equipment service problems, and interfere with normal activities in the space. The stand-alone equipment is generally of lower quality, has a more limited useful life, and, in the case of room air conditioners and other unitary equipment, is deficient in humidification capability and variable fresh air control. In some cases, it may be difficult or impossible to provide outside air for ventilation to stand-alone units because of their remote location from an outside air source.

Stand-alone units require either air- or water-cooled condensers. They can be moved from one location to another if changes in building use require it. It is often simpler to relocate stand-alone units than to modify extensive duct and piping systems. Stand-alone units are generally noisier than central systems.

A closed-loop water-to-air heat pump system (see Chapter 7) contains individual refrigeration compressors, and heat is transferred from units in the cooling mode to the closed loop, making it available to units requiring heating. While the coefficient of performance (COP) of the individual heat pump may not be as high as that of central equipment, the closed-loop heat pump system may be more efficient on a seasonal basis. Such systems often require a supplemental boiler to supply heat when heating demand exceeds coincidental heat rejection from the units in the cooling mode.

2.5 IMPACT OF COMPUTERS

2.5.1 Background

The availability of low-cost personal computers has significantly changed the way in which practicing engineers design HVAC systems. Computers are routinely used for peak-load calculations, energy analyses, duct and pipe sizing, preparation of drawings, cost estimates, spreadsheet analyses, as well as for office accounting and word processing of correspondence, reports, and specifications. Some large offices have workstations and plotters for computer-aided drafting, but most offices still prepare drawings by hand, using copy machine reproduced decals for schedules, legends, and standard details.

The HVAC designer is primarily interested in peak-load calculations to obtain equipment sizes. Most HVAC systems are selected based on the past experience of the designers and the lowest first cost, not on energy studies with life-cycle cost analysis of alternative systems. The emphasis in design offices is on production of drawings and specifications within a budgeted fee, not on elaborate studies.

2.5.2 History

In the early 1960s, practicing HVAC design engineers developed small computer programs to ease the burden of manual calculations for such tasks as determining shadowing due to window overhangs to establish the areas exposed to solar loads at various times throughout the year. During this period, researchers developed programs to analyze dynamic heat transfer through building materials and room energy balances. The first generally accepted peak heating and cooling calculation (HCC) program was developed by practicing mechanical engineers in the mid-1960s

(APEC 1967). The program was based on the equivalent temperature method (ASHRAE 1967, 1972) and calculated design-day space/zone loads, supply air quantities, and total building loads imposed on the central heating and cooling equipment.

In 1967, ASHRAE formed a special Task Group on Energy Requirements to establish procedures for calculating heating and cooling loads for energy calculations by computer. Algorithms for building heat transfer, including energy balance and weighting factor techniques, and procedures for simulating the performance of HVAC system components were published (ASHRAE 1969a, 1969b). The first generation of loads/energy programs was developed by the U.S. National Bureau of Standards (now the National Institute of Standards and Technology), using the energy balance method (NBSLD) (Kusuda 1974), and the U.S. Post Office, using the weighting factor method (Lokmanhekim *et al.* 1971).

Publication of these public domain programs and the oil embargo in the early 1970s stimulated ASHRAE researchers and others in the scientific community. In the mid-1970s, the U.S. Army Construction Engineering Research Laboratory combined NBSLD with HVAC systems and central plant equipment to form the complete BLAST energy analysis program (Hittle 1977). At about the same time, the National Aeronautic and Space Administration (NASA) upgraded the Post Office program to NECAP (Henninger 1975). In 1976, the Lawrence Berkeley Laboratory (LBL), Los Alamos Scientific Laboratory (LASL), and Argonne National Laboratory (ANL), with funding from the California Energy Commission (CEC) and the U.S. Energy Research and Development Administration (ERDA), joined with other consultants to significantly upgrade NECAP and introduced the Building Design Language (BDL). This modified NECAP program was then renamed CAL-ERDA. After receiving the final product from LBL, CEC adopted the program as the official energy program for California and renamed it CAL/CON. At the same time, ERDA was changed to the Department of Energy (DOE), and the same program was renamed DOE-1. With continued DOE funding, it was upgraded to DOE-2. During the same period, numerous computer programs were developed in the private sector with proprietary methodologies that did not always follow ASHRAE procedures (ASHRAE 1986).

In 1977, the cooling load temperature differences (CLTD) and cooling load factors (CLF) were developed to simplify the calculation procedures and still account for radiant heat transfers and thermal storage. The results were published in the ASHRAE *Handbook* (1977), and a separate load calculation manual using these factors was published (Rudoy 1979). The HVAC industry generally accepted this methodology, and numerous load/energy programs were then developed by the private sector. Programs are currently available that use both the equivalent temperature differential and the CLTD/CLF methodologies. The numerical values for the

CLTD and CLF factors were upgraded through a series of ASHRAE research projects (ASHRAE 1988). Values for these factors covering a large variety of conditions are available in computer disk format (ASHRAE 1992).

2.5.3 Computer Programs

Design-day heating and cooling loads are required to (1) determine peak hour sensible and latent heat loads in each thermal zone, (2) establish supply air quantities, (3) select cooling and heating coils, and (4) obtain building diversified loads (also called block loads) for sizing the equipment. To determine the zone peak loads, perform calculations for each hour of the year to be sure that each thermal zone is exposed to the greatest impact due to the weather (temperature, humidity, solar) and internal loads (people, lights, equipment). Coincident zone loads add up to the diversified load imposed on the central heating and cooling equipment.

In most design offices, PC-based programs that calculate design-day heating and cooling loads are routinely used. The programs are an important production tool and, for maximum use, must have the capacity to handle a large number of thermal zones. Other multiparameter calculations, such as the sizing of duct and piping networks and the analysis of sprinkler loops, can also be handled by available PC programs.

On the other hand, energy calculation programs are used primarily for energy code compliance and, in most offices, are only infrequently used for comparative detailed energy studies. Remember that internal load schedules for peak-load calculations are not the same as those used in annual energy simulations! Energy calculations require average use schedules, while peak-load calculations require maximum load schedules to be certain that the HVAC system will meet all the design-day space loads. A typical office building load schedule is shown in Table 2-3.

Table 2-3 Typical Office Building Load Profile

	Percent of Peak		
Hour	**Lights**	**Receptacles**	**Occupancy**
1	5	0	0
2	5	0	0
3	5	0	0
4	5	0	0
5	5	0	0
6	5	0	0
7	10	0	0
8	90	55	100
9	90	55	100
10	95	50	100
11	95	55	80
12	95	90	40
13	80	60	80
14	80	80	100
15	90	70	100
16	90	75	100
17	95	30	30
18	80	30	10
19	70	50	10
20	60	5	10
21	40	0	0
22	30	0	0
23	20	0	0
24	20	0	0

2.6 EFFECTS OF ALTITUDE

Standard psychrometric charts and performance data published by manufacturers generally assume equipment operation at sea level. When the project is located at a significantly higher altitude, allowances must be made. Factors by which the usual (sea level) data must be multiplied when operating at higher altitudes are summarized in Table 2-4. For items not listed, consult the appropriate sources, such as the *Engineering Guide for Altitude Effects* (Carrier).

2.7 CODES AND STANDARDS

All mechanical design is subject to codes. These are laws or ordinances or other types of regulations that specify certain aspects of the design and construction of buildings. All states in the United States and all Canadian provinces have building codes, which are state or provincial laws pertaining to buildings. Many U.S. municipalities, particularly the larger ones, have promulgated local building codes,which are generally stricter than the state codes over which they take precedence. Most U.S. state and local building codes follow one of four major model codes: Building Officials and Code Administrators International (BOCA), International Conference of Building Officials (ICBO), Southern Building Code Congress International (SBCCI), and Uniform Building Codes (UBS). State and local building codes affect HVAC design in several ways. They may specify minimum efficiencies for equipment, minimum ventilation rates, and various measures for fire and smoke control. Buildings of the federal government are not subject to state or local codes, and the designer must follow applicable regulations issued by the General Services Administration (GSA) or the responsible federal department or agency. Where state or local codes do not provide any guidance, use ASHRAE or other suitable standards or guidelines for good design practice. Also consult the chapter Codes and Standards in the back of each volume of the ASHRAE *Handbook*.

Table 2-4 Altitude Correction Factors

	Altitude, ft			
Item	2500	5000	7500	10,000
Compressors	1.0	1.0	1.0	1.0
Condensers, air-cooled	0.95	0.90	0.85	0.80
Condensers, evaporative	1.00	1.01	1.02	1.03
Condensing units (air-cooled)	0.98	0.97	0.95	0.93
Chillers	1.0	1.0	1.0	1.0
Induction room terminals (chilled water)	0.93	0.86	0.80	0.74
Fan-coil units				
Total capacity (*SHF = .40 – .95)	0.97	0.95	0.93	0.91
Sensible capacity (SHF = .40 – .95)	0.92	0.85	0.78	0.71
Total capacity (SHF = .95 – 1.00)	0.93	0.86	0.79	0.73
Packaged air-conditioning units, air-cooled condenser				
Total capacity (SHF = .40 – .95)	0.98	0.96	0.94	0.92
Sensible capacity (SHF = .40 – .95)	0.92	0.85	0.78	0.71
Total capacity (SHF = .95 – 1.00)	0.96	0.82	0.88	0.84

*SHF = Sensible Heat Factor = (Sensible Heat)/(Sensible + latent heat)
Table excerpted by permission from *Engineering Guide for Altitude Effects* (Carrier).

Of special importance to HVAC design is a class of codes and standards pertaining to energy conservation that has been created since the mid-1970s. Foremost among these are ASHRAE standards of the 90 and 100 series. ASHRAE *Standard* 90 was first issued in 1975 and has been revised several times. In its current form (*Standard* 90.1-1989), the standard specifies three ways in which a building may be designed. Under the prescriptive method, the designer follows a clearly defined methodology using prescribed performance for mechanical equipment, lighting, and wall and window insulation. Alternately, the system performance method is based on annual rather than seasonal energy use and allows the designer greater freedom in the choice of components. If the designer wants to use an even more innovative design approach, the building energy cost budget method is indicated. In this method, the designer must simulate the energy cost performance of the proposed building and compare it to the performance of the same building designed by the prescriptive method. Actual energy costs or utility tariffs in force at the location of the proposed building must be used in the calculations. If the annual energy cost of the proposed building is no larger than that of the building designed by the prescriptive or the system performance method, the building is deemed to comply with the standard. The standard encourages the use of the last method; however, this requires much more effort than the other two methods.

Although written as a guideline to good design and not as a legal document, ASHRAE *Standard* 90, or parts of it, has been incorporated by reference into the building codes of all states. Since this standard is updated more frequently than state legislation, ascertain for each project whether a particular state code refers to the latest version or to earlier versions of the standard.

The state with the strictest code affecting HVAC design is California. Its Energy Commission has issued the California Energy Code, which, while following the general outline of ASHRAE *Standard* 90, has more severe requirements. Based on history, one might speculate that the rest of the United States will follow California's lead in this respect. The California Energy Code specifies two ways of meeting its requirements—energy budget and custom budget. The energy budget ($Btu/ft^2 \cdot yr$), based on the performance of prototypical buildings, is specified depending on climate, building function, and the ratio of floor area to building perimeter. In the custom budget method, the energy use of the proposed building is compared to the energy use of that building modified to include prescribed conservation features. If the energy use of the proposed building is no greater than that of the building with the prescribed features, it is deemed to be in compliance with the standard. Certain alternative calculation methods (ACM) must be approved by the commission before the results can be used to demonstrate code compliance. Criteria for the approval of ACMs have been issued by the commission.

On the whole, Canadian building codes and design practices are more energy conserving than those in the United States. This may be due to the more severe climate of that country and/or to the fact that its federal and provincial building agencies have greater influence on building design and construction than their counterparts in the United States.

Other codes of major importance to the HVAC designer are the National Fire Codes—NFPA 90A, Installation of Air Conditioning and Ventilation Systems, and NFPA 92A, Smoke Control Systems. Follow their provisions where local requirements are less stringent or do not exist.

The greatest impact of ASHRAE *Standard* 55-1981, Thermal Environmental Conditions for Human Occupancy, and ASHRAE *Standard* 62, Ventilation for Acceptable Indoor Air Quality, is on minimum ventilation requirements. The provisions of these standards are being incorporated into state and local building codes.

2.8 ECONOMIC CONSIDERATIONS

2.8.1 Overview

No engineering design can be independent of economics. The type of economic analysis to be used, however, strongly depends on the orientation of the client, whose criteria on that matter must be ascertained before the engineer begins an economic analysis.

A developer who puts up a speculative building is primarily concerned with first cost, and concern for operating costs may vary from minor to none. On the other hand, an institutional client who expects to own and occupy the building during its entire useful life is frequently willing to accept additional first costs if these result in operating cost savings. On many projects, the U.S. government requires a life-cycle cost analysis (which covers capital, operating, and maintenance costs and includes the effect of interest), while industrial or commercial clients may want to know the rate of return on their investment (ROI).

In addition to determining the client's preferred method of analysis, try to ascertain the client's usual or preferred financial assumptions, such as projected rates of inflation, discount rates, fuel cost escalation rates, and similar data. Using the methods and data generally used by the client for financial projections makes your economic analysis more believable to the client, and it avoids subsequent criticisms and objections relative to their use.

Regardless of the method of financial analysis used, annual costs for each air-conditioning system under consideration must be determined. Items whose costs should be considered are listed in Chapter 33, Owning and Operating Costs, of ASHRAE *Applications* (1991). For a realistic analysis, the costs of maintenance and repairs, which are difficult to obtain, should also be included in the economic analysis. Frequently, manufacturers either do not have the information or are reluctant to divulge it. Potential sources for this information are operators of equipment similar or identical to that being designed, operators of other buildings occupied by the client, and Building Owners and Managers Association (BOMA) publications. (The Hartford Steam Boiler Inspection and Insurance Company, 1 State Street, Hartford, CT 06102 has also collected much useful data related to equipment failure.)

2.8.2 Methods of Economic Analysis

Payback Period. This method is sometimes used to determine whether a particular addition to first cost is warranted by savings in operating costs. The first year's annual savings are divided into the additional first cost to obtain the payback period (in years) to recoup the additional investment. While the method is very simple, it ignores the time value of money (interest or discount). It should be used only for periods not exceeding three to five years. It can be modified by discounting savings occurring in future years (see below).

Discounted Cash Flow. Revenues (or savings) and costs are calculated separately for each year over the assumed lifetime of the building or piece of equipment. They are then discounted and summed to a specified year, usually either the first or the last year of the analysis period. The discount factor takes into account the time value of money. If all costs are referred to the first year, the discount factor for the rth year is $1/(1 + i)_r$, where i is the discount rate in decimal form (a 6% rate would be 0.06). If costs are referred to the last year, the discount factor for the rth year is $(1 + i)_{n-r}$, where n is the number of years over which the analysis extends.

Net Present Value (NPV). The NPV is the difference between the present value of revenues and the present value of costs. It is the sum of all annual discounted cash flows referred to in the first year of the analysis. The higher the NPV, the more desirable the project, subject to the initial cash limitations of the investor.

Life-Cycle Cost (LCC). This is the discounted cash flow method including first cost, operating cost, maintenance cost, and any salvage value, usually referred to the last year of the analysis. Life-cycle costing is required on many federal government projects.

Profitability Index. The profitability index is defined as the ratio of the net benefit to the net cost. It can be expressed as (NPV + C)/C, where C is the total initial investment. This index normalizes the total amount of benefits to a single unit of invested capital. It is a useful concept for the choice among different projects when the amount of available capital for investment is limited.

Internal Rate of Return (IRR). The IRR is the discount rate that makes the NPV equal to zero. Therefore, it must be obtained by iterative calculations once the cash flow stream has been identified. Because the IRR is a percentage number, it is intuitively easy to communicate to a broad variety of clients. An investor may have a "hurdle rate," which is the minimum acceptable rate of return for a project. If the IRR is higher than that, the project will be undertaken. If it is not, the investor will balk.

After-Tax Analysis. An after-tax analysis includes the effect of taxes, particularly income taxes, on the financial aspects of the project. These have to be calculated for each year before discounting, as illustrated in a table in Chapter 33, Owning and Operating Costs, of ASHRAE *Applications* (1991).

Levelized Cost. This method is used only by public utilities, since their rates are set based on an allowable return on investment. It does not apply to the private sector.

Economic analyses may be performed either in current (nominal) dollars or in constant dollars (where the effects of inflation are removed). Constant dollar analyses are generally easier to use and, for that reason, are preferable. Many economists also believe that they yield a truer picture of the financial viability of a project. However, since certain tax items, such as depreciation, are always given in current dollars, after-tax financial analyses must be conducted in current dollars.

In private investment analysis, the discount rate is generally based on the investor's cost of capital or required rate of return. This rate may be increased to account for technological uncertainties or perceived risks. In public investment analysis, the discount rate should be at least equal to the cost of borrowing money on the part of the government agency plus an amount representing the public's lost opportunity cost (by the purchasers of the government bonds).

2.8.3 Rentable Area

The rentable area of a tenant-occupied building is the basis for determining income. Full-service leases including utility costs are based on this parameter. The rentable area may be defined in the lease, sometimes by reference to a standard, such as that of the BOMA. Strangely, many leases fail to define the term at all, and it becomes defined by the established practice in an individual building. Significant disparity can occur among buildings in the same region. For an office building, the definition will vary somewhat depending on whether a floor is leased to single or to multiple tenants. For single tenants, the rentable area is often measured from the inside surface of the exterior walls less any areas connecting with and serving other floors, such as stairs, elevator shafts, ducts, and pipe shafts. Toilet rooms, mechanical equipment rooms serving the floor, janitorial closets, electrical closets, and column spaces are included in the rentable area. For multiple tenants on one floor, the term usually excludes public corridors, lobbies, toilet rooms, mechanical equipment rooms, etc.

2.8.4 Space Utilization

The ratio of net rentable space to gross building area is an important indicator of the profitability of a building. Building construction costs are related to the gross area, whereas income potential relates to the rentable area. The ratio tends to be higher for larger buildings—it usually varies from 90 to 75%. This ratio can be significant in the economic evaluation of air-conditioning alternatives. The air-conditioned area—which relates most closely to mechanical equipment initial and operating costs—may be similar to the single-tenant rentable area defined previously. Construction costs, including mechanical costs, however, are reported on the basis of gross building area. Therefore, if such cost information is to be used for budgeting initial cost, gross area should be used to create a total construction model. On the other hand, operating costs may be more realistically based on rentable area.

To some degree, the concept and design of a building's air conditioning can influence the ratio of rentable area to gross area. While equipment located on the floor under windows may occupy otherwise usable space, this space is almost always included in the rentable area. Thus, no reduction in income results from the use of under-window units. If air-conditioning equipment to serve each floor is located in a mechanical equipment room on that floor, such space may also be included in the net rentable area. Compare this concept to a central system serving an entire building from a rooftop penthouse through supply and return duct shaftways. Due to the method by which rentable area is defined, the penthouse and the shaftway space may be nonrevenue-producing. Thus, paradoxically, even though more total building area and perceptually more valuable space may be used by locating equipment on the tenant floor than by putting it in a rooftop penthouse and using duct risers, the decentralized arrangement could result in more revenue and a better net return for the construction cost investment.

2.8.5 Public Utility Tariffs (Rate Schedules)

The design engineer must be familiar with the details of the tariffs of the local electric utility (and the gas utility if gas service is contemplated for the building). This is necessary for the detailed analysis of operating costs, and it enables the designer to look for ways of taking advantage of utility rates to decrease these costs. Sometimes, special rates can be negotiated with the local utility if special conditions exist.

Most utility tariffs for nonresidential buildings consist of a *customer charge*, a *demand charge*, and an *energy charge*. Few residential tariffs include demand charges, primarily because of the high cost of demand meters, but this may change as less expensive meters and different methods of reading them become available. The customer charge is a flat monthly charge. The energy charge is a charge per kilowatt hour (or Btu or cubic foot of gas) used. It may be a flat charge or a decreasing block charge in which the cost per energy unit decreases as monthly use increases, or an increasing block charge in which the opposite occurs. Some tariffs make the energy charge a function, among other variables, of the monthly demand.

The demand charge is based on the highest monthly demand that the customer makes on the utility network (electric kW or gas Btu/h), usually measured at 15- or 30-min intervals. Some utilities calculate their monthly charges on a billing demand that is never less than a certain fraction of the highest demand during the previous 12 months (ratchet clause). Thus, a high air-conditioning demand during a particularly hot summer day may raise utility costs for that customer for an entire year. It is thus beneficial to investigate methods of avoiding simultaneous operation of high-demand equipment, if that is feasible. Thermal storage (see Section 9.2) and control strategies (see Section 9.3 and Chapter 10) are methods of accomplishing this; they may also contribute to a reduction in time-of-use charges.

Time-of-use charges, in which demand and/or energy rates depend on the time of day during which they are made, have recently become more prevalent; some utilities have even started to introduce them for their residential customers. Generally, they are higher during periods of heavy use (on-peak, daytime) and lower during periods of light use (off-peak, nighttime). Some utilities reduce or waive their demand charges during off-peak periods. These periods may be nights, weekends, or even entire seasons of low demand, such as the summer months for gas use. Thus, time-of-use or other charges may differ from summer to winter.

2.9 REFERENCES

APEC. 1967. HCC—Heating/cooling load calculation program. Dayton, OH: Automated Procedures for Engineering Consultants, Inc.

ASHRAE. 1967. *Handbook of Fundamentals.*

ASHRAE. 1969a. *Procedure for determining heating and cooling loads for computerized energy calculations: Algorithms for building heat transfer sub-routines.*

ASHRAE. 1969b. *Procedure for simulating the performance of components and systems for energy calculations.*

ASHRAE. 1972. *Handbook of Fundamentals.*

ASHRAE. 1977. *Handbook of Fundamentals.*

ASHRAE. 1981. ASHRAE *Standard* 55-1981, Thermal environmental conditions for human occupancy.

ASHRAE. 1986a. *A bibliography on available computer programs in the area of heating, ventilating, air conditioning and refrigeration,* 2nd ed.

ASHRAE. 1986b. *Bin and degree hour weather data for simplified energy calculations.*

ASHRAE. 1988. Development of expanded wall, roof, and zone classification for cooling load calculation methods. ASHRAE *Transactions* 94(2)1: 716-777.

ASHRAE. 1989a. ASHRAE *Standard* 62-1989, Ventilation for acceptable indoor air quality.

ASHRAE. 1989b. *ASHRAE pocket guide for air conditioning, heating, ventilation, and refrigeration.*

ASHRAE. 1989c. Commissioning of HVAC Systems. *Guideline* 1-1989.

ASHRAE. 1989d. *Handbook—Fundamentals.*

ASHRAE. 1991. *Handbook—Applications.*

ASHRAE. 1992. *Handbook—Systems and Equipment.*

Engineering guide for altitude effects. Syracuse, NY: Carrier Corporation.

Henninger, R.H., ed. 1975. NECAP, NASA's energy-cost analysis program. Washington, D.C.: National Aeronautics and Space Administration.

Hittle, D. 1977. The building loads analysis and system thermodynamics program, BLAST. Champaign, IL: U.S. Army Construction Engineering Research Laboratory.

Klote, J. and J. Milke. 1992. *Design of smoke management systems*. Atlanta: American Society of Heating, Refrigerating and Air-Conditioning Engineers, Inc.

Kusuda, T. 1974. NBSLD, computer program for heating and cooling loads in buildings. Washington, D.C.: National Bureau of Standards.

Lokmanhekim, M., *et. al.* 1971. Computer program for analysis of energy utilization in postal facilities, vols. I, II, III.

McQuiston, F.C. and J.D. Spitler. 1992. *Cooling and heating load calculation manual*, 2nd ed. Atlanta: American Society of Heating, Refrigerating and Air-Conditioning Engineers, Inc.

NFPA. 1989. Standard for the installation of air conditioning and ventilating systems. NFPA *Standard* 90A-89. Quincy, MD: National Fire Protection Association.

NFPA. 1991a. Code for safety to life from fire in buildings and structures. NFPA *Standard* 101-91. Quincy, MD: National Fire Protection Association.

NFPA. 1991b. Standard on fire protection for laboratories using chemicals. NFPA *Standard* 45-91. Quincy, MD: National Fire Protection Association.

Rudoy, W. 1979. *Cooling and heating load calculation manual*. Atlanta: American Society of Heating, Refrigerating and Air-Conditioning Engineers, Inc.

USAF. 1988. *Engineering weather data*, AFM 88-29. Washington, D.C.: U.S. Government Printing Office.

Chapter 3 OCCUPANT COMFORT

3.1 OVERVIEW

One of the first steps in designing an air-conditioning system is to establish comfort criteria for the intended service. Such criteria should include space temperature and humidity, ventilation rate, and sound level. The selection of these design parameters will be influenced by a number of conditions: the ages and activities of the occupants, the occupant density, and the contaminants present in the space. Even the physical character of the space can have some bearing on occupant comfort. For example, surface temperatures of walls and floors could affect human comfort and influence the design space temperature.

The designer must also consider economic parameters. A balance frequently must be sought between the optimum environmental conditions and system performance capabilities on the one hand and the initial and operating cost targets on the other. If that is not enough, also carefully consider the construction and operating complexity of the system concepts. Design objectives will have but a slim chance of being realized if system design features reach beyond the capabilities or understanding of installing contractors and operating staff (assuming there is to be an operating staff). This point can hardly be overemphasized and deserves serious client and contractor discussion at an early stage.

3.2 THERMAL COMFORT

3.2.1 Thermal Environmental Conditions

Satisfaction with the thermal environment is based on a complex subjective response to several interacting variables. The design, construction, and use of an occupied space, as well as the design, construction, and operation of its heating and air-conditioning systems, will determine the degree of satisfaction with the thermal environment. Not all individuals perceive a given thermal environment with the same degree of acceptability. The perception of comfort relates to an individual's physical condition, body heat exchange with the surroundings, and physiological characteristics. The heat exchange between the individual and the surroundings is influenced by:

- Dry-bulb temperature
- Relative humidity
- Thermal radiation
- Air movement
- Amount of clothing
- Activity level
- Direct contact with surfaces not at body temperature

While ideal thermal conditions are complicated to define for any one individual in a particular setting, ASHRAE *Standard* 55-1992 (see below) specifies conditions likely to be thermally acceptable to at least 80% of the adult occupants in a space. A more complete explanation of the condition, factors affecting comfort, and the methods for their determination and measurement can be found in that ASHRAE standard and in Chapter 8, Physiological Principles for Comfort and Health, in ASHRAE *Fundamentals* (1989a).

3.2.2 Temperature and Humidity

The design space temperature and humidity for both heating and cooling seasons should be based on the ASHRAE comfort chart for most applications. This revised chart (Figure 3-1) appears in ASHRAE *Standard* 55-1992, Thermal Environmental Conditions for Human Occupancy (ASHRAE 1992). This comfort zone is defined for people in winter and summer clothing engaged in primarily sedentary activities (≤ 1.2 met). If the designer selects values from the center of the comfort envelopes, temperatures of 77°F for summer and 72°F for winter are the likely choices, with relative humidity values of 45 and 50%, respectively. On the other hand, many designers frequently choose 75°F for cooling and 72°F for heating. ASHRAE *Standard* 90.1-1989, Energy Conservation in New Building Design (ASHRAE 1989b), recommends 78°F for summer and 72°F for winter. Some discomfort in the cause of energy conservation may be tolerable, especially if the design load peaks are of short duration. For spaces with peak cooling loads of long duration (*e.g.*, interior offices), however, continued operation at 78°F may well overtax the occupants'

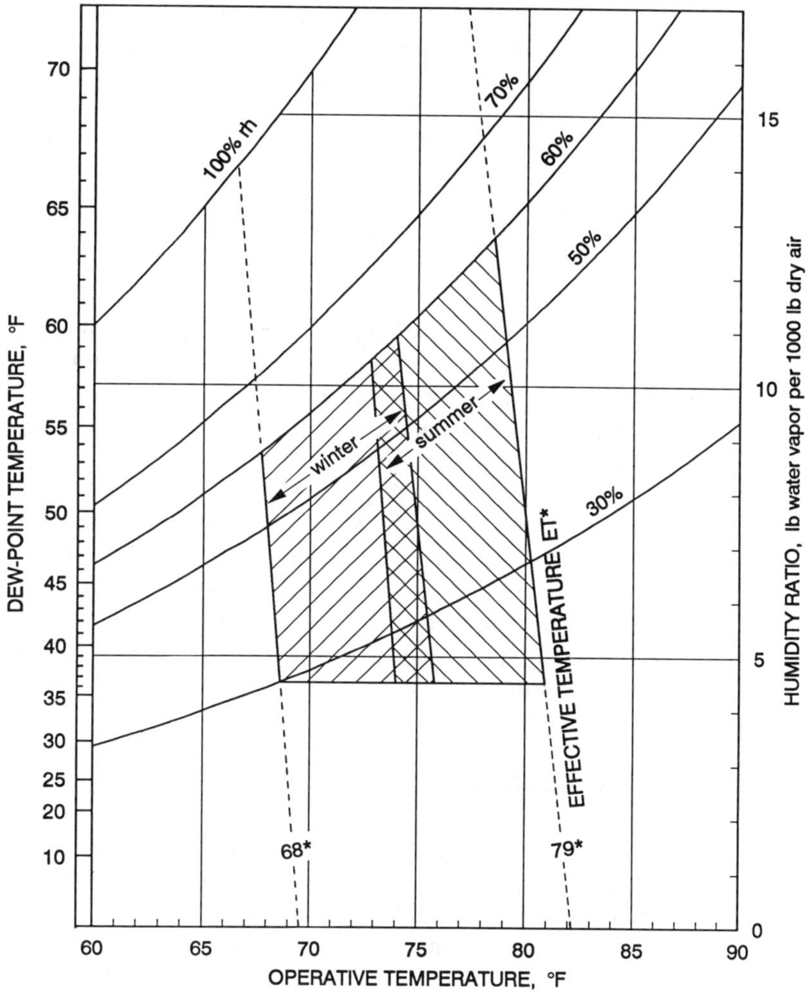

Fig. 3-1 Acceptable Ranges of Effective Temperature and Humidity for Human Comfort

sense of duty regarding saving energy, impairing their productivity, and inducing the sick building syndrome (SBS).

As the comfort chart indicates, within the range of 30 to 60%, relative humidity does not have a significant bearing on human comfort in most applications as long as the space dry-bulb temperature is in the comfort range. Relative humidity, though, does affect odor perceptibility and respiratory health. Because of these considerations, 40 to 50% rh is the preferred design range. However, maintaining humidity within this range during winter is complicated by (1) energy considerations, (2) the almost certain risk of condensation on windows and window frames during cold weather, and (3) the need to provide and maintain humidifying equipment incorporated in the air-conditioning system. The economic value of winter humidity control for occupant well-being is not always appreciated. However, significantly reduced absenteeism among children, office workers, and army recruits as a result of winter humidification has been reported (Green 1979, 1982). Where winter humidification is provided for comfort, a minimum relative humidity of 30% is generally acceptable.

If a higher level of humidity is acceptable under summer conditions, considerable energy savings can be realized, as shown in Figure 3-2 (Dubin and Long 1978). To determine an approximate value of the energy used for dehumidification at a constant 78°F dry-bulb temperature, enter the annual wet-bulb degree-hours above 66°F at the bottom left, intersect this value with the indoor relative humidity chosen, then draw a vertical line to the weekly hours of cooling system operation and read the energy used (in million Btu per 1,000 cfm) on the upper-left scale. Repeating this procedure for a different value of relative humidity yields the energy savings obtainable by raising relative humidity. However, be cautious about choosing excessively high humidities. Computer rooms, particularly their printers, and drafting rooms are two examples of applications for which relative humidity in excess of 50 or 55% is undesirable or unacceptable.

3.2.3 Zoning Considerations

Even when appropriate design temperatures are selected, they are achievable only at the space control instruments and then only with variations set by the control band of that instrumentation (see Chapter 10). Furthermore, the larger the control zone (and the distance from occupant to controller) and the more diverse the thermal load characteristics within the zone, the more the selected parameters of occupant comfort will vary.

The obvious desirability of individual occupant control of the temperature is compromised by the physical arrangement of the space, the mobility of the occupants, the inherent system capabilities, and the higher cost of providing a temperature control zone for each person. The degree of compromise is an important design concern. A single thermostat to control space temperature throughout a commercial building is not appropriate and will guarantee a low level of occupant satisfaction. This is especially true if the space is compartmented, as in an office building or a hotel.

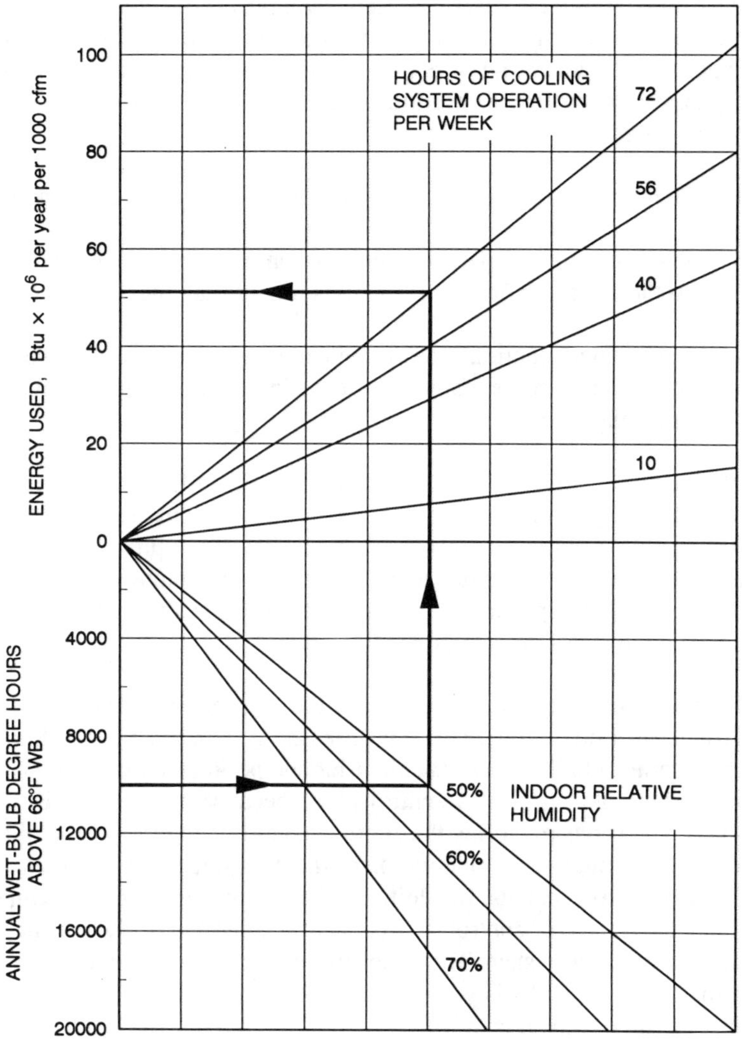

Fig. 3-2 **Summer Energy Requirements for Dehumidification** (Dubin and Long 1978)

Each compartmented space can become a control zone if provided with a thermostat (or manual) control and a terminal device that can be regulated to produce the proper capacity of heating and/or cooling to maintain the desired room temperature. If separate systems provide heating and cooling, one of the following control methods may be used.

- Thermostat control of both heating and cooling in a coordinated fashion.
- Two independent space temperature controls, although this method risks control "fighting" if the heating setpoint is higher than the cooling setpoint.
- Control one function at each small zone, while controlling the other function is controlled to serve many such zones in common, such as all those sharing the same exposure.

The last alternative is often installed in office buildings. For example, supply air terminals may be provided for each office module, but the perimeter heating system for an entire building, or for separate exposures of a building, may be controlled from a common point. Attempting to both cool and heat a number of perimeter rooms with different solar exposures and potentially diverse thermal characteristics from a single control device is not recommended.

Whenever possible, rooms that require the same environmental conditions should be grouped in a single zone. This simplifies the system and makes it less costly to install and operate. On the other hand, interior and perimeter spaces should almost always be on separate systems or control zones because of the disparity of their thermal loads. Decisions regarding the zoning capability and the cost implications must be resolved through the collaboration of the design team, the owners, and, if feasible, the prospective tenants. In addition to defining the size of the temperature control zone, the degree of control desired has considerable impact on the selection of the air-conditioning system. The rapidity of change in space temperature—as can occur with on/off control—is yet another consideration of system and control concepts bearing on occupant comfort.

3.3 VENTILATION

In contrast to common usage, in ASHRAE terminology *ventilation* usually refers only to the outdoor air introduced into the conditioned space, not to the total amount of air supplied.

When an amount of ventilation air is specified, it refers to the outdoor air portion only—except in special cases defined in Section 6.1.3.2 of ASHRAE *Standard* 62-

1989 (1989c). The total amount of air supplied to a space is generally determined by the cooling requirements of that space, subject to the minimum ventilation air specified in the applicable code for the particular use of the space.

3.3.1 Ventilation Techniques

The purpose of ventilation is to introduce contaminant-free air into an occupied space at a rate and manner sufficient to dilute or remove contaminants generated within the space to a satisfactory degree considering occupant comfort and health. Buildings are usually ventilated by supplying cleaned outdoor air through the air-conditioning system. Different systems have different capabilities in meeting the ventilation requirements.

Natural ventilation through open windows or through infiltration is the method for many residences and other buildings. However, such ventilation is variable and, to a large measure, unquantifiable and uncontrollable. Nevertheless, some air-conditioning systems in moderate climates are designed to depend on natural ventilation. However, mechanical ventilation should be used, unless the adequacy of this method is historically or otherwise demonstrable.

Local and general exhaust systems usually complement the ventilation supply system to contain and remove certain sources of contamination, such as toilet exhaust. It was once the practice to supply sufficient outdoor air through the air-conditioning system to make up the air exhausted plus an additional amount to offset infiltration by pressurization. Energy considerations now generally dictate that the ventilation air supply not be increased above the amount needed for dilution of contaminants except to balance the exhaust airflow. Infiltration is more effectively limited by the tightness of the building envelope.

Ventilation is not the only means of limiting contaminant levels, and it should not be considered a cure-all. Source control, where practicable, is more effective. Building materials, such as carpets and wall coverings, should be selected for low emission of volatile organic compounds. Physical containment or segregation of emission sources may be appropriate. Indeed, control of some sources may be beyond the capabilities of even a well-designed ventilation system. Therefore, the objectives and capabilities of the proposed ventilation system should be understood by all who are concerned with the construction and use of the building.

3.3.2 Indoor Air Contaminants

To better understand the methods for controlling indoor air contamination, the designer should have some background concerning the nature of such contaminants. Indoor air contaminants can be solid or liquid particles, gases, or vapors. Some can be irritants or odiferous, thus affecting occupant comfort. The same contaminants at higher concentrations, as well as others of which occupants may be unaware, can be health risks. People vary in their sensitivity to contaminants. Even minute concentrations of certain fungi and other impurities can cause serious discomfort and impairment of sensitive individuals while not affecting most occupants. Standards for vapors and gases specify a quantity of pollutant per unit volume in parts per million (ppm) of air. Standards for particles often specify the mass concentration of particles expressed as micrograms per cubic meter ($\mu g/m^3$). They include all particle sizes or the total suspended particulate concentration (TSP). Large particles are filtered by the nasal passages and cause no adverse physiological response unless they are allergenic or pathogenic. Smaller respirable suspended particles (RSP) are important because they can lodge in the lungs. Respirable particles range in size up to 5 µm. Particles of specific interest include:

- Respirable particulates as a group
- Tobacco smoke (solid and liquid droplets), which also contains many gases
- Asbestos fibers
- Allergens (pollen, fungi, mold spores, and insect feces and parts)
- Pathogens (bacteria and viruses), which are almost always contained in or on other particulate matter

Vapors and gases of interest include:

- Carbon dioxide (CO_2)
- Carbon monoxide (CO)
- Radon (decay products become attached to solids)
- Formaldehyde (HCHO)
- Other volatile organic compounds (VOC)

Some contaminants, such as sulfur dioxide (SO_2), are brought in with outside air by mechanical ventilation or by uncontrolled infiltration. Many of the contaminants, such as nitrogen oxides and carbon monoxide, found in outdoor air may have indoor sources as well. However, most indoor pollutants emanate from inside sources. People are sources of CO_2, biomatter, and other contaminants characterized as "body odors." People's activities (smoking, cleaning, cooking, gluing, and refinishing furniture) also cause pollution, and building materials and finishes can "outgas" pollutants. Furnishings, business machines, and appliances, particularly unvented or poorly vented wood- and fossil-fueled heaters and ranges, can be contaminant sources. The soil surrounding a building can be a source of radon and pesticides that enter the building through cracks or drains or by diffusion. Heating, ventilating, and air-conditioning systems, drains, plumbing systems, and poor construction or maintenance practices can have "environmental niches" where pathogenic or allergenic organisms can collect and multiply to be reintroduced into the air.

An additional complicating factor in the buildup of contaminants is the variation in dilution rates and effectiveness of the ventilation delivery systems often found within buildings. Concentrations vary spatially as well as over time. These variations add further nonuniformity to the pollutant concentration. ASHRAE *Standard* 62 (1989b) specifies maximum concentration levels of common indoor contaminants. It also sets forth acceptable quality parameters for outdoor air used for building ventilation. If the outdoor air source exceeds the contaminant parameters, it must be cleaned or purified prior to introduction into occupied spaces.

3.3.3 Determination of Ventilation Rate

ASHRAE *Standard* 62 provides the designer with a means of determining ventilation rates needed to achieve acceptable indoor air quality, which is defined as

"air in which there are no known contaminants at harmful concentrations . . . and with which a substantial majority (80% or more) of the people exposed do not express dissatisfaction."

Two procedures for determining the required ventilation rate are offered to the designer in the standard—the ventilation rate procedure and the indoor air quality (IAQ) procedure.

The ventilation rate procedure sets forth prescriptive rates, usually on the basis of cfm of outdoor air per occupant, for an array of applications. Unless unusual pollutants are present, these rates are intended to produce acceptable IAQ. The basis for most of the rates specified is an underlying minimum of 15 cfm per occupant as a means of controlling CO_2 to a concentration of 1000 ppm. While CO_2 per se is not a contaminant of concern at this low concentration, it is an easily measurable surrogate for other contaminants, such as body odor vapors and tobacco smoke. The indoor air quality procedure offers an analytical alternative, allowing the designer to determine the ventilation rate based on knowledge of the contaminants being generated within the space and the capability of the ventilation air supply to limit them to acceptable levels.

While ASHRAE *Standard* 62 contains guidelines concerning maximum concentration of some pollutants, "threshold limit values" for industrial applications are published by the American Conference of Governmental Industrial Hygienists (ACGIH 1992). Frequently, local building and occupational codes prescribe threshold limit values and ventilation rates. When they are more stringent than current ASHRAE and ACGIH recommendations, they must be followed unless waived by the governing authority; when they are less stringent, the ASHRAE (or ACGIH) guidelines are recommended since they generally represent a more recent consensus of expert opinion on the subject and may have legal weight.

3.3.4 Air Cleaning

The selection of air filters, gas or vapor removers, or other methods of air contaminant reduction influences the quality of the ventilation air supply, the dilution capability of the total air supplied to the occupied space, the concentration of room air contaminants, and the resultant occupant health and comfort. Filtration above 50% efficiency (according to the ASHRAE dust-spot determination) must frequently be installed if the particulate limits established by both codes (ASHRAE and ACGIH) and ASHRAE *Standard* 62 for ventilation air are to be met. Filters with efficiencies above 80% (dust spot) can be effective in removing a significant portion of respirable particulate contaminants, including tobacco smoke.

When high-efficiency supply air filters are installed, the total supply airflow helps control the particulate concentrations within the space. Consequently, all-air constant-volume systems and, to a lesser extent, variable-volume systems, if equipped with such filters, produce respirable particulate concentrations in the space environment that are lower than those achievable by systems in which the supply airflow rate is limited to the ventilation rate (*e.g.*, air-and-water systems).

Certain applications demand removal of gaseous contaminants present in the outdoor air or produced within the conditioned space. Harmful gases and vapors can be removed by adsorption or oxidization. Activated carbon is the absorptive material most commonly used in HVAC systems. Potassium permanganate impregnated into the carbon or an alumina base is used to oxidize certain chemicals for which carbon has limited effectiveness. Before being incorporated into the system design, the efficacy limitation, bed depth requirements, and maintenance procedures should be thoughtfully evaluated in light of the need.

Air washers (see Section 5.1.3) are also effective in reducing SO_2 and other acid-forming gases. But such scrubbers need continuous maintenance to keep the recirculated water from becoming highly corrosive and its reservoir a breeding ground for biological contaminants. Rare-book rooms and valuable display or storage areas in museums and archival depositories are candidates for gas removal provisions, but the users should understand that careful maintenance is required for effective performance. Notwithstanding such gas removal provisions, the ventilation rate should be maintained at no less than 15 cfm per person (except where occupancy is variable, as permitted by ASHRAE *Standard* 62-1989) to limit the carbon dioxide concentration.

3.4 ROOM AIR DISTRIBUTION

3.4.1 Predicting Thermal Comfort

ASHRAE *Standard* 55 (1992), Thermal Environmental Conditions for Human Occupancy, states "there is no minimum air movement that is necessary for thermal comfort," although this is contrary to the experience of many building operators (see Section 3.4.2). The standard further prescribes a maximum rate of air movement of 30 fpm for winter and 50 fpm for summer. Those values essentially agree with the data found in Chapter 31, Space Air Diffusion, in ASHRAE *Fundamentals* (1989a).

That chapter describes different methods of room air distribution. It also presents a space air distribution performance index (ADPI), which permits predicting the degree of comfort that can be achieved with a given supply air distribution design. The higher the index, the more uniform the conditions of comfort as determined by temperature variation from a control value and air velocities at various locations in the occupied zone. Using the guidance of that chapter, you can develop air distribution layouts that are likely to maintain a high level of thermal comfort. Note that the ADPI concept is based on research using a space (or control) temperature of 76°F. Therefore, the predictions of success apply only if the average space temperature is close to that value. For instance, if a room is at 68°F, it is likely to be considered uncomfortably cool even though the ADPI is quite high.

3.4.2 Minimum Air Motion

Low air velocity affects the ability to maintain uniformity of a comfortable temperature throughout the occupied zone and the dilution of contaminants generated within that zone. Occupant comfort has been reported to suffer as a consequence of low total supply airflow in the space, even when the space temperature is well inside the comfort envelope. Because of this, many designers have adopted a minimum total supply airflow of 0.6 to 0.8 cfm/ft^2 for office applications. These values are based on an all-air system with conventional supply outlets. They can be reduced when outlets with high induction ratios are employed, since they increase the average room air motion.

Minimum room air circulation is particularly important in the case of VAV systems when the conditioned air supply is excessively reduced in response to reductions in the sensible cooling load. Fan-powered terminals can be used to maintain a minimum circulation rate. When using outlets with horizontal discharge patterns, consider the following points:

1. An outlet with a relatively low throw coefficient will produce a smaller absolute change in throw values with variation in volume and thus tends to minimize the change in air motion within the occupied space due to changes in airstream pattern.

2. Choose outlets for relatively small quantities of air. This will make absolute values of throw vary least with variations in outlet flow rate. If the application requires modular outlet arrangements for occupancy flexibility, as with diffusers in combination with light troffers or with air-bar ceiling suspensions, the number of outlets does not need to be increased.

3. All outlets have a tolerance in throw and pressure drop that permits their use without concern for small volume reductions. However, modular ceiling slot-type diffusers have the most desirable characteristics for large volume reductions. To have maximum leeway for airflow reduction, select diffusers for full volume at full capacity and maximum permissible outlet velocity.

In cold-air systems, air is supplied at low temperatures, *e.g.*, less than 50°F, to reduce both capital and operating costs compared to systems using conventional supply air temperatures (see Section 9.3.8). Here again, special care in selecting supply air terminals ensures adequate air circulation and avoids occupant exposure to cold airstreams.

Another cause of occupants' complaint, not always coincident with uncomfortable space temperature, is shoulder-high partitioning. In such offices, occupants may spend long periods sitting inside a box that is fully enclosed except at the top. The ventilation effectiveness of an otherwise excellent space air distribution system can be defeated by such partitioning. The solution that will yield adequate air motion in the occupied zone (the box) to some extent rests with the design of the partitioning system. Nonetheless, prime responsibility for the environmental conditions within such cubicles rests with you as the HVAC designer. Maintaining a good rate of air circulation within the total space will be a partially mitigating factor.

Air distribution in specific occupancies, such as theaters, hospital operating rooms, laboratories with exhaust hoods, and clean rooms, requires special analysis of both the air supply and return or exhaust distribution systems.

Air-and-water systems inherently supply less air to the occupied space than all-air systems. Air circulation is enhanced by the terminal unit if it is an induction or fan-coil unit. Radiant panel systems usually do not contribute to air circulation.

3.4.3 Jet Stream Air Distribution

Some all-air systems have air jets that sweep over the occupant's workstation. Flow and direction are manually adjustable to provide individual control. This refined form of spot cooling has the advantages of reduced cooling load and total airflow. As of 1990, this application was not common in buildings in North America.

3.4.4 Return Air

As a rule, the provisions made for return air do not materially influence space comfort. An exception is the placement of return air grilles too close to supply air terminals, which may short-circuit space air. Except in unusual circumstances, return air finds its way back to the air-handling unit. Recommended face velocities at the return air inlet are given in Table 5, Chapter 32, Duct Design, in ASHRAE *Fundamentals* (1989a).

3.5 NOISE AND VIBRATION

Air conditioning was once known as the "unseen servant." To offer occupants the complete panoply of comfort, it must be the "unheard servant" as well. This is true except in those cases where air-conditioning noise is discreetly blended to produce a pleasantly balanced level over the audible frequency range and becomes a welcome mask for other disagreeable or distracting noises.

An initial step in design is to establish space noise criteria. These criteria should be communicated to the client early in the design process. All noise-generating sources within the air-conditioning system must be identified, and their magnitude must be controlled. Methods for accomplishing this are described in Section 2.4.3.

3.6 REFERENCES

ACGIH. 1992. *Threshold limit values and biological exposure indices for 1992-1993*. Updated annually. Cincinnati, OH: American Conference of Governmental Industrial Hygienists.

ASHRAE. 1989a. *Handbook—Fundamentals.*

ASHRAE. 1989b. Energy conservation in new building design. *Standard* 90.1-1989.

ASHRAE. 1989c. Ventilation for acceptable indoor air quality. *Standard* 62-1989.

ASHRAE. 1992. Thermal environmental conditions for human occupancy. *Standard* 55-1992.

Dubin, F.S. and C.G. Long, Jr. 1978. *Energy conservation standards for building design, construction and operation.* New York: McGraw-Hill, 164.

Green, G.H. 1979. The effect of indoor relative humidity on colds. ASHRAE *Transactions* 85(1).

Green, G.H. 1982. The positive and negative effects of building humidification. ASHRAE *Transactions* 88(1): 1049.

Chapter 4 LOAD CALCULATIONS

4.1 DEFINITIONS

The following terms are commonly used to describe heating and cooling loads.

Instantaneous load occurs during a defined time step, usually one hour.

Dynamic load is a load the intensity of which varies in time; it is usually measured in small time steps, such as seconds or minutes.

Coincident load is a load that occurs at the same time as another load, such as the wet-bulb temperature that is coincident with an extreme dry-bulb temperature.

Peak load is the largest load occurring in a space, a zone, or an entire building. In the latter case, it is also the maximum simultaneous or coincident load.

Diversified load is the portion of the sum of the peak loads that is coincident. Load diversity accounts for the fact that the peak loads of different building zones do not occur simultaneously. Therefore, the building peak load is generally smaller than the sum of the zone peak loads.

Block load is the diversified load that occurs in a portion of a building or in the entire building. It is the maximum load that the equipment (air handler, chiller, pump, fan) actually sees. It is sometimes called the *refrigeration load*.

4.2 OUTDOOR AND INDOOR DESIGN CONDITIONS

Before performing heating and cooling load calculations, both outdoor and indoor design conditions must be defined. Outdoor conditions can be obtained from Chapter 24, Weather Data, in ASHRAE *Fundamentals* (1993). Typical design conditions recommended by ASHRAE *Standard* 90.1 are the 97.5 and 2.5% values listed in that chapter. For special projects where precise control of indoor temperature and humidity is required, the 99 and 1% values should be used.

Indoor design conditions are governed either by occupant comfort conditions or by special requirements for the materials or processes housed in the space. In most buildings, such as offices and residences, indoor comfort is the only requirement, and small fluctuations in both temperature and humidity are not objectionable, as illustrated by Figure 3-1. In other occupancies, however, more precise control of temperature and humidity is required; refer to the appropriate chapter in ASHRAE *Applications* (1991). ASHRAE *Standard* 90.1 specifies 72°F for winter and 78°F for summer indoor design conditions. Special state energy codes and particular owner requirements may also affect the establishment of indoor design conditions.

Heating and cooling sensible and latent loads depend on heat transfer through the building envelope; solar heat gain through windows and skylights; infiltration through openings in the building envelope; internal heat gains due to lighting, people, and equipment in the conditioned space; and outside air for ventilation and building pressurization. They are described in detail in Chapter 25 and Chapter 26 in ASHRAE *Fundamentals* (1993). Typically, the heating load calculation does not include the heat gains to the space, unless they occur during occupied periods. Even then, credits are taken only for an estimated portion of lighting, occupancy, and equipment gains but not for solar gains.

4.3 EXTERNAL LOADS

External loads are highly variable, both by season and by time of day. They cause significant changes in the heating and cooling requirements, not only of the exterior building spaces, but of the total building cooling plant.

4.3.1 Conduction through Building Envelope

Most new buildings are well insulated, and conduction loads through walls and roofs are small compared to those through windows and skylights. New glass and plastic products, however, are improving the performance of these envelope openings, and you should obtain the technical data directly from the manufacturers, remembering to make allowances for conduction through the metal mullions.

4.3.2 Solar Heat Gain through Glazing

The sun is often the major cooling load, and the solar heat gain is highly variable with time. Therefore, carefully analyze the heat gain through windows, skylights, and doors. Facade self-shadowing, adjacent building shadowing, and ground reflections from water, snow, and sand must be considered. Heavily glazed areas must be analyzed for occupant comfort near the windows (especially in corner offices). The cool supply air must enter the space in a manner that will offset the warmed glass. Also, consider providing exterior or interior shading devices to avoid direct radiant transfer to the occupants.

4.3.3 Ventilation Load

The outside air ventilation load does not have a direct impact on the conditioned space, but it imposes a load on the HVAC equipment. It is introduced through the HVAC system and adds a load (sensible and latent) to the heating and cooling coils, thus affecting the necessary humidification. The amount of ventilation depends on the occupancy and function of each space. Refer to ASHRAE *Standard* 62 (1989) for recommended ventilation rates and the requirements of the local building, mechanical, and energy codes.

4.3.4 Infiltration Load

Most commercial buildings have inoperable sash windows and are pressurized by the HVAC system to reduce the amount of outside air infiltration. It is generally assumed that a pressurized building prevents infiltration (although infiltration will occur over the lower third of a building taller than 80 ft) and, when the system is shut off, the infiltration can be calculated by the crack method. In addition, the highest heating load often occurs in the early morning (2 to 3 A.M.) when buildings are not occupied. It may also occur during occupied hours if the ventilation load, present during these periods only, predominates. In entrance lobbies, determine the infiltration rate based on door-opening rates and pressure differentials due to wind, temperature, and stack effect [see Chapter 23, Infiltration and Ventilation, in ASHRAE *Fundamentals* (1993)].

4.4 INTERNAL LOADS

While external loads can be heat gains or heat losses, internal loads (except for duct losses) are always heat gains.

4.4.1 Heat Gains from Occupants and Their Equipment

Heat gains from lights, people, and equipment (*e.g.*, computers, copy machines, and process equipment) must be determined. The lighting level and fixture type are usually determined by the electrical engineer, and the HVAC engineer must determine the amount of heat gain to the space. For example, in office buildings, the radiant and convective heat losses from lighting fixtures must be quantified to determine how much can be captured by upward flow of the air around the fixture into the hung ceiling and how much is radiated to the floor to be added to the space load for air volume calculations. Heat gains from people are, of course, activity related: athletes in a gymnasium release eight times the amount of heat released by a seated audience (see the tables in Chapter 8, Physiological Principles, Comfort, and Health, and Chapter 26, Air-Conditioning Cooling Load, in ASHRAE *Fundamentals* (1993). The number of people in a space is obtained by seat counts, owner estimates, and the design engineer's experience. For example, how many people are crowded around the dice tables in a gambling casino?

The heat gains from equipment in offices are developed from simple uniform allowances per square foot, but in other occupancies, estimating these gains is more complex. For example, in computer rooms, research laboratories, and processing plants, the heat released from equipment creates most of the cooling load. In a complex laboratory or manufacturing facility, it is important to understand the operation of each piece of equipment, its connected load, sensible and latent heat release to the space, and the length of its operating cycle. For example, an electric oven will use the maximum power only during start-up periods, but, once it reaches its set temperature, it will use only what is required to make up for the heat loss to its surroundings. The heat gain to the space, therefore, is the heat loss from the oven, not the connected load or the power required during start-up. Where several pieces of equipment are in one zone, you must determine the use diversity (*i.e.*, how many units will be operating at the same time) and, with assistance from the owners and users (or from experience), develop a schedule of use to establish the maximum heat gain to the space. Numerous other miscellaneous loads in special occupancies cannot be ignored, such as cooling benefits from open freezers in markets or heat and humidity gains from indoor swimming pools and water fountains.

4.4.2 Heat Gains from Fans and Pumps

Supply and return air fans add heat to the airstream, depending on static pressure and fan/motor efficiencies. If the motor is located in the airstream, the heat released due to motor inefficiency must be included in the load calculations. It is the quantity q_{sf} in Figure 6-1, Chapter 6.

$$\text{fan BHP} = 0.000157 \times \text{cfm} \times TP/\text{efficiency}$$

$$\Delta t = (\text{thermal input})/(1.1 \times \text{cfm}) = \text{fan BHP} \times 2545 \text{ Btu/hp}/(1.1 \text{ cfm})$$

$$\Delta t = 0.000157 \times 2545\ TP/(1.1 \text{ efficiency})$$

$$\Delta t = 0.363\ TP/(\text{motor efficiency} \times \text{fan efficiency} \times \text{drive efficiency})$$

This heat gain is approximately equal to 0.7°F per 1 in. of water static pressure. When the motor is mounted outside the airstream, $\Delta t = 0.363\ TP/(\text{fan efficiency})$, which is approximately 15% less.

The only portion of fan energy that is not converted to a rise in air temperature is the velocity pressure, which is the energy required to increase the air velocity at the fan. The heat of friction from the gradual pressure drop in any duct causes no temperature rise along the duct because this heat is equal to and nullified by the expansion cooling effect of the static air pressure reduction.

If the supply fan blows air through the coil, all fan energy except that from velocity pressure is absorbed in the cooling coil and does not affect the air supply load. You must, however, increase the supply air capacity to make up for the fan energy gain.

In all-water and air-and-water systems, include the heat added by the circulating pumps to the chilled water, which can be assumed to be approximately 1.5% of the building load.

4.4.3 Duct Losses

Chapter 32, Duct Design, in ASHRAE *Fundamentals* (1993) describes methods for calculating heat gains, which are approximately equal to 3% of total system air volume. Greater leakage may occur, however, and could affect the design calculations (see quantity q_{sd} in Figure 6-1, Chapter 6). It is customary to specify leakage testing for high-pressure systems but not for low-pressure systems. Leakage in untaped low-pressure duct systems can run from 10 to 20% of total air volume. Low-pressure lighting troffer connections can leak as much as 35% of terminal air volume unless taped and sealed. Leakage does not affect the overall system loads unless it occurs within unconditioned spaces, but it always affects supply air quantities. Leakage through zone control dampers affects supply air quantities for multizone and dual-duct systems, as explained in Chapter 6. Suitable rules for duct design are given in Section 5.7.

4.5 CALCULATION METHODS

A wide variety of heating and cooling load calculation procedures are available to the designer. Most of the procedures are based on ASHRAE publications with simplifications and adjustments for specific applications. For example, load-dominated buildings with small envelopes are often calculated manually for a single design-day peak hour with the addition of a safety factor when selecting the equipment, while loads on large, multizoned buildings are calculated by computer.

Do not readily accept the results of computer calculations. Until the 1960s, all HVAC calculations were performed manually. A good understanding of building behavior was gained through repeated manual calculations. Although computers have shortened calculation time and improved analytical techniques, review manual calculation methods first. Carefully review the computer program documentation to determine what calculation procedures are used, compare the results with manual calculations, and conduct sample tests to confirm that the program delivers what it claims.

The software vendor does not accept responsibility for the correctness of the calculations or the use of the program.

Most computer programs only calculate cooling loads for 12 daylight hours, 1 day a month for 6 months of the year, because the position of the sun in the spring is the same as in the fall. The user is required to input the monthly high and low temperatures, and the program then generates the daily temperature profiles using the ASHRAE typical cooling design-day temperature curve. The summer outdoor design temperatures may be obtained from Chapter 24, Weather Data, in ASHRAE *Fundamentals* (1993), but the outdoor design temperatures for other months have to be obtained from local weather statistics. To eliminate the weather data problem, some programs provide 12 monthly typical-day weather profiles for most major cities in the United States and foreign countries, allow user modifications for specific locations, and then calculate the cooling load for all 24 h of the day.

The largest cooling loads in conventional residential/office buildings are due to solar radiation through glass and internal loads. In many special-purpose buildings (factories, shopping malls), internal loads predominate. Inputs affecting these loads must therefore be developed with care. They include window and skylight sizes, properties of glazing materials, interior and exterior shading devices, lights, sensible and latent loads due to people and equipment, and operating schedules. Some programs completely ignore exterior facade shadowing, most simulate overhangs and side fins, but only a few account for adjacent building shadowing. Bear in mind that connected electrical loads are not hourly demand loads, and most equipment does not operate at full load during the building's operating hours.

4.6 COMPUTER INPUTS AND OUTPUTS

The accuracy of the computer outputs depends on the user's inputs and the programmed calculation procedures. The inputs include not only the typical architectural parameters of areas and building materials but also indoor design conditions, occupant ventilation requirements, internal loads, and their schedules of use.

4.6.1 Computer Inputs

All computer programs require inputs to perform the necessary calculations; some are mandatory and others are not. If the user does not provide data for nonmandatory inputs, the built-in program default values are used. As the designer, you must understand and accept the appropriateness of the default values or override them by inputing correct values. Most of the load programs require user inputs of the basic properties of construction types. When the proposed construction is not defined in Chapter 26, Air-Conditioning Cooling Load, in ASHRAE *Fundamentals* or in other publications, you may either use data from a similar type of construction or develop the required input data by hand or by computer. For this purpose, you may use component properties listed in that chapter, in Chapter 22, Thermal and Vapor Transmission Data, or in other publications. Some programs allow such data to be input by reference to a name.

In commercial buildings, internal loads are a large portion of the cooling load. Most programs allow input of these loads in actual quantity per square foot of floor area. If the load is specified on a square-foot basis, the program will use the zone area to calculate the load. Be sure to enter the 24-h maximum use schedules for people, lights, and equipment so that the load diversity is included in the HVAC system and central plant calculations. Remember that maximum use schedules are not the same as average use schedules. Maximum schedules are used to determine peak cooling loads in each space, while average schedules are used for energy calculations. If the use schedules are not available, assume a constant load profile and add an experience diversity factor to the system and central plant load calculations. Energy calculations, unlike load calculations, are significantly affected by equipment control schedules.

Most programs use the air change method to calculate the infiltration load. Input the number of air changes per hour, and the program multiplies this value by the input zone area and height. In specifying the zone height, distinguish between the zone's floor-to-ceiling height and the building's floor-to-floor height. The zone height must correspond to the air change rate for the same height. Most programs allow the user to specify the ventilation load in cubic feet per minute (cfm), cfm per square foot, or cfm per person. You may also input the zone exhaust in cfm. The program compares the ventilation rate with the exhaust rate and selects the larger of the two as the outside air requirement.

Select the HVAC system and zones that will be served by that system. For each system, input the maximum and minimum supply air temperatures or the cfm supplied to each zone. If the supply air temperatures are input, the program will determine the zone cfm based on the calculated zone sensible heat. After the supply and outside air quantities have been determined, the program will calculate the fan heat and perform the necessary psychrometric analysis.

4.6.2 Computer Outputs

All computer programs provide a summary of peak sensible and latent cooling loads and sensible heating loads for each thermal zone. The cooling load is itemized by roof conduction, wall conduction, floor conduction, glass conduction, glass solar transmission, people, lights, equipment, and infiltration. The time and date when the peak cooling loads occur are usually given, and some programs also indicate the outdoor conditions at time of peak. Most programs also provide psychrometric data for each HVAC system, including state points for room air dry-bulb temperature and humidity, outside and supply air cfm, outside and mixed-air dry- and wet-bulb temperatures, and cooling coil entering and leaving conditions. These data are necessary for the final selection of heating and cooling coils.

Do not assume that the computer output is correct because frequently there are input errors, and there can also be errors in the computer program itself. The outputs must be studied in detail to determine their correctness, and any inconsistencies must be resolved. It is very important to check some of the outputs with hand calculations and review the Btu/ft^2 and cfm/ft^2 results for reasonableness. Also, realize that the program output cannot be used directly if the zone requires close control of temperature and humidity, and that additional psychrometric analysis is required to establish the final HVAC system design in such cases.

Chapter 5 COMPONENTS

Components of air-conditioning systems are treated several chapters in the *Systems and Equipment* volume of the ASHRAE *Handbook* (1992). That volume should be your primary reference because this manual covers only a few items of particular importance for system design.

5.1 COOLING EQUIPMENT

5.1.1 Compression Refrigeration

The prevalent method of air conditioning uses the thermodynamic cycle of compression refrigeration. For small units, reciprocating or rotary compressors are used; medium-sized units frequently employ screw compressors; and large units use centrifugal compressors. Reciprocating compressors for air conditioning require approximately 1 kW/ton, while centrifugal units can require much less.

The centrifugal compression system for building cooling usually consists of one or more systems that either directly cool air (DX = direct expansion systems) or water (chiller), which is then conducted to the conditioned spaces where it cools air. Each system is composed of a centrifugal compressor driven by a prime mover (usually an electric motor using as little as 0.65 kW/ton) through a speed-increasing gear. The compressor takes its suction from a heat exchanger (evaporator) and discharges it into a second heat exchanger (condenser). Condensed liquid passes from the condenser to the evaporator through a high-pressure float drainer (receiver) and a throttling valve, thus providing a closed refrigerant circuit.

A refrigeration system operating as a chiller works against *thermal lift*, the difference in temperatures between return warm water and chilled supply water. This is usually 10 to 15°F but may be as high as 20°F or even 24°F in special cases. Most buildings operate with a 42 to 48°F chilled-water supply. Raising the chilled-water temperature 1°F saves approximately 1.75% of energy use. The advantages of using ultra-low chilled-water temperatures are discussed in Section 9.3.8.

Split systems are DX systems in which the compressor and condenser are placed at different locations. (As a rule, the compressor is located near the condenser.) The upper limit for total distance between them is 300 ft, but the only true limit is based on careful design of the refrigerant piping. The compressor discharge line should be sized for a gas velocity of no less than 1000 feet per minute (fpm) under minimum load with upward gas flow and no less than 500 fpm with horizontal flow. The maximum recommended flow velocity under full load is 4000 fpm because higher velocities generally cause excessive pressure drop and noise.

The upper limit for vertical lift (difference in elevation) is 60 ft. When the net lift is no more than 8 ft, no U-traps are required to ensure return of oil to the compressor (Figure 5-1). However, in these cases, the pipe must pitch away from the compressor at least 0.5 in. for every 10 ft of horizontal run. Suction and hot gas risers that exceed 8 ft of net lift require a trap at the base of the riser and an additional trap at each 25-ft interval of net lift (Figure 5-2). If the compressor/evaporator is low, the hot discharge gas must have double suction risers to capture the oil in a *P-trap* and lift it to the condenser for eventual return to the compressor. In this case, always use an oil separator directly behind the compressor discharge valve and an inverted P-trap entering the condenser with a liquid shutoff valve leaving the condenser. A P-trap consists of three line-size elbows arranged to trap oil in two of the elbows before the horizontal run. A smaller diameter bypass permits gas to continue flowing until sufficient pressure is built up to propel the trapped oil upward to the condenser.

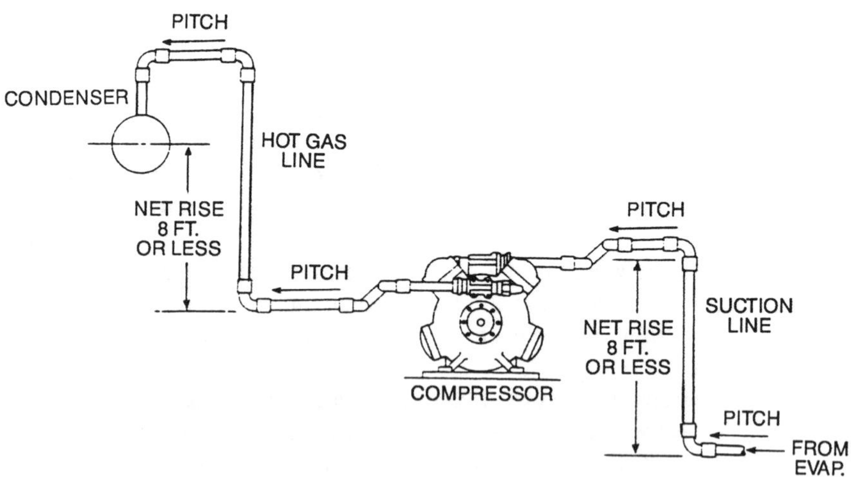

Fig. 5-1 Refrigerant Line for Low Lift

For prudent design, always include a receiver between the condenser and the evaporator. This is a horizontal tank that allows separation of refrigerant gas and liquid; it serves as a reservoir for the refrigerant and accommodates changes in inventory in the refrigerant loop under different operating conditions.

Receivers should be designed for storing the entire refrigerant charge when the system is shut down for the winter or when high-side components are removed for repair. This avoids the need for venting refrigerant, a practice no longer permissible because of its harmful effect on the stratospheric ozone layer. Size the receiver to be no more than 80% full, allowing 20% for expansion.

The *Uniform Building Code* (ICBO 1991) specifies minimum ventilation requirements in the compressor room for dilution if refrigerant is lost. The code also prescribes the proper size for the mechanical room exhaust fan and limitations on the auxiliary equipment that may be installed in that room. Follow local codes if they are more restrictive than national or regional codes. It is recommended that two "break-glass stations" be provided, one at each exit door, to start the exhaust fan in case of an emergency. Be sure to provide sufficient makeup air, which need not be conditioned.

Predominant refrigerants have been R-11, R-12, R-113, R-114, and R-115, depending on system size; however, all these refrigerants are scheduled to be phased out in

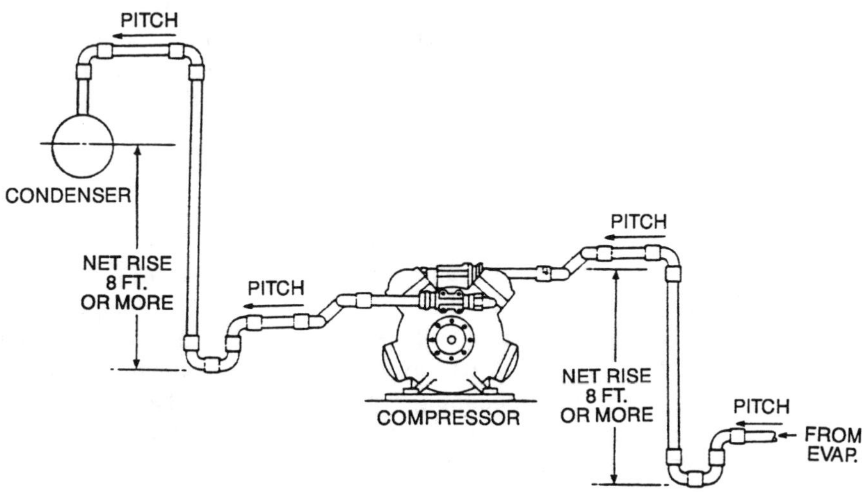

Fig. 5-2 Refrigerant Line for High Lift

accordance with the Montreal Protocol. This international treaty, signed in 1987 and ratified since then by most industrial nations including the United States and Canada, aims to reduce the use of chlorofluorocarbons (CFCs) in order to protect the ozone layer of the stratosphere. Among possible alternative refrigerants are R-22 for small or medium-sized systems, R-123 to replace R-11, and R-134a to replace R-12. Many chemical companies and producers of compressors and cooling systems are performing extensive research and development to find and test alternative refrigerants. Keep informed on current progress in that area.

5.1.2 Absorption Refrigeration

The absorption refrigeration cycle is mostly employed in large units. Often the initial cost of the two types of refrigeration systems are within 15 to 20% of each other, the compression cycle being less, except in rare circumstances where a gas utility offers a partial subsidy as an incentive. Due to the greater heat rejection of absorption systems (generally 31,000 Btu/ton) compared to compression systems (15,000 Btu/ton or less), cooling towers, condenser water pumps, and other auxiliaries must be larger and are therefore more expensive for absorption systems than for compression systems. Hence, any economic comparison depends greatly on relative energy costs. Absorption systems can be direct-fired (when a heat source is directly applied to the machine) or indirect-fired by steam or hot water generated by gas or oil from another source. They also require a few electric auxiliaries. Since most air-conditioning ton-hours occur in the summer months and many gas utilities often have a cheap rate during that season, absorption may be economically attractive under those conditions and should be evaluated. Also, absorption machines do not use any CFCs, which avoids the restrictions placed on those chemicals.

5.1.3 Evaporative Cooling

The application of evaporative cooling equipment for use with air-conditioning systems has decreased since the 1950s, when the emphasis on initial cost overwhelmed consideration of operating cost. Building developers were writing leases that simply passed on any increases in operating cost to the tenant. However, depending on the building's operating hours and location, evaporative cooling can be applied effectively (for example, in the southwestern United States, where low outside wet-bulb temperatures are prevalent, or in other locations where wet-bulb temperatures are below 55 to 60°F for a substantial number of hours, or the building is to be air conditioned at night, and there is a fairly wide diurnal swing).

Two general types of equipment are available for evaporative cooling: spray-type air washers and wetted media-type coolers. In all cases of evaporative cooling, provide for 100% outside air and exhaust capability; therefore, the cost, space, filter capacity, etc., must be weighed against the energy savings of fewer hours of operation of the chiller system.

5.1.3.1 Spray-Type Air Washers

Spray-type air washers are versatile in that they can use chilled water from a refrigeration system when the outside wet-bulb temperature is high, but they can also perform without chilled water when that temperature is suitable to condition the air without refrigeration.

Two types of spray air washers are available. One type is the "full open" spray washer. It is generally either 7 or 11 ft long in the direction of airflow, depending on whether it has one or more spray banks. Air passes through the air washer at about 500 fpm and passes through a spray bank of water. The direct contact of air with the water droplets provides a more efficient heat exchange than air that simply passes over finned coils. Depending on air velocity and spray density (gpm/ft^2), the adiabatic efficiency (the ratio of actual dry-bulb depression to the difference between entering air dry-bulb and spray water temperature) can be as high as 98%.

Another type of air washer is the "short coil" washer. It consists of a conventional extended finned coil bank placed in a casing with a deep water pan and an opposing spray bank. The short coil washer has a much lower spray density than the spray washer and only requires 4 ft in the direction of airflow. When the outside air wet-bulb temperature is not conducive to evaporative cooling, the finned chilled water coils do the job. When the outside air wet bulb is sufficiently low, the chilled water coils are shut off and the spray of water cascading over the extended coil fins provides the cooling. The closed circuit chilled water of the short coil washer requires less chilled water pumping energy, but, because of lower spray capacity, it has an adiabatic efficiency of only 50%, thereby requiring more hours of active refrigeration operation.

Water reservoirs require constant maintenance to prevent corrosion and a buildup of bacteria and fungi that would otherwise be dispersed into the atmosphere during the evaporative cooling process. When this is not done, air washers can become a potential source of biological contaminants and a cause of the sick building syndrome.

5.1.3.2 Wetted Media-Type Coolers

Because wetted media coolers use nonrefrigerated water, they are adaptable to air-conditioning systems only in those parts of the southwestern United States where very low ambient wet-bulb temperatures prevail year-round. In a climate like that of

Phoenix, Arizona, a dual system of air conditioning may be indicated under certain conditions—wetted media for 10 months of the year and compression refrigeration during the 2-month humid season. The extremely low operating cost of the wetted media cooler justifies the higher initial investment.

The systems described thus far are *direct evaporative coolers* because water is injected into the primary airstream, which is cooled adiabatically. In *indirect evaporative coolers*, secondary air (outdoor air or exhaust air) on one side of a heat exchanger is cooled evaporatively. The primary (supply) air is on the other side of the heat exchanger and is indirectly cooled through the heat exchanger walls; thus, no moisture is added to the primary air whose enthalpy is decreased. The leaving dry-bulb temperature of the primary air can be no lower than the entering wet-bulb temperature of the secondary air. One example of an indirect evaporative cooler used as a precooler is shown in Figure 5-3. Other examples are given in Chapter 19, Evaporative Air Cooling, in ASHRAE *Systems and Equipment* (1992). That chapter also covers recent technological improvements, which broaden the scope of application of evaporative cooling through the use of two-stage indirect/direct evaporative cooling and even three-stage cooling.

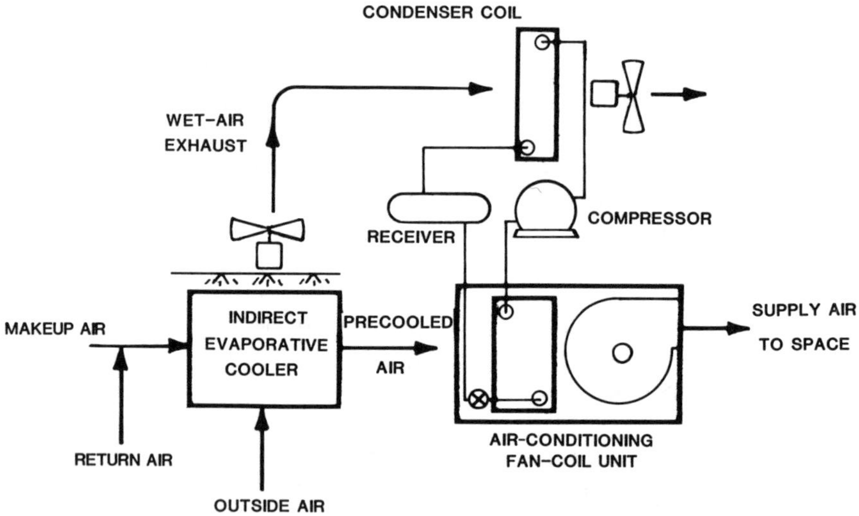

Fig. 5-3 Indirect Evaporative Cooler Used as Precooler

5.1.4 Desiccant Systems

In desiccant systems, air is dried by exposure to desiccants, *i.e.*, liquids or solids that have a high affinity for water (hygroscopic substances). The desiccants are dried (regenerated) by heating. Such systems are discussed in Section 9.1 and in Appendix E.

5.1.5 Prime Movers

The predominant prime mover for compressors is the induction electric motor. Where local electric utility tariffs incorporate a penalty for a low power factor, the addition of capacitors or a change to synchronous electric motor drive may be cost-effective.

Where gas is cheap (due to summer utility rates or for other reasons), a combination of gas- or steam turbine-driven centrifugal compression and absorption is sometimes used, particularly for large buildings. This scheme is referred to as a *piggyback system,* in which steam is generated in a boiler and first sent to a back-pressure turbine, which drives a centrifugal compressor. The exhaust from the steam turbine becomes the thermal input to an absorption chiller. When a gas turbine drives the compressor, its exhaust gases generate steam for the absorption unit in a heat recovery boiler. As a simple rule, the steam-turbine-driven compressor is selected to have one-half the capacity of the absorption unit.

Example. For a total air-conditioning load of 900 tons, select a 300-ton centrifugal chiller driven by a back-pressure turbine with a steam rate of 36 lb/hp; couple this with a single-effect, low-pressure, 600-ton absorption chiller having a steam rate of 18 lb/ton. The result is 900 tons using 10,800 lb of steam per hour or 12 lb/ton. This compares with 18 lb/ton for a straight absorption system.

Where district steam is available in the summer at low rates, steam-driven turbines are sometimes used as prime movers for large centrifugal chillers. Here again, the selection of steam versus electric drive depends primarily on the relative utility costs, and a careful analysis should be made to determine the relative merits of the prime movers. On rare occasions and for very large buildings, gas turbines have been used as prime movers for centrifugal compressors, either with or without waste heat boilers.

5.1.6 Variable-Speed Drives

The use of variable-speed drives for compressors, air handlers, and pumps has increased because they have become available at reasonable cost. These drives are solid-state devices that were, at one time, add-ons to conventional equipment. They are now frequently part of the factory-supplied equipment for air handlers and

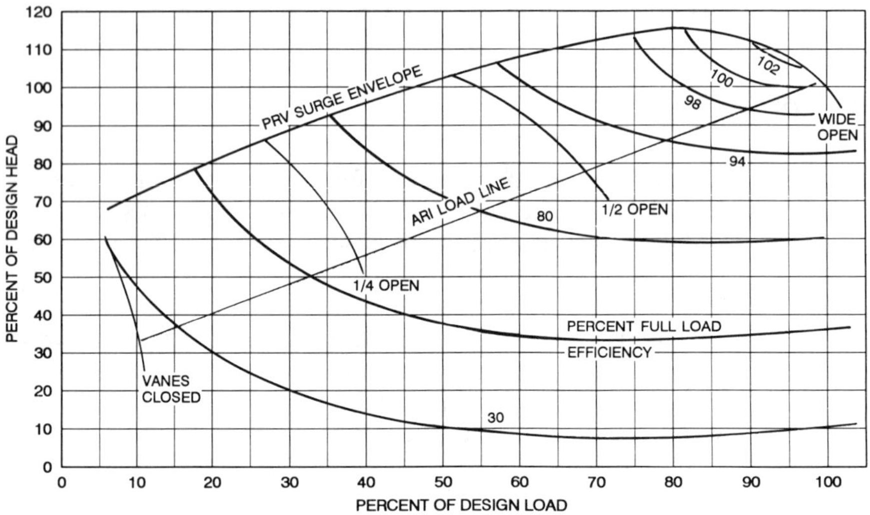

Fig. 5-4 Constant Speed Compressor Performance under Pre-Rotation Vane Control (Courtesy of *Heating/Piping/Air Conditioning* 1984)

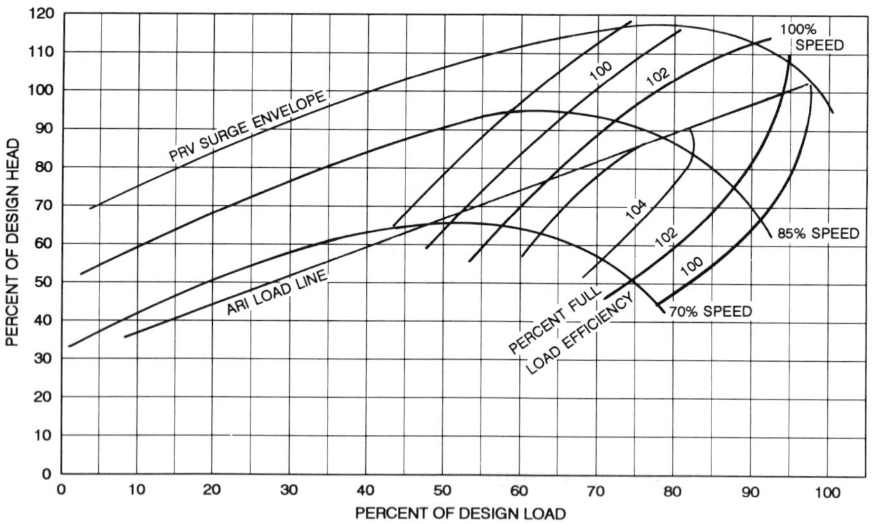

Fig. 5-5 Variable-Speed Compressor Performance (Courtesy of *Heating/ Piping/Air Conditioning* 1984)

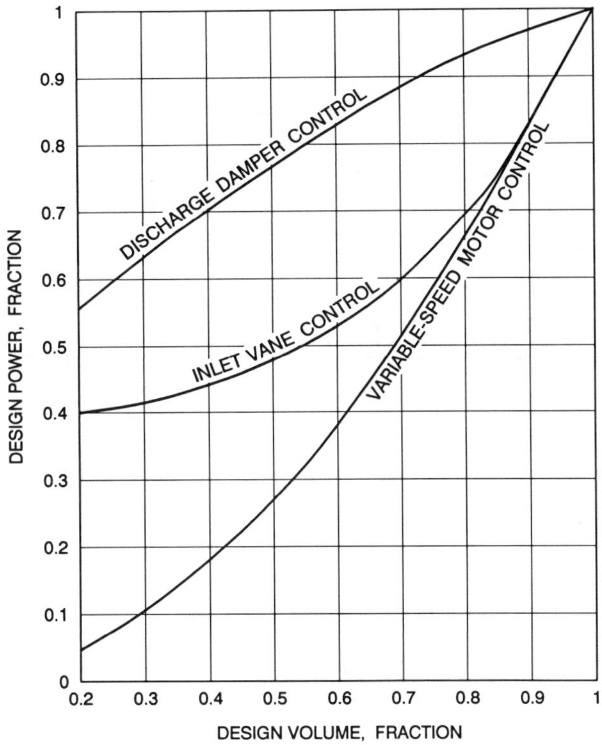

Fig. 5-6 **Fan Power versus Volume Characteristics for Different Control Methods** (Courtesy of *Heating/Piping/Air Conditioning* 1984)

pumps, and one manufacturer offers a variable-speed compressor. Variable-speed drives save energy whenever electric motors run at less than full power. Since most HVAC equipment rarely runs at full power, large energy savings occur with these drives. Typical centrifugal compressor performance curves under pre-rotation vane control and under variable-speed control are shown in Figures 5-4 and 5-5, respectively (*Heating/Piping/Air Conditioning* 1984). Power demands of fans under different methods of capacity control are compared in Figure 5-6.

Of the two methods of controlling fan capacity—inlet vane and variable-speed—the inlet vane choice usually results in an installation of lower first cost. When dropping the flow rate from full capacity, inlet vanes are initially efficient, but, as they progressively close, the reduction in flow rate is attributable more to throttling than to imparting initial spin to the air entering the impeller. Variable-speed control maintains a high efficiency over the entire operating range and is therefore more energy efficient (Figure 5-6).

Variable-speed controls for centrifugal pumps have characteristics similar to those shown in Figure 5-6. System implications of variable-speed pumping are discussed in Section 8.8. Additional information is contained in Chapter 35, Compressors, and Chapter 36, Condensers, in ASHRAE *Systems and Equipment* (1992).

Like all solid-state devices, variable-speed drives are sensitive to differences in phase loads. Therefore, ensure that the phase differences are no more than 10% on circuits incorporating such devices.

5.2 HEATING EQUIPMENT

5.2.1 Boilers

Four principal types of boilers are commercially available to designers of heating plants for large buildings:

Type	Fuel
Sectional cast iron	Oil, gas, or dual fuel
Water tube	Oil, gas, or dual fuel
Fire tube	Oil, gas, or dual fuel
Electric resistance	Electricity

In today's economy, oil, gas, and electricity are the fuels commonly used in new buildings. On rare occasions, coal, wood chips, or refuse are used. The principal reasons for selecting boilers for heating are (1) to provide a plant at a central location that can distribute a secondary heating medium (hot water or steam) throughout a building or a multibuilding complex, (2) to take advantage of the diversity of otherwise individual zone loads, and (3) to centralize maintenance. The choice of fuel should be made on the basis of life-cycle cost, not on first cost alone. With the unsettled world situation affecting the price of fuel, you should, if the budget permits, consider dual-fuel burners (oil or gas) when selecting the most cost-effective of the three principal fuels. When recommending dual-fuel burners, instruct the owner/operator to exercise the alternate fuel two or three times each season to maintain readiness.

If the boiler plant is to be a heat source for a heat pump system, as in the case of a closed loop system, heat pump sizing is important. In spite of the increased first cost, design the plant with a heat exchanger interposed between the boiler and the heat pump system's closed loop. This is a safety precaution, because if the heat pump closed-loop is directly circulated through the boiler and the boiler control fails, the closed loop temperature could rise to a temperature where the refrigerant relief valves in all the heat pump units would pop open—not only creating a great maintenance expense but, more important, causing the heating system to fail and putting the terminal units out of service for a considerable period.

In selecting the type of boiler, consider these factors. Consider installing a sectional cast iron boiler if unskilled operators are probable. Cast iron is least susceptible to oxidation due to poor water treatment. Its principal hazard would be thermal shock. If heavy oil is going to be the principal fuel, consider a fire tube boiler because it is easier to clear the tubes of soot (by rodding them), whereas the water tube boiler requires soot blowers in the exhaust gas stream. However, they are being outlawed in many communities. In choosing fuel burners, consider pressure-atomizing burners up to 500,000 Btu/h. Above this level, investigate air-atomizing, flame retention burners. They become favorable at 1-million Btu/h and above.

If electric boilers are selected, do not consider the low first cost only. Electric resistance boilers vary in cost with the power density of elements (50 W/in^2 is preferable to 100 W/in^2), and a large number of control steps is also desirable.

5.2.2 Furnaces

Furnaces are seldom considered in the design of large office buildings, hospitals, and similar buildings. However, they are often used in small office buildings, large shopping malls, and industrial plants.

5.2.3 Electric Resistance Heating

Heating by centralized electric boilers is discussed in the previous section on boilers. Many designers do not choose centralized direct electric heating with large electric duct heaters because of the lack of localized control and load diversity. If direct electric heating (electric resistance elements in the airstream) has any merit, it is the flexibility it offers in locating small units (baseboard convectors, unit heaters, or terminal duct heaters) in individual spaces and controlling them separately. In the case of terminal duct heaters, provide flow sensors to disconnect the electric duct heaters whenever airflow ceases in order to avoid a fire hazard.

While, at first glance, electric resistance heating may not appear to be economical, carefully examine the local electric utility tariffs to determine whether this first impression is justified. Many utilities offer extremely attractive incentives to all-electric buildings, and additional incentives for thermal storage (Section 9.2) are frequently available. Since the first cost of electric resistance heating is usually the lowest alternative, this option deserves a careful economic analysis. A guide to electric utility tariffs is given in Section 2.8.5.

5.2.4 Electric Heat Pumps

Heat pumps are increasingly being used in building air conditioning, partly because of the evolution of energy conservation codes and their influence on architectural design. In the quest to reduce energy use, architects are reducing glass areas in walls and replacing them with better insulated opaque walls. Remaining glass areas are often double-glazed to reduce winter heat loss; and, whereas transmission heat gain, solar heat gain, and internal heat gains are additive in determining the building's cooling load, internal heat gains during occupied hours are subtracted from heat loss in winter. Consequently, when the transmission loss is greatly reduced by added insulation in the skin, the building's heating-to-cooling ratio is materially reduced—often to the point where no heating is needed during occupied periods. Closed loop heat pump systems (see Chapter 7) are frequently used advantageously under those conditions.

In large buildings, the perimeter-to-floor area ratio is relatively small compared to that in small buildings. Consequently, internal gains attributable to lights, computers, occupants, and other heat sources are high compared to transmission losses. Also, during unoccupied periods, the electrical demand attributable to these heat sources is now available for increased heating of the perimeter by heat pumps. This so-called *umbrella of demand* is the principal reason why most heat pumps are electrically driven and supplemented with electric resistance heat.

Heat pumps need not provide 100% of the heating requirements of a building. Frequently, it is more practical to evaluate those areas easily served with heat pumps and to supplement the heat pump with supplemental loop heat or point-of-use heat from another medium (electric resistance, gas) for special areas such as entrance vestibules, etc. Another heat source to be considered for heat pumps is the building exhaust air after it has first been exploited in a runaround system coupled with outside air intakes. These and other heat pump uses are described in Chapter 8, Applied Heat Pump and Heat Recovery Systems, in ASHRAE *Systems and Equipment* (1992). Packaged terminal air conditioners, often employed for cooling core areas of large buildings, and water-source heat pumps, used in closed loops (see Chapter 7), are described in the same volume.

5.3 HEAT TRANSFER EQUIPMENT

5.3.1 Condensers and Evaporators

These types of heat exchangers are described in more detail in Chapter 36, Condensers, in ASHRAE *Systems and Equipment* (1992). Only a few of the most important aspects relating to system design are mentioned here.

In building air conditioning, shell-and-tube condensers are the most prevalent type. Generally, these are available in one, two, three, or four passes on the water tube side, depending on the flow of condenser water versus pressure drop. Frequently, you can choose between two- and three-pass or three- and four-pass with little difference in condenser performance and, by doing so, you can select a water piping arrangement most suitable to the equipment room layout. Remember that large buildings generally require large-diameter condenser water mains, which are both expensive and occupy a large amount of space. As a first cut, design heat exchangers with a 3 to 5°F approach temperature. Shell-and-tube condensers require periodic cleaning of the inside of the tubes to maintain performance. Therefore, on large condensers, consider marine-type water boxes. They permit removal of condenser heads for tube cleaning without dismantling the condenser water piping.

In most urban atmospheres, condenser water circulated over open cooling towers can rapidly build up scale on condenser tubes. It is inconvenient to shut down cooling systems during the summer months, yet tube scale buildup occurs in late summer just when full capacity is needed. Consequently, consider specifying centrifugal chilling systems with a fouling factor of 0.001 rather than the commercial factor of 0.0005, because a 0.0005 fouling factor can be built up during the first weeks of the cooling season.

Most manufacturers of centrifugal or absorption chillers offer more than one condenser size. Due to competition, manufacturers are inclined to consider lowest first cost in their initial product offer. Such offers most often include a condenser that is marginal. Any condenser with limited condenser surface is apt not to meet the full-load capacity rating, which most likely will be needed at the same time the condenser tubes are fouled. Today's large shell-and-tube condensers are manufactured using extruded finned tubes of nearly the same ratio of primary (tube) and secondary (fin) surface. As a general rule, provide at least 6 ft^2 of total external surface (primary plus secondary) per ton of full-load rating for conservative sizing of the condenser. This added condenser surface will pay dividends in two ways: longer tube-cleaning intervals (tube cleaning is labor intensive and therefore expensive) and lower condensing temperatures with attendant reduced energy use. In direct expansion systems, estimate about 1/15 hp per ton of refrigeration for the condenser fan.

> **Raising the evaporator temperature 1°F decreases energy use by approximately 1.75%. Lowering the condensing temperature 1°F produces approximately one-half of this saving.**

Life-cycle costing must be considered when specifying the size of evaporator and condenser surfaces. Absorption systems are more sensitive to condenser tube fouling than compression systems, and you should either devote extra care to either design for easy cleaning or allow for large fouling factors in the analysis. Many systems can be retrofitted with on-line tube-cleaning equipment for low fouling factors. Some manufacturers offer such devices, which maintain a low fouling factor through the entire operating period.

5.3.2 Double-Bundle Heat Exchangers

In recent years, double-bundle condensers for large heat pump systems and heat recovery systems have been used more frequently. As the name implies, two water tube bundles are contained in a single shell. The high-pressure refrigerant vapor discharged into the condenser shell from the compressor surrounds, and is in contact with, both tube bundles. The condensing temperature (vapor pressure) is then based on the number of active tubes. For example, when a double-bundle condenser provides hot water (100 to 110°F) for low-density fin-type baseboard or unit heaters or for reheat hot water, the water is circulated through the winter bundle of the condenser, where it picks up the heat rejected from the condenser. When the evaporator-compressor portion of the system is called on to provide more cooling than can be absorbed by the reheat hot water, the summer bundle is activated by circulating the cooling tower water through it and rejecting the excess heat through the cooling tower. Close control is provided by bypassing a varied quantity of water

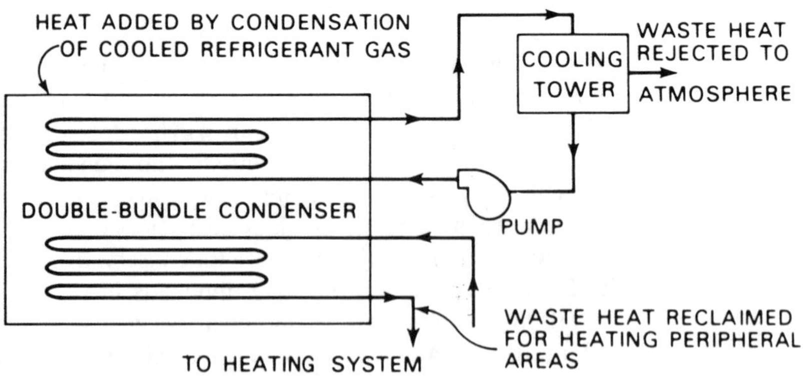

Fig. 5-7 Double-Bundle Condenser (Griffin 1974)

around the cooling tower as needed to sustain the temperature of the reheat hot water leaving the winter bundle (see Figure 5-7).

Double-bundle evaporators have been employed only rarely. In the case of a well-water heat pump providing chilled water in summer and hot water in winter, a double-bundle evaporator and a double-bundle condenser can be used. In summer, the chilled water is circulated through the summer bundle of the evaporator and the well water is circulated through the summer bundle of the condenser to reject the heat. Conversely, in winter, the hot water is circulated through the winter bundle of the condenser and the well water is circulated through the winter bundle of the evaporator to absorb heat from the well water.

Heat exchangers are often used in place of double-bundle condensers. However, this is less efficient because it adds the penalty of an additional temperature difference. In lieu of a double-bundle condenser, an auxiliary condenser is added side by side to the main condenser to act as the winter bundle. With this arrangement, care should be taken in piping from the compressor discharge into the condenser so that the superheat is picked up by the lead condenser, which should always be the winter bundle.

5.3.3 Plate Heat Exchangers

The plate-type heat exchanger can transfer heat from one liquid to another at high efficiency (2 to 3°F approach). It consists of a series of gasketed, embossed metal plates (usually stainless steel) held between two rigid end frames by threaded rods and spanner nuts. The alternate plates create a heat exchanger with a cold medium on one side and a warm medium on the other side of the plate, creating counterflow in very narrow passages. Heat transfer rates are very high due to the large surface area that even a small plate heat exchanger provides and due to turbulence created by embossing the plates with corrugations, which develop the flow into a thin turbulent film that eliminates any stagnant areas. Because of the turbulence created and the high velocity of the liquid film, dirt deposits are minimized so that cleaning outages are reduced. When cleaning is required, the exchanger is simply opened by unscrewing the spanner nuts and spreading the plates to permit easy wash-down by a high-pressure hose accompanied by brushing of plate surfaces if needed. Pressure drops are comparable to those of shell-and-tube heat exchangers because of the many parallel paths existing in plate heat exchangers.

The application of plate heat exchangers to air conditioning is gaining in use, including:

- Cooling chilled water directly with cooling tower water in winter or whenever the temperature of the tower water is less than that of the chilled water return (see Section 9.3.9)

- Using them in place of double-bundle condensers (or double-bundle evaporators) in many heat pump applications

- Heat exchange in a solar collector or other applications requiring heat exchange between glycol and water

- Heat exchange between thermal storage systems (hot or cold) and main hot water or chilled water streams

5.3.4 Cooling Coils

Cooling coils are usually rated within the following limits:

Entering air dry-bulb temperature	65 to 100°F
Entering air wet-bulb temperature	60 to 85°F
Air face velocity	300 to 800 fpm, may be as low as 200 fpm and as high as 1500 fpm)
Refrigerant saturation temperature at coil outlet	30 to 50°F
Refrigerant superheat at coil outlet	6°F or more
Refrigerant entering temperature	35° to 65°F
Chilled-water quantity	1.2 to 2.4 gpm per ton, equivalent to a rise in water temperature of 10 to 20°F)
Chilled-water velocity	3 to 8 fps
Chilled-water pressure drop	< 20 ft of water

However, for special applications, the range in the above design variables may be exceeded [see ASHRAE *Systems and Equipment* (1992)].

5.3.5 Cooling Towers

Cooling towers are used to reject condenser water heat to the ambient air. For large building air conditioning, the vertical counterflow tower is most prevalent. It is equipped with either induced draft or forced draft fans. Since condenser water systems are open systems in which air is continuously in contact with the water, they require a somewhat different approach to pump selection and pipe sizing than do closed heating and cooling systems. The contact with air introduces impurities, which can result in continuous scaling and corrosion. Therefore, aging of the piping must be anticipated, resulting in an increased pressure drop with time and higher than usual fouling factors. Condenser water systems without water-regulating valves do not necessarily require a strainer. If a cooling tower is used, strainers provided in the tower basin or at the pump will usually be adequate. Also:

- Size the tower to avoid windage drift of tower carryover, which might fall on adjacent buildings or parked cars (where cooling tower water can damage paint finishes)

- Provide water treatment and makeup systems, including adding water meters to permit deduction from any sanitary sewer charges that would be based on water intake

- Winterize the tower to allow for ready availability in the event of a sudden change in the weather

Do not:

- Locate cooling towers near fresh air intakes to prevent excessive entering air wet-bulb temperatures and potential bacterial contamination

- Locate cooling towers near the building facade to avoid staining it and to prevent noise entry into occupied spaces

Typical values are per ton of refrigeration and include:

- 2.8 gpm condenser water for vapor-compression systems

- 3.5 gpm condenser water for absorption systems

- 300 cfm outside air

- 1/15 fan hp for induced draft (draw-through) towers

- 1/7 fan hp for forced draft (blow-through) towers

Space requirements include:

- 1 ft^2 per 500 gross ft^2 of building area (8-ft-high towers)
- 1 ft^2 per 400 gross ft^2 of building area (higher towers)

Installation of piping from the tower sump to the pump requires caution. Place the sump level above the top of the pump casing to provide positive prime. Pitch all piping up either to the tower or to the pump suction to eliminate air pockets. Suction strainers should be equipped with inlet and outlet gages to indicate when cleaning is required. Normally, piping is sized to yield water velocities between 5 and 12 fps. Use friction factors for aged or old pipes.

In multiple cooling tower installations, ascertain that sufficient net positive suction head (NPSH) (see Section 5.4.2) for the condenser water pump is provided, and design the connections so that the flow through each cooling tower has the same pressure drop. Install equalizing lines of proper size between towers; otherwise, one tower will fill and the other one will overflow. Winterize the tower piping if necessary or provide for drain-down when pumps shut off. In the latter case, the isolation valves must be located in a heated space. Provide valving or sump pump isolation so that any cell can be serviced while the others are active.

5.3.6 Evaporative Condensers

Evaporative condensers are used to reject heat directly from refrigerant to the ambient by means of the evaporative action of water. Water is sprayed on the condenser coils, and its heat of evaporation, combined with its sensible heat, cools the liquid inside the coils. The water is usually recirculated. An evaporative condenser uses much less makeup water than the cooling tower for a conventional water-cooled condenser (approximately 1.6 to 2.0 gph/ton by evaporation plus 0.5 to 0.6 gph/ton for bleed water). Evaporative condensers are used primarily to cool refrigerant. In such applications, they should be located as close as possible to the refrigeration machinery to keep the refrigerant lines short. Another use is to cool the loop water in closed loop water-source heat pump systems (see Chapter 7).

Figure 5-8 shows two evaporative condenser types using city, well, or river makeup water. Piping should be sized in accordance with the principles outlined in Section 5.3.5 with flow velocities of 5 to 10 fps. Where city water is used, a pump is usually not required since the water arrives under pressure. For well or river water, pumps may be necessary, in which case the procedure for pump sizing generally would be that indicated below for cooling tower systems. For practical purposes, the velocity head is insignificant and can usually be ignored in these systems. Since evaporative

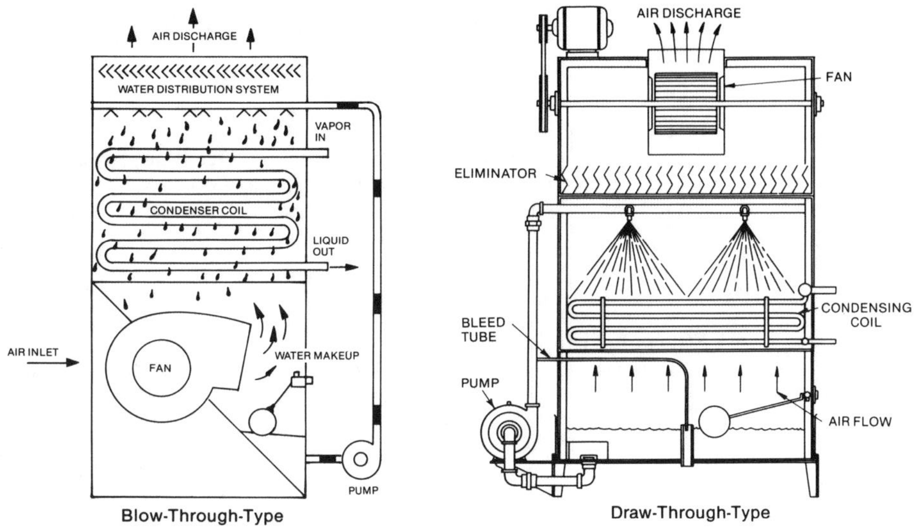

Fig. 5-8 Evaporative Condensers

condensers have little storage capacity, liquid receivers (see Section 5.1.1) should be installed downstream of such condensers to hold the refrigerant until it is required by the system.

5.4 PUMPS

5.4.1 General Characteristics

The different types of centrifugal pumps and their applications to HVAC systems are treated in ASHRAE *Systems and Equipment* (1992). Pumps of that type are most prevalent in HVAC systems. Table 5-1 shows some of the effects on pump performance of variations in system or operating parameters. The use of pump curves for system design is explained in Section 8.8.

Table 5-1 Characteristics of Common Types of Pumps

	Positive Displacement Pumps			Centrifugal Pumps	
	Rotary	Piston	Radial	Mixed Flow	Axial Flow
Flow:	Even	Pulsating	Even	Even	Even
Effect of increasing head—					
On flow:	Negligible decrease		Decrease	Decrease	Decrease
On bhp:	Increase	Increase	Decrease	Small decrease to large increase	Large increase
Effect of closing discharge valve—					
On pressure:	Can destruct unless relief valve is used		Up to 30%	Considerable increase	Large increase
On bhp:	Increases to destruction		Decrease 50 to 60%	10% decrease to 80% increase	Increase 80 to 150%

5.4.2 Net Positive Suction Head

To eliminate cavitation, a certain minimum net positive suction head (NPSH) must be maintained at the inlet side of a pump. The required NSPH for a specific pump is available from the manufacturer, either from catalog data or on request. Although usually given as a single number, it increases with flow. The required NPSH can be considered to be the pressure required to overcome pump inlet losses and to keep water flowing into the pump without the occurrence of vapor bubbles, which cause cavitation. As a result of design and physical limitations, a piping system will produce an available NPSH. For satisfactory pump operation, the available NPSH must always exceed the required NPSH; if it does not, bubbles and pockets of vapor will form in the pump. The results will be reduction in capacity, loss of efficiency, noise, vibration, and cavitation. The available NPSH in a specific system is expressed by:

$$\text{Available NPSH} = p_a + p_s + V^2/2g - h_{vpa} - h_f \quad (1)$$

where

available NPSHA	=	net positive suction head available
p_a	=	atmospheric pressure at elevation of installation (34 ft at sea level)
p_s	=	water pressure at pump centerline
$V^2/2g$	=	velocity pressure at pump centerline
h_{vpa}	=	absolute vapor pressure at pumping temperature
h_f	=	friction and entrance head losses in the suction piping

All the above are expressed in feet of water (gage).

Net positive suction head is not normally a concern in closed systems. It also is not ordinarily a factor in open systems unless hot fluids are pumped, the suction lift is large, or there is considerable friction in the pump suction pipe. Insufficient available NPSH can occur because of undersized piping, too many fittings, if a valve in the suction is throttled, or if the mesh of a strainer on the suction side of the pump becomes clogged.

Example. The pump in Figure 5-9 has a NPSH of 22.4 ft of water. The pressure loss of the suction piping is 4.6 ft of water, and *L* equals 12 ft. Determine the additional lift available from the pump.

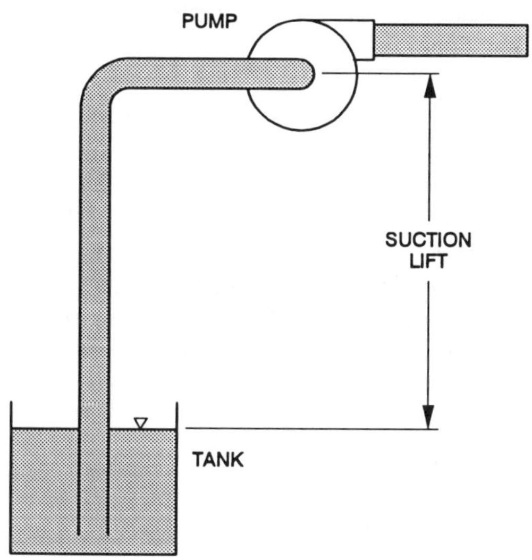

Fig. 5-9 Pump with Suction Lift

Solution: The additional lift equals the NPSH minus the suction head and piping pressure losses. Additional lift = 22.4 – 4.6 – 12 = 5.8 ft of water.

5.5 VALVES

5.5.1 Types

The following valves are used in HVAC systems.

Ball and butterfly valves operate from the fully open to the fully closed position with a 90° turn. They can be used for throttling and offer very low resistance to flow when fully open.

Balancing valves provide flow measurement and adjustments. *Check* valves prevent flow reversal. They can be spring-loaded to close before an actual reversal of flow has occurred.

Gate valves are used to open or close fully and are not used for throttling. They have low resistance to flow when fully open.

Globe valves are used for throttling and have high resistance to flow.

Pressure-reducing valves maintain a constant downstream pressure under varying flow conditions.

Pressure-relief valves open when the system reaches a preset level to prevent overpressuring a system.

Control valves provide automatic operation by means of an actuator mounted to the valve body to provide a push-pull or a rotary motion to the valve stem. Control valves are available with electric, hydraulic, thermostatic, or self-operated actuators. A control valve provided with an automatic controller and sensor becomes part of an automated control system (see Chapter 10).

5.5.2 Characteristics

The internal design of a valve determines the characteristics of the percent flow versus the percent stroke as the valve is positioned from open to closed. Three typical characteristics—quick opening, linear, and equal percentage—are shown in Figure 5-10 at constant pressure drop. The *quick-opening* characteristic is usually applied

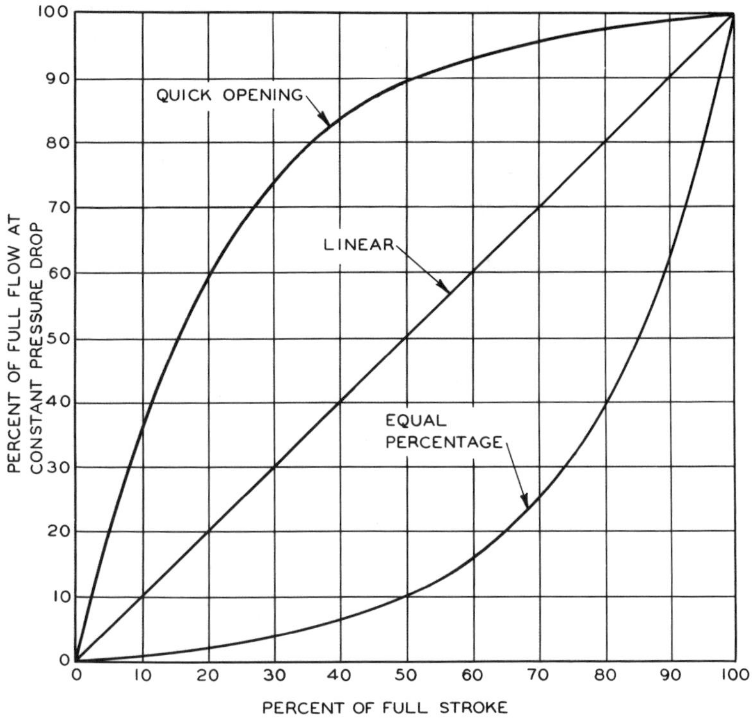

Fig. 5-10 Typical Valve Flow Characteristics

where the process requires nearly on-off operation. For example, a preheat coil in the outside air duct may be operated by a quick-opening valve, the controller of which has its sensing bulb in the outside air.

A *linear* valve gives a uniform percentage change in flow for each percentage change in valve position. A typical application may be on a steam-heating coil operated by a controller whose sensor is on the discharge side of the coil. Since the coil output is uniform with steam flow, the linear valve provides a uniform change in temperature as the proportional controller increases or decreases the valve position to match the load to maintain its temperature setpoint.

An *equal-percentage* valve gives a very gradual change in flow for each percentage change in valve position near the closed position and a very rapid increase in flow near the fully open position. An example is the application to the control of a hydronic heating or cooling coil. The nonlinear coil heat transfer characteristic (Figure 5-11a) is counteracted by the equal-percentage valve characteristic (Figure 5-11b), resulting in a nearly linear coil output of flow versus valve position (Figure

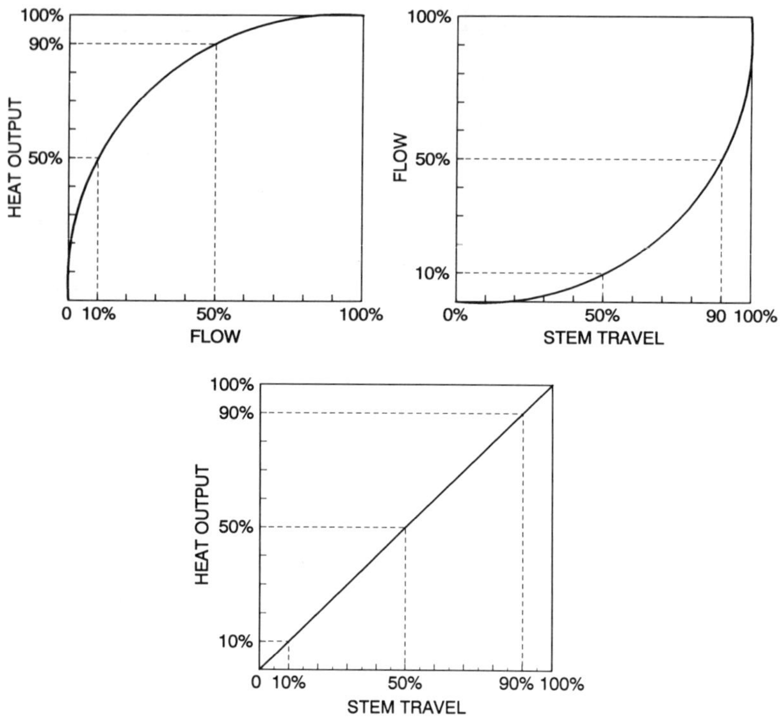

Fig. 5-11 Combination of Characteristics of Cooling (or Heating) Coil and Equal-Percentage Control Valve

5-11c) to assist the controller in matching the flow to the load and to minimize coil discharge fluctuations.

5.5.3 Sizing

Selection of the size of the control valves should be made by determining the C_v valve coefficient required to provide the design flow conditions at full load, *i.e.*, in the wide-open position. Manufacturers will furnish values of the C_v coefficient for each valve size. By definition, the C_v coefficient is the flow (in gpm) through a wide-open valve at a 1-psi pressure drop across the valve. For water, this reduces to

$$\text{gpm} = C_v\,(\Delta p)^{0.5}$$

For steam, this formula becomes

$$W_s = 2.1\,C_v\,[(p_1 + p_2)\,(p_1 - p_2)]^{0.5} \tag{2}$$

where

C_v = required valve coefficient
p_1 = valve inlet pressure, psia
p_2 = valve outlet pressure, psia
W_s = full-load steam flow rate, lb/h
Δp = pressure drop across valve

5.5.4 Applications

Two-way valves vary the amount of total volume of water flowing in a heat transfer coil (Figure 5-12). Uniform total flow volume can be maintained by varying the volume of water flowing through the coil and allowing a bypass flow around the coil using a *three-way mixing* valve located on the leaving side of the coil (Figure 5-13). A more expensive *three-way diverting* valve, located on the entering side of the coil, varies the volume of water flowing through the coil and diverts flow around the coil as the coil flow is decreased.

The application of valves to air-and-water systems is treated in Chapter 7; the application to all-water systems is treated in Section 8.7.

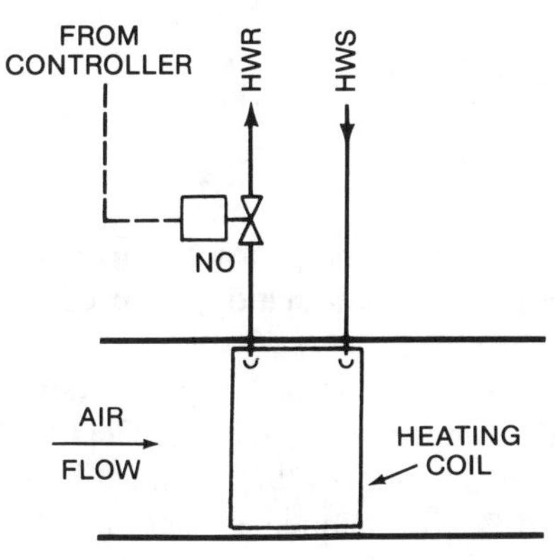

Fig. 5-12 **Two-Way Valve Controlling Hot Water Circuit**

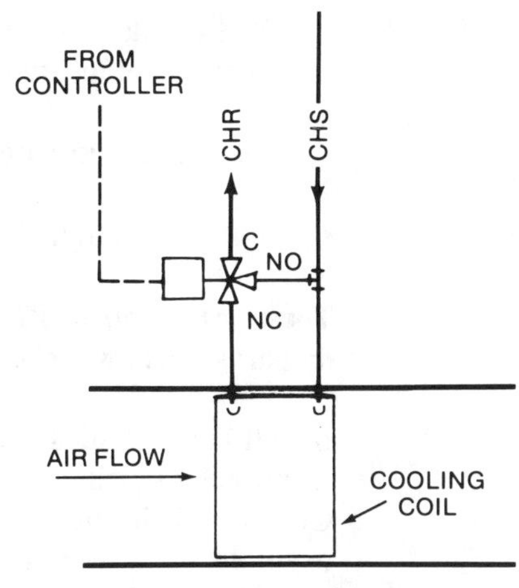

Fig. 5-13 **Three-Way Valve Controlling Chilled-Water Circuit**

5.6 PIPING

The design of piping systems is described in Chapter 34, Pipe Sizing, in ASHRAE *Fundamentals* (1993), while physical pipe sizes are given in ASHRAE *Systems and Equipment* (1992). Assume a value of $C = 140$ in the Hazen-Williams equation as used in the *Handbook*. The usual values are between 130 and 148. Friction loss in hydronic systems is usually between 1 and 4 ft per 100 ft of pipe, with an average of 2.5 ft per 100 ft. Typically, add 50% to the friction loss in piping to account for fittings. The following rule is often used to account for the pressure drop in elbows: the equivalent straight length of pipe (in feet) for an elbow equals twice the nominal pipe diameter (in inches). Thus, a 1-in. elbow causes the same friction loss as 2 ft of straight 1-in. pipe. Pumping power is obtained from

$$\text{BHP} = (\text{gpm}) \times (\text{head of water, ft})/[3960 \times (\text{pump efficiency})] \quad (3)$$

with pump efficiencies ranging from 0.50 to 0.87.

For calculating the required size of compression/expansion tanks, use the table in Chapter 12, Hydronic Heating and Cooling System Design, in ASHRAE *Systems and Equipment* (1992), which lists volumes in gallons per foot of pipe. The water volumes in chillers, coils, heat exchangers, etc. must be added to those volumes.

Approximate sizes for air control expansion tanks are

- 16% of total water volume for chilled water
- 20% of total water volume for heating and hot water under standard conditions of sea level location, 30 to 40 ft static head
- 12 psig fill pressure, 30 psig relief valve
- 200°F hot water, 44°F chilled water

These sizes can usually be cut in half if compression tanks, which contain a diaphragm and work on the principle of air elimination, are used.

If antifreeze liquids are used, the reduced thermal capacity (specific heat) and increased viscosity of these liquids compared to water [see Chapter 18, Secondary Coolants (Brines), in ASHRAE *Fundamentals* (1989a)] must be considered. For example, the flow resistance of a 30% glycol aqueous solution is 15% greater than that of water. Tests on drag-reducing agents have recently been performed, so look for new developments in that area.

The following velocities are recommended for closed (pressurized) systems:

Service	Velocity Range (fps)
Pump discharge	8 to 12
Pump suction	4 to 7
Drain line	4 to 7
Header	4 to 15
Riser	3 to 10
General service	5 to 10
City water	3 to 7

Maximum recommended water velocities to minimize erosion are as follows:

Normal Annual Operating Hours	Maximum Recommended Water Velocity (fps)
1500	12
2000	11.5
3000	11
4000	10
5000	9
8000	8

The pressure drop calculations for a typical 10,000-ft^2 low-rise office building may be as follows. They are based on manufacturers' data, except where noted.

Pressurized System

Item	Pressure Drop, ft of water
Boiler (assume 20 gpm)	1.0
Pump tree	6.0 (from calculations)
Strainer (in tree)	8.0
Supply piping	12.0 (from calculations, assume 400 ft at 3 ft/100 ft)
Return piping	12.0 (from calculations, assume 400 ft at 3 ft/100 ft)
3-way valve	12.0 (assume 50% of loop)
Total	51.0

Heating hot water pump BHP = (20 gpm)(51 ft)/(3960 × 0.75) = 0.34 BHP

Open System

Item	Pressure Drop, ft of water
Cooling tower return pipe	1.5 (from calculations)
Indoor retention tank	0.5
Basket strainer (3.5 psi)	7.0
Pump tree	6.0 (from calculations)
Chiller	6.0
Cooling tower supply pipe	1.5 (from calculations)
3-way modulating valve	3.0
Cooling tower nozzle	12.0
Lift to tower nozzle	12.0 (from building plans)
Total	49.5

Condenser water pump BHP = (3 gpm/ton)(49.5 ft)/(3960 × 0.75)
= 0.05 BHP/ton

5.6.1 Air Control and Venting

If air and other gases are not eliminated from the hydronic piping circuit, they may cause "air binding" in the terminal heat transfer elements and noise and/or reduction in flow in the piping circuit. Piping can be run level, provided flow velocities in excess of 1.5 fps are maintained. Otherwise, pitch the piping up in the direction of flow to a high point containing an air vent or a runout up to a room terminal unit. Vent high points in piping systems and terminal units with manual or automatic air vents. As automatic air vents may malfunction, provide valves at each vent to permit service without draining the system. Pipe the discharge of each vent to a point where water can be wasted into a drain or container to prevent damage to the surroundings. If a standard compression or expansion tank is used, free air contained in the circulating water can be removed from the piping circuit and trapped in the expansion tank by air-separation devices. If a diaphragm-type tank is used, vent all air from the system.

5.6.2 Drains, Shutoffs, and Strainers

Equip all low points with drains. Provide separate shutoff and drain valves for individual equipment so that the entire system does not have to be drained for service of a particular item. Use strainers where necessary to protect equipment. Size strainers placed in the pump suction large enough to avoid cavitation. Large separating chambers, which serve as main air-venting points and dirt strainers ahead of pumps, are available. Automatic control valves or spray nozzles operating with small clearances or openings require protection from pipe scale, gravel, welding slag, etc., which may readily pass through the pump and its protective separator. Individual fine mesh strainers are often required ahead of each control valve.

5.6.3 Thermometers and Gages

Include thermometers or thermometer wells to assist the system operator and the test and balance technician and for use in troubleshooting. Install permanent thermometers with correct scale range and separable sockets at all points where temperature readings are regularly needed. Thermometer wells should be installed where readings will be needed only during start-up and balancing. Install gage cocks at points where pressure readings will be required. Note that gages permanently installed in the system will deteriorate due to vibration and pulsation, and they cannot be counted on to provide reliable readings over the long haul. A single gage connected to both supply and return piping with appropriate valving will permit checking the pressure differential, thus providing some of the benefits of calibration.

5.6.4 Flexible Connectors

Flexible connectors at pumps and other machinery help reduce pipe vibration. Vibration can be transmitted, however, through the water across a flexible connection and thereby reduce the effectiveness of the connector. Yet flexible connectors prevent damage caused by misalignment and thermal expansion/contraction of equipment piping.

5.7 DUCTS

Recommended and maximum air velocities for ducts are given in Table 5-2. Three methods for the design of ducts are in common use—equal friction, static regain, and the T-method. Refer to Chapter 32, Duct Design, in ASHRAE *Fundamentals* (1993), in which all three methods are explained in detail.

Table 5-2 Maximum and Recommended Air Velocities in Ducts

	Recommended Velocities, fpm		
Designation	**Residences**	**Schools, Theaters, Public Buildings**	**Industrial Buildings**
Outdoor air intakes[a]	500	500	500
Filters[a]	250	300	350
Heating coils[a,b]	450	500	600
Cooling coils[a]	450	500	600
Air washers[a]	500	500	500
Fan outlets	1000 – 1600	1300 – 2000	1600 – 2400
Main ducts[b]	700 – 900	1000 – 1300	1200 – 1800
Branch ducts[b]	600	600 – 900	800 – 1000
Branch risers[b]	500	600 – 700	800
	Maximum Velocities, fpm		
Outdoor air intakes[a]	800	900	1200
Filters[a]	300	350	350
Heating coils[a,b]	500	600	700
Cooling coils[a]	450	500	600
Air washers[a]	500	500	500
Fan outlets	1700	1500 – 2200	1700 – 2800
Main ducts[b]	800 – 1200	1100 – 1600	1300 – 2200
Branch ducts[b]	700 – 1000	800 – 1300	1000 – 1800
Branch risers[b]	650 – 800	800 – 1200	1000 – 1600

[a]These velocities are for total face area, not the net free area; other velocities in the table are for net free area.
[b]For low-velocity systems only.

The Duct Design chapter (ASHRAE 1989b) and the HVAC *Systems Duct Design Manual* (SMACNA 1990) contain extensive tabulations of pressure losses in ducts. Manufacturers' data are also freely available. Roughness factors for other than smooth sheet metal ducts can also be found in these references. Internal linings may add 25 to 40% to air resistance, while flexible ducts may add 50%. The latter require special attention so that they remain round and do not collapse or become crushed. Fan power is obtained from

BHP = (cfm)[static pressure + velocity pressure (in. of water)]/4100 (4)

Recommended cfm per square foot for different conditions are given in ASHRAE *Pocket Guide* (1993). Minimum ventilation (outside air) requirements are given in ASHRAE *Standard* 62 or in local codes, which, if they exist, supersede the standard. For supply air systems, use approximate pressure losses of

- 0.08 in. of water per 100 linear feet of duct for quiet areas
- 0.10 in. of water per 100 linear feet of duct for ordinary areas
- 0.15 in. of water per 100 linear feet of duct for factory areas

The pressure loss calculations for a typical 10,000-ft^2 low-rise office building may appear as follows. They are based on manufacturers' data or on the SMACNA *HVAC Systems Duct Design Manual* (1990) except where noted.

Supply Air Fan

Item	Pressure Drop, in. of water
Outside air louver	0.05
Mixed-air damper plenum	0.05
Preheat coil	0.13
Filters (dirty)	0.75
Cooling coil (wet)	1.30
Heating coil	0.13
Fan inlet	0.50
VAV inlet vanes	0.75
Fan outlet	0.50
Primary ductwork	2.00 (from calculations)
Terminal box	0.20 (from calculations)
Air diffuser	0.10
Total	6.66

Return Air Fan

Item	Pressure Drop, in. of water
Room/plenum/ceiling	0.04
Plenum drop	0.05
Duct inlet	0.03
Fan inlet	0.30
Ductwork	0.50 (from calculations)
Fan inlet vanes	0.50
Fan outlet	0.10
Total	1.52

Exhaust Air Fan

Item	Pressure Drop, in. of water
Exhaust register	0.10
Fire damper	0.04
Ductwork	0.50 (from calculations)
Total	0.64

A prudent designer will add a safety factor to these values to be sure that adequate fan power has been provided for the system.

Makeup air must equal exhaust air plus exfiltration at all times.

The exhaust air fan will cause outdoor air to infiltrate unless outdoor air fans supply a quantity of outdoor air equal to the exhausted plus exfiltrated air. Ignoring this fact may lead to frozen coils or sprinkler piping in cold climates. Infiltration due to the stack effect alone may require insulation of noncirculating sprinkler piping within 15 ft of the building envelope on lower floors of high-rise buildings. Buildings should be pressurized to reduce or prevent infiltration. Therefore, it is customary to supply approximately 5 to 10% more outside air than exhaust air to make up for exfiltration.

The static pressure control needs to reference outside air conditions. The outside air sensor must not be sensitive to weather or wind. Also, use a large chamber to dampen sudden variations in its readings.

Standard louvers, hoods, and direct openings are usually sized at 700 to 1500 fpm through the free area of the opening. Specify stormproof louvers approximately 50% larger than standard louvers to minimize the resistance of the extra rain breaks in the blades. Install flashings at the outside wall, and weep holes or a floor drain to carry away rain or melted snow entering the intake. In cold regions, a snow baffle may be required to direct fine snow particles to a low-velocity area below the dampers. At maximum airflow, overall resistance of the louver, dampers, and outside air intake duct should approximately equal the resistance of the return air systems.

Construct relief openings in large buildings similarly to outdoor air intakes but equip them with motorized or building pressure-operating backdraft dampers to prevent reversal of airflow caused by high wind pressures or building stack action when the automatic dampers are open. Treat self-acting dampers to prevent rattling.

5.8 REFERENCES

ASHRAE. 1992. *Handbook—Systems and Equipment.*

ASHRAE. 1993. *ASHRAE Pocket guide for air conditioning, heating, ventilation, and refrigeration.*

ASHRAE. 1993. *Handbook—Fundamentals.*

Griffin, C.W. 1974. *Energy conservation in buildings.* Alexandria, VA: Construction Specifications Institute.

Heating/Piping/Air Conditioning. 1984. May, p. 85.

ICBO. 1991. Uniform Building Code. Whittier, CA: International Conference of Building Officials.

SMACNA. 1990. *HVAC Systems duct design manual*. Chantilly, VA: Sheet Metal and Air Conditioning Contractors' National Association, Inc.

Trane. 1977. *Refrigeration manual*. LaCrosse, WI: The Trane Company.

Chapter 6 ALL-AIR SYSTEMS

6.1 INTRODUCTION

All-air systems provide the entire sensible and latent cooling capacity through the cold air supplied to the conditioned space. No additional cooling is required at the zone. Heating may be accomplished by the same airstream, either in the central system or in the zone. Here a *zone* is an area controlled by a thermostat, while *room* refers to a partitioned area that may or may not have a separate thermostat. The psychrometrics of air-conditioning systems are treated in detail in this chapter. Much of this discussion is also applicable to the systems discussed in Chapters 7 and 8.

It is assumed that you are familiar with Chapter 2, All-Air Systems, in ASHRAE *Systems and Equipment* (1992), which gives a general description of these systems. To conserve space, material in that volume is generally not repeated here.

6.2 ADVANTAGES AND DISADVANTAGES OF ALL-AIR SYSTEMS

Among the advantages of all-air systems are:

- Major equipment is centrally located, which allows for maintenance in unoccupied areas.
- There is no drain piping or power wiring in occupied areas.
- The systems are well suited to air-side economizer, heat recovery, winter humidification, and large outside air requirements.
- They are the best choice for close control of zone temperature and humidity.
- There is simple seasonal changeover.
- The systems permit simultaneous heating and cooling in different zones.

Among their disadvantages are:

- Duct space requirements may add to building height.
- Air balancing is difficult.
- It is difficult to provide full comfort in locations with low outdoor temperatures where air is used for perimeter heating.
- Accessibility to terminals requires close coordination between mechanical, architectural, and structural designers.

6.3 SYSTEM CONCEPTS AND BASIC PSYCHROMETRICS

Procedures for the calculation of air-conditioning cooling loads are presented in Chapter 6, Psychrometrics, in ASHRAE *Fundamentals* (1993) and in Chapter 4 of this manual. Proper representation of each load on the psychrometric chart will help you understand the interaction of these loads and their effect on systems design. Figure 6-1A shows a representative single-duct, single-zone, all-air system. Heat flows q and airflows are indicated by arrows, and temperatures are indicated by t. The subscripts in the sequence of the airflow are:

R = room
rp = return plenum
rd = return duct
o = outside air
cc = cooling coil
hc = heating coil
rf = return air fan
sf = supply fan
s = supply

Figure 6-1B shows the corresponding psychrometric chart. All pertinent points are identified by the same subscripts. The room sensible and latent loads are denoted by q_{SR} and q_{LR}, respectively, and the outside air sensible and latent loads are denoted by q_{So} and q_{Lo}, respectively. The cooling load q_{cc} is the difference in enthalpies between states m and cc.

Note that the cooling coil discharge air draws heat from the supply air fan and the supply air ducts, accounting for the difference in dry-bulb temperatures between points cc and s in Figure 6-1B before entering the room. Room sensible and latent loads due to occupants, lights, machinery, solar radiation, transmission, etc. are

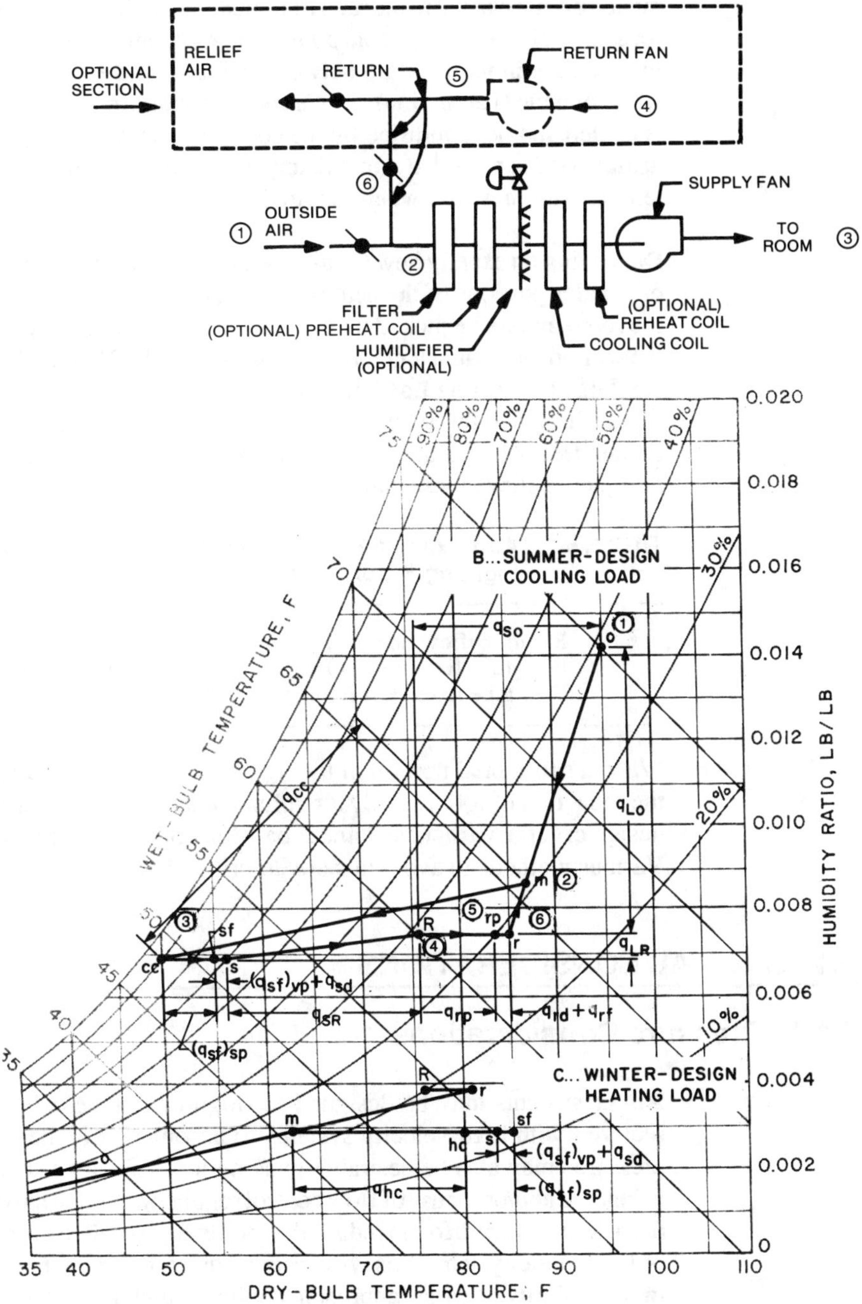

Fig. 6-1 **Single-Duct, Single-Zone Schematic Diagram and Psychrometric Chart**

picked up and carried to the return air plenum. Additional heat may be picked up from recessed ceiling lights, floors above, the roof, and the return air fan, accounting for the increase in temperature between points R and r. Some of the air is exhausted, while outside (ventilation) air o is taken in, resulting in a mixed airstream m, which is cooled and dehumidified by the cooling coils, producing the state of air at cc. A heating coil is provided immediately downstream of the cooling coil to raise the air temperature in winter when required.

Depending on exactly how the air is supplied to and exhausted from the room, larger or smaller portions of the heat from lights must be added to the other heat loads in the room or may be picked up in the return air plenum. This is explained in detail in the section on plenum temperatures, in Chapter 26, Nonresidential Air-Conditioning Cooling and Heating Load, in ASHRAE *Fundamentals* (1993). When return air is routed through light fixtures (troffers), the approximate fractions of heat removed by the airstream under different conditions are shown in Table 6-1, although some designers believe these values are high.

TABLE 6-1 Approximate Heat Removed (Percent of Total Power Input) from Lighting Fixtures by Air

Return airflow rate, cfm	10	20	30	40
Four-tube fixtures, 4 ft long	38	55	60	65
Two-tube fixtures, 4 ft long	43	58	65	75

When a substantial portion of the lighting heat is removed by the return airstream, the reduction of room supply air volume may fall short of the minimum ventilation design criteria, particularly under part-load operation of variable-volume systems. Various methods to avoid this are described later.

6.4 SPECIAL CONSIDERATIONS

6.4.1 Pressure Considerations

All-air systems may be designed for high or low velocity. Higher pressures are required with high-velocity systems to account for higher friction losses. Higher pressures may also be desirable in some low-velocity systems for greater ease of balancing and/or for use of flow control regulators, which have a substantial pressure drop. Make every effort to reduce the total energy needed to produce the required rate of flow in the system. This will ensure a quieter system, reduce duct leakages, and, in the majority of cases, approach maximum owning and operating economy. Duct design methods are described in Section 5.7, which also presents maximum and recommended velocities for high- and low-pressure systems.

The chief justification for high duct velocity is space limitation. Most installations have points where ducts must be restricted in size. As a result, the air velocity in the duct must be high. As soon as these points are passed and space becomes less critical, the lower duct velocities should be restored to reduce friction rates, then gradually decreased toward the far end of the duct system. A substantial saving in duct friction is usually achieved by carefully studying the duct mains adjacent to the machinery room. The connections from plenums to main supply ducts should be as streamlined as possible to reduce the loss at the entrance to the high-velocity duct. In practice, this is usually accomplished by using a conical-shaped fitting between the high-velocity ductwork and the plenum. In many cases, the space conditions at the machinery room are not critical, and a large duct can be used in this area. The duct velocity can then be stepped up gradually when the point of maximum space restriction is reached.

Return air ducts should be sized for low velocities. One way of simplifying the arrangement of the system is to use hung ceilings or corridors for return plenums and to collect all return air at central points on each floor. In some localities, there are code restrictions to this method.

One way of ensuring a quiet installation is to have some length of lined ductwork on the leaving side of the terminal unit when system layout permits. The lined ductwork, especially when it contains one or two elbows, is a very effective sound attenuator. It will provide a necessary safety factor in acoustical design if noise regeneration occurs in the air distribution system because of poorly constructed ducts, fittings, or tap-offs. Another method is to use sound attenuators, which avoids potential contamination from fiberglass duct lining. Noise and vibration control is treated in more detail in Section 2.4.3.

6.4.2 Temperature-Volume Considerations

The volume of air required for any area can be governed by the peak room sensible heat load, by the dehumidification or humidification (latent) load, or by ventilation requirements. (*Note*: In HVAC terminology, "ventilation" always refers to outdoor air alone, not to total air.) Ventilation requirements are rarely the governing condition. However, they may govern under minimum supply air conditions in variable-air-volume (VAV) systems (see Section 6.7) and in rooms with high occupancy rates or high exhaust air requirements.

Although load requirements, ventilation or air movement standards, and equipment limitations may predetermine air volume and temperature, the system designer still has some latitude in selecting the temperature-volume relationship. Weigh the economics of low initial volume against possible distribution, odor, or air movement problems. Low air volume, usually accompanied by low supply air temperature, requires increased refrigeration capacity and increases the operating cost of refrigeration, although not the combined operating costs of refrigeration and air-handling systems. Low temperatures may cause drafts and condensation on grille surfaces. For occupant comfort, it is undesirable to inject air into a room that is more than 20°F colder than the room air temperature. When the supply air is colder than that, for example in cold air systems (see Section 9.3.8), room air must be induced and mixed with the supply air inside the supply terminal before being injected into the room.

6.5 USEFUL EQUATIONS FOR AIRSIDE SYSTEM DESIGN

6.5.1 Basic Calculations

1. Sensible Heat

$$q_S = \text{(Specific heat) (Specific volume) (Mass flow rate) (Temperature difference)}$$

$$q_S = (0.244\ \text{Btu/lb}\cdot°\text{F})\ (0.075\ \text{lb/ft}^3)\ (60\ \text{min/h})\ (\text{cfm})\ \Delta t$$

$$q_S = 1.1\ (\text{cfm})\ \Delta t\ (\text{Btu/h}) \quad (1)$$

where

q_S = sensible heat corresponding to change in dry-bulb temperature (Δt) for given volumetric flow of standard air, Btu/h

cfm = airflow rate of standard air, cubic feet per minute

While the specific heat of air is 0.24 Btu/lb, the factor 0.244 is the specific heat of air with an average of 1% moisture.

2. Latent Heat

$$q_L = \text{(Latent heat of vaporization) (Air mass flow) (Change in humidity ratio)}$$

$$q_L = (1076\ \text{Btu/lb water})(0.075\ \text{lb air/ft}^3)(60\ \text{min/h})(\Delta W)$$

$$q_L = 4840(\text{cfm})\Delta W = 0.69(\text{cfm})\Delta G \quad (2)$$

where

q_L	=	latent heat corresponding to the change in absolute humidity ratio
ΔW	=	pound of water per pound of dry air
ΔG	=	grains of water per pound of dry air for given volumetric flow of standard air, where 7000 grains equal 1 pound

3. Total Heat

$$q_T = \text{(Air mass flow)(Change in enthalpy)}$$

$$q_T = (0.075\ \text{lb/ft}^3)(60\ \text{min/h})(\text{cfm})\Delta h$$

$$q_T = 4.5(\text{cfm})\Delta h = q_S + q_L \tag{3}$$

where

q_T	=	total heat corresponding to change in enthalpy Δh for given volumetric flow of standard air, Btu/h

4. Air Mixtures

When the flow rate cfm_1 at temperature t_1 and humidity ratio W_1 mixes with cfm_2 at $t_2 < t_1$ and $W_2 < W_1$ to yield a mixed condition of cfm_m at t_m and W_m, then

$$\text{cfm}_m = \text{cfm}_1 + \text{cfm}_2 \tag{4}$$

$$t_m = t_2 + (t_1 - t_2)\ (\text{cfm}_1/\text{cfm}_m) \tag{5a}$$

or

$$t_m = t_1 - (t_1 - t_2)\ (\text{cfm}_1/\text{cfm}_m) \tag{5b}$$

$$W_m = W_2 + (W_1 - W_2)\ (\text{cfm}_1/\text{cfm}_m) \tag{6a}$$

or

$$W_m = W_1 - (W_1 - W_2)\ (\text{cfm}_2/\text{cfm}_m) \tag{6b}$$

In the abovementioned equations, t applies to either the dry- or the wet-bulb temperature. Such mixtures, plotted on a standard psychrometric chart, fall on a straight line drawn between states 1 and 2, with the distances from these points inversely proportional to the flow rates.

6.5.2 Fan Laws

The fan laws predict the effect on fan performance of changing such quantities as the conditions of the air, fan operating speed, or fan size. The form of the fan laws given here is one of several variations in which they can be expressed. The five fan laws are:

1. Variation in fan speed—constant density of air, constant system (no changes in ductwork or damper positions)
   Q varies as rpm
   SP varies as rpm^2
   VP varies as rpm^2
   P varies as rpm^3

where

Q	=	flow rate of air
SP	=	static pressure
VP	=	velocity pressure
P	=	fan power

2. Variation in fan size—similar fan proportions, constant rpm, constant air density
   Q varies as D^3
   SP varies as D^2
   P varies as D^5

where

D	=	diameter of fan wheel

3. Variation in air density—constant system, constant fan size, constant fan speed
   Q remains constant
   SP varies as air density
   P varies as air density

4. Variation in air density—constant static pressure, constant system, constant fan size, variable rpm
   Q varies as (air density)$^{-0.5}$
   SP is constant
   rpm varies as (air density)$^{-0.5}$
   P varies as (air density)$^{-0.5}$

5. Variation in air density—constant mass flow of air, constant system, constant fan size, variable rpm
   Q varies as the reciprocal of air density
   SP varies as the reciprocal of air density
   rpm varies as the reciprocal of air density
   P varies as (air density)$^{-2}$

Typical applications of the fan laws:

Fan law 1. Required increase in rpm to increase the rate of airflow a specified amount.

Fan law 2. Size of fan to achieve a greater flow rate but with the operating speed remaining constant.

Fan law 3. Change in static pressure and power if air density changes but fan speed remains constant.

Fan law 4. Speed change necessary to maintain a certain static pressure if air density changes.

Fan law 5. Amount rpm must be changed when air density changes if a fixed mass rate of airflow is to be maintained (for example, to preserve a certain change in temperature across a heat exchanger with a fixed rate of heat addition).

You will most likely use fan laws 1 and 5 in design. If the testing and balancing operation shows that a 15% greater flow rate is required, fan law 1 indicates not only how much the fan speed must be increased, but what the consequences are on the power required of the fan motor. Fan law 5 may be required when the air handled by the fan and duct system is at an appreciably different temperature from the one for which the ratings are available. This fan law may be necessary when designing systems to operate at high altitudes and thus lower atmospheric pressures than at sea level. Fan law 3 may give an order-of-magnitude estimate of the change in fan size needed to achieve a revised flow rate, but realistically it would probably be better to consult the fan manufacturer's catalog for such information.

Example 1. A fan installed in a new system operates at 800 rpm and provides a measured flow rate of 4100 cfm working against a static pressure of 2.8 in. of water. The catalog rating of the fan at these conditions specifies a power requirement of 4.5 hp, and a 5-hp motor has been selected. Instead of the actual flow rate of 4100 cfm, 4500 cfm is desired. What are the consequences of increasing the fan speed to achieve 4500 cfm?

Solution: Fan law 1 states that, with constant air density and system, Q varies linearly with rpm. The fan speed must, therefore, be changed to

$$800(4500/4100) = 878 \text{ rpm, or by a ratio of } 1.0975$$

The new static pressure will be 2.8 $(1.0975)^2 = 3.37$ in. of water.

The revised power requirement will be $4.5(1.0975)^3 = 5.95$ hp, which will probably necessitate changing the 5-hp motor to a 7.5-hp motor. Figure 6-2 shows the original and revised operating conditions.

Example 2. Measurements made on a newly installed air-handling system were: 1200-rpm fan speed, 4500 cfm, 1.9 in. of water static pressure, and 3-hp shaft power when the air temperature was 70°F and the pressure was standard atmospheric. In actual use, the fan and duct system are to handle 150°F air. If the fan speed remains at 1200 rpm, what will be (a) the volume rate of airflow, (b) the static pressure, and (c) the power requirement?

Solution: Fan law 3 applies because the air density changes, but the fan speed, fan size, and system remain constant.

a. The volume rate of flow remains constant at 4500 cfm. Incidentally, the mass rate of flow decreases because the air density at 150°F is 0.065 lb/ft^3 in comparison to 0.0749 lb/ft^3 originally.

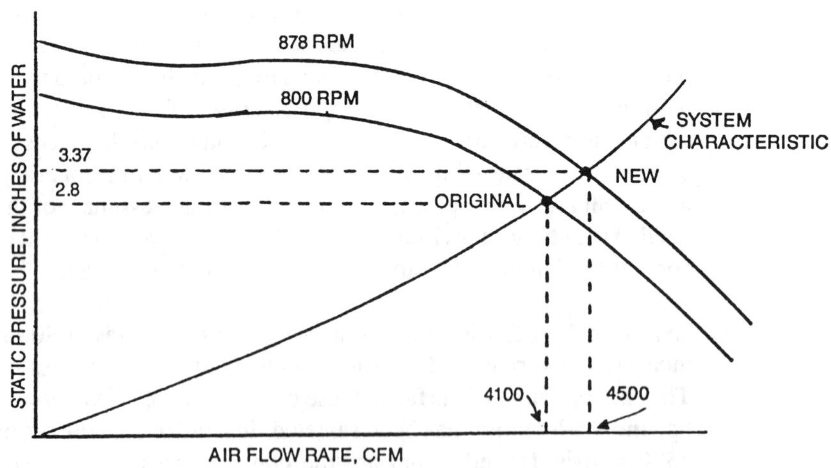

Fig. 6-2 Original and Revised Operating Conditions for Example 1

b. The static pressure varies in proportion to the air density, so $SP = (1.9)(0.065/0.0749) = 1.65$ in. of water.

c. The power varies in proportion to the air density, so $P = (3 \text{ hp})(0.065/0.0749) = 2.61$ hp.

In the selection and operation of a combination of supply and return air fans, an underlying objective is to regulate the air pressure in the occupied spaces so that it is close to outdoor atmospheric pressure and perhaps just slightly higher. This objective becomes a special challenge in VAV systems (Section 6.7), where the duct-damper characteristics are constantly changing as the zone loads change.

6.6 SINGLE-ZONE SYSTEMS

The simplest form of the all-air system is a single refrigeration machine serving a single temperature control zone (Figure 6-1). The unit may be installed within or remote from the space it serves and may operate with or without distributing ductwork. Properly designed systems can maintain temperature and humidity closely and efficiently and can be shut down when desired without affecting the operation of adjacent areas. They are energy efficient, easy to control, and easily adaptable to economizer cycles (see Section 9.3.7). Their disadvantage is that they respond to only one set of space conditions. Therefore, their use is limited to situations where variations occur approximately uniformly throughout the zone served or where the load is stable. Single-zone systems are applicable to small department stores, small individual shops in a shopping center, individual classrooms for a small school, computer rooms, hospital operating rooms, and large open areas, such as gymnasiums. For example, a rooftop unit, complete with refrigeration system, serving an individual space is considered a single-zone system. The refrigeration system, however, may be remote and may serve several single-zone units in a larger installation. A return fan is necessary only if 100% outdoor air is used for cooling, and then it may be eliminated if air can be relieved from the space with very little pressure loss through the relief system. A system using multiple fan-coil units is a collection of single-zone systems put together to control different zones.

Units that incorporate all the components of an air-conditioning system except ductwork and condenser water piping are referred to as *self-contained* units because they contain a refrigeration compressor, a direct expansion coil, the refrigeration piping, a fan and motor, and filter and controls, all in one casing. In many cases, a heating coil is also included. Sometimes the units are attached to ductwork, but frequently they are used with a free-discharge grille. These units are manufactured with either a water-cooled or an air-cooled condenser and with or without gas heating units.

Frequently the fan and evaporator section of these self-contained units is mounted in or next to the conditioned space, and the compressor-condenser unit is mounted on the roof or some other remote location (see Section 6.11). In this case, refrigerant piping must be installed between the evaporator and the compressor, in an arrangement referred to as a *split system.*

A single-zone system can be controlled by varying the quantity or the temperature of the supply air or by providing reheat, or face and bypass dampers, or by a combination of these. Single-duct systems without reheat offer cooling flexibility but cannot control summer humidity independent of temperature requirements. Single-duct systems with reheat provide flexibility for both temperature and humidity control; the cooling coil cools the air to the desired humidity level, and the reheat coil raises the dry-bulb temperature to the desired value. However, ASHRAE *Standard* 90.1, Section 9.5.1, severely restricts the application of reheat, limiting this option to special cases. If high humidity and low dry-bulb temperatures are desired, a humidifier may have to be included in the system.

Appendix B contains sample design calculations for a single-zone system.

6.7 VARIABLE-AIR-VOLUME (VAV) SYSTEMS

6.7.1 System Concepts

In constant-air-volume systems, cooling or heating is accomplished by varying the supply air temperature and keeping the volume of the supply air constant. Variable-air-volume (VAV) systems accomplish cooling/heating by keeping the air temperature constant and varying the volume of the air supply.

6.7.1.1 Simple VAV

This system applies to cooling-only service with no requirement for simultaneous heating and cooling in different zones; a typical application is the interior of an office building. To permit system volume variations without fan volume variation, install a fan bypass. Vary zone air volume while keeping fan and system volume substantially constant by dumping excess air into a return air ceiling plenum or directly into the return air duct system. Dumping cold air into the return air plenum, however, wastes energy and can cause overcooling under low load conditions due to radiation from the cool ceiling surface to the zone below. Dumping can also cause a shortage of system volume if it is used for system balancing as well as temperature control.

6.7.1.2 VAV Reheat or VAV Dual Duct

By sequencing reheat or blending in each zone after throttling the zone's cold air supply, full heating/cooling flexibility can be achieved much more energy-efficiently than with comparable constant-volume systems. This can be applied to both interior and perimeter spaces. You can keep system volume and/or fan volume constant by fan bypass or return air dumping, as for the simple VAV system just described.

6.7.1.3 VAV with Independent Perimeter System

All-air cooling and heating can be accomplished by a VAV system serving interior spaces in connection with a constant-volume perimeter system. The variable-volume system provides cooling year-round, taking care of all variations in all zone internal heat gains as well as skin solar gains. The perimeter system uses an outdoor/indoor temperature scheduled, constant-volume air supply, which simply offsets the skin transmission gains or losses. The perimeter system requires no individual zone control (except for operating economy) and no outdoor air since the VAV system takes care of each zone's load variations (except transmission through the skin) and all outdoor air ventilation requirements.

If a hydronic perimeter heating system is provided, the VAV system accomplishes all cooling in all zones with air year round, while the perimeter heating system offsets the winter transmission heat losses but not the summer transmission heat gains.

6.7.1.4 VAV with Constant Zone Volume

Individual zone fans may be used to maintain minimum or constant supply air to the zone while the system primary air fed to the zone is throttled. Terminals in these systems are commonly referred to as *fan-powered terminals*. The load is satisfied by recirculating return air, thus keeping the sum of the throttled system air and the recirculated return air substantially constant. This technique is particularly useful for zones with large variations of internal loads, such as conference rooms, and it may be combined with terminal reheat. Fan-powered terminals can be used to ensure good air circulation in occupied spaces during periods of reduced cooling load.

6.7.1.5 VAV with Economizer Cycle

When the enthalpy of the outside air is lower than that of the return air, chiller power can be reduced by taking in more outside air than required for ventilation and dumping the excess return air. Under favorable conditions, all of the return can be dumped and replaced by outside air. This mode of operation is called an *economizer cycle* (see Section 9.3.7). While this cycle requires large outside air intakes and exhausts, it improves the economy of operation except in areas such as the southeastern United States, where these favorable conditions occur so rarely that the additional first cost of providing for economizer operation is not justified (Mutammara and Hittle 1990).

6.7.1.6 Applications

Variable-air-volume systems are easy to control, are highly energy efficient, allow fairly good room control, and are easily adaptable to economizer cycles. A potential drawback includes the possibility of poor ventilation, particularly under low zone loads. They are suitable for offices, classrooms, and many other applications, and are currently the system of choice for most commercial and institutional buildings in spite of the fact that humidity control under widely varying latent loads is difficult, as illustrated in Figure 6-3 and in Appendix C.

6.7.2 Design Considerations

6.7.2.1 Psychrometric Analysis

A typical arrangement of the components for a variable-volume system with separate perimeter heating is shown in Figure 6-3, with typical performance for maximum and partial load for both summer and winter operation. For cooling-only applications, air is delivered to the space at the fixed supply temperature to offset maximum space heat gains. The space thermostat simply reduces the volume of supply air as the load falls off. Where some warm-up is desired after system shutdown, a heating coil may be installed and manually or automatically actuated for a short period only. When system air is throttled while outdoor air volume is kept constant, tempering outdoor air may be necessary at cold ambient conditions in winter. Equations (1) through (6) for fan gains, ceiling temperatures, and air mixtures apply. Zone ductwork and variable-volume terminals must be sized for the governing factor, either sensible, latent, or ventilation loads.

The designer must determine maximum system volume from either (1) coincident system (block) load or (2) block sensible load plus all air requirements from governing latent or ventilation loads that exceed the volume required in any zone to satisfy the sensible load in that zone when the entire system reaches its peak sensible load. Thus, if a system peaks at 4 P.M. and an east exposure's air requirement from sensible loads is less than the volume required by its latent or ventilation loads at that time, the latter volume must be used.

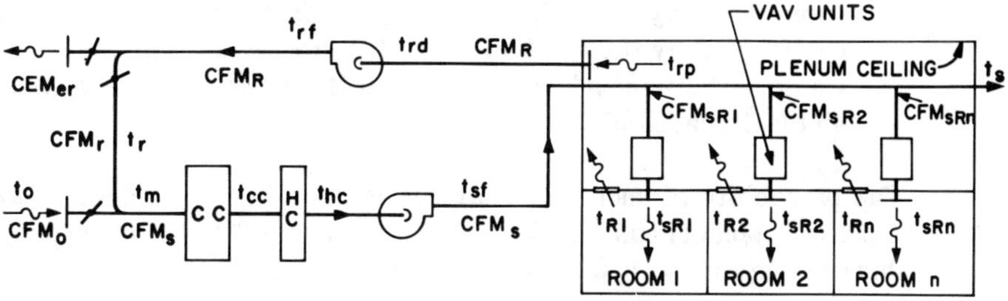

A... TRUE VAV SYSTEM CYCLE DIAGRAM

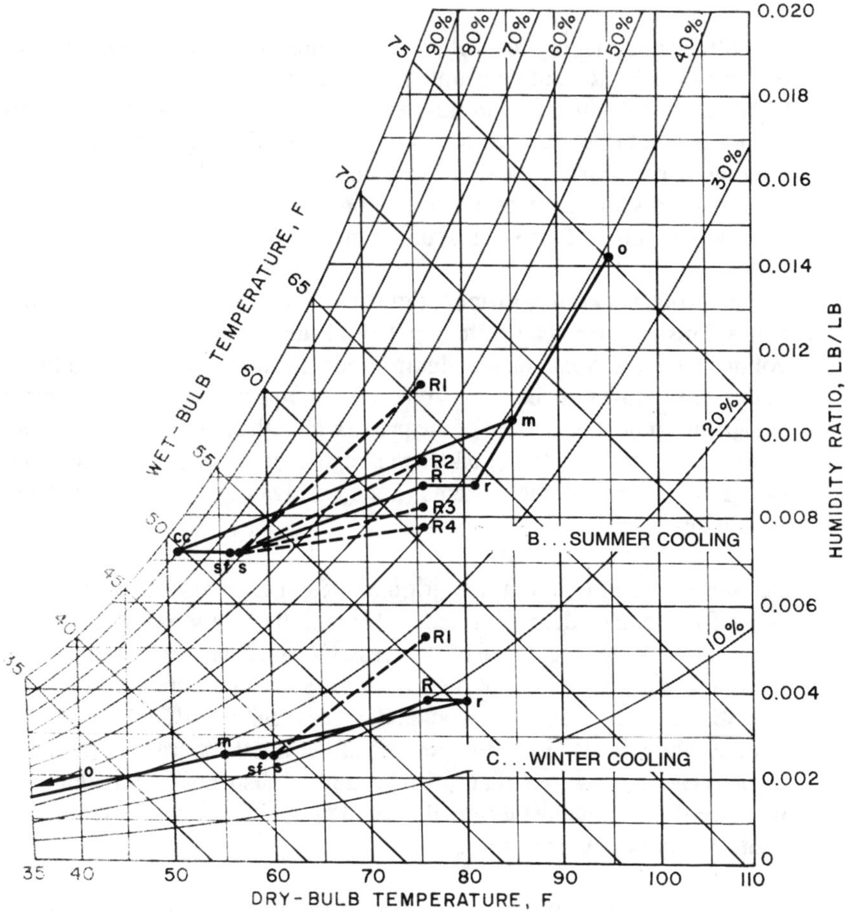

Fig. 6-3 Variable-Air-Volume Cycle—Schematic Diagram and Psychrometric Chart

> **Be aware that if building occupants set their thermostats below the values used in your design, more supply air is required than the amount determined by the block load. Under those conditions, the occupants would suffer from a lack of supply air if the system was designed based on block loads only.**

The probability of such an occurrence is quite high—ASHRAE *Standard* 90.1 specifies an indoor summer design temperature of 78°F, and many people find that this setting provides insufficient cooling comfort.

> **Whenever supply air volume is throttled at a given dew-point condition, its dehumidification capacity decreases.**

The effect of reducing air volume on dehumidification capacity is illustrated in Figure 6-3; here, R1 and R4 denote zones with high and low latent loads. Spaces with high latent loads must be carefully examined on the psychrometric chart before the proper air volume can be determined. Some designers prefer to design zones governed by latent or ventilation loads as constant-volume zones because the thermostat does not respond to variations in either of those loads. This ensures a sufficient amount of air volume under all conditions.

As each zone peaks at a different time of day, it borrows the extra air from off-peak zones. This transfer of air from low- to high-load zones occurs only in true variable-volume systems. Systems that dump excess air into a return air ceiling or bypass it to the return duct may be using zone VAV temperature control principles, but their design air volume is the sum of the peaks which they circulate on a constant-volume basis. These systems do not take advantage of smaller fans and duct mains than are used in constant-volume systems (a major advantage of VAV systems), and they should be avoided.

Be sure to examine minimum air quantities in each zone and in the system. First, check for adequate dehumidification [Equation (2)] and ventilation (Section 3.3.3); second, check the increase in back pressure resulting from terminal throttling and its effect on fans and ducts and their need for static pressure control; third, check for potential noise problems in terminal devices from throttling. If, at minimum system volume, the increase in static pressure at the fan discharge and at the last terminal device is not so large as to cause fan instability, distribution imbalance, or excessive pressure drop in the last terminal, volume regulation and/or system static pressure control may not be necessary.

Distinguish between the minimum throttling ratio (ratio of throttled flow to full flow) in any zone and that in the system. The ratio can be as little as 0.25 in a zone with a peak of, for example, 3 cfm/ft^2, without necessarily violating ventilation criteria, while the system ratio might never go below 0.5. The lower the system design static pressure, the smaller the duct network, and the higher the throttling ratio, the less need there is for system static control. Static pressure controls can be used to maintain fan stability and distribution balance, to improve operating costs, and to stabilize system sound characteristics.

Minimum ventilation air quantities can be maintained by

- Raising supply air temperatures on a system basis
- Raising supply air temperatures on a reheated zone-by-zone basis
- Auxiliary heat in the room independent of the air system

Full-zone air circulation can be maintained by

- Individual zone fan recirculation with blending of varying amounts of supply with room air or with ceiling plenum air
- Ceiling induction unit recirculation

Oversized VAV terminals tend to become noisy because they must always be throttled. You can replace low-frequency noise, which is difficult to attenuate, with high-frequency noise, which is easier to attenuate, by installing a piece of perforated metal upstream of the box, sized to reduce the pressure at the terminal at full flow by the amount needed to provide better acoustics. The multiple holes cause low frequencies to be "end reflected" and energy to be dissipated in forcing air through the multihole orifice. Unfortunately, the following relationship applies

$$\mathrm{cfm}_2/\mathrm{cfm}_1 = (P_2/P_1)^{0.5}$$

which means that the orifice will exhibit a parabolic control function. Thus, it is unsound practice to install manual control dampers in VAV systems.

When radiation is provided for perimeter heat (see Section 6.7.1.3), size it to offset transmission losses. Heat is supplied by radiation in proportion to outdoor temperature, and the VAV system usually provides all outdoor air requirements for interior as well as perimeter spaces. If it is assumed that radiation exactly offsets transmission loss, interior loads will cause the variable-volume terminal to remain open, providing ventilation requirements and air movement for comfort and odor control. The same effect may be produced when a perimeter or skin air system is used to offset transmission loads.

The solid lines in Figure 6-3b represent average system conditions, while dotted lines represent individual room conditions. The psychrometric chart does not show variations in airflow; any drop in room sensible load with no change in latent load appears as a steeper internal SHF (sensible heat factor) line *s*-*R*1 due to the reduced air volume. A typical winter cooling condition for severe climates is also shown in Figure 6-3c. The moisture level is reduced due to mixing of return air with low dew-point outdoor air. Supply air temperature at *s* is maintained by controlling outdoor air dampers and return air dampers to produce the mixed-air state *m*. Some preheating may be needed to maintain the state, which would further reduce the space moisture level. Control of temperature is maintained at each individual control zone in the system by reducing zone airflow. Again, moisture level will rise as sensible loads are reduced.

All the preceding considerations apply when a separate perimeter radiation system handles the perimeter transmission and infiltration heating loads, while a common air system serving both interior and perimeter zones with the same temperature throughout handles cooling.

6.7.2.2 Air Skin System

A special perimeter system satisfies all conduction loads from the envelope, in effect thermally separating the envelope from the rest of the building. It can be a hot-water (radiation) or a separate air system along the perimeter skin, although the former is preferred in cold regions. If the interior system is sized for 100% of the solar load on a year-round basis, air volume requirements for the air skin system are usually governed by heating needs. A typical perimeter system temperature schedule without solar cooling capability might be as follows:

Outdoor air temperature, °F	95	75	0
Perimeter system supply air temperature, °F	65	75	120

Design concerns of the air skin system are the same as for the constant-volume, single-duct system given in Section 6.6. Heating requirements normally govern air volume requirements unless very high discharge temperatures are used. Outlets should be located to "wash" the glass areas by perimeter air. They should be modular for partitioning flexibility, and they should be coordinated with the interior VAV system's distribution terminals for maximum comfort at low-volume VAV operation and low ambient temperatures. Sometimes, perimeter air can be added to the VAV air downstream of the VAV terminal, thus saving a set of perimeter slot diffusers.

Examine the justification and practicality of zoning perimeter radiation or air skin system exposures on individual temperature schedules in view of the uncertainty of factors that limit their usefulness: (1) moving shadows from neighboring buildings that vary the solar load, (2) nonuniform glass areas, and (3) varying positions of movable shading devices. The purpose of such zoning is either to conserve energy by avoiding excess heat supply to areas with high solar or internal loads or, in the case of VAV, as a sequencing temperature control method to keep VAV throttling at some low limit by supplying excess heat to areas with a low cooling load. In either case, a blanket approach to large and sometimes heterogeneous areas is ineffective. Some form of modular control on the perimeter air supply in each room, even if manual, may be preferable to automatic control.

If the perimeter system handles only recirculated air with a separate central air handler, the heat of lights and the return fan heat are available as free heat. This avoids the wasteful cooling and subsequent reheating of room return air that would be required if a common air-handling system were used for both interior and perimeter systems.

6.7.2.3 Air Volume

In VAV systems, pay particular attention to proper evaluation of the fraction of total heat from lights picked up by the return air in a hung ceiling. If, at design cooling loads, room sensible heat has been reduced so much by recessed or return air lighting troffers that the design airflow approaches the minimum criteria, no leeway will be available for variable-volume reduction at part loads. For this reason, return air troffers can be a disadvantage in variable-volume systems. If net room loads are not determined with this in mind, a VAV system will be overaired, resulting in partial throttling at full load.

From a practical standpoint, if a variable-volume device were to close completely at a satisfied thermostat setting, the condition would be similar to that prevailing in millions of residential areas with no apparent discomfort—although contrary to most ventilation codes for commercial structures. Except for special applications, such a condition need never exist. On the other hand, drafts, which are a frequent problem in constant-volume systems, can be nonexistent in variable-volume systems (when proper distribution is used) except at full-load cooling conditions. Environmental research indicates a 2°F mean radiant temperature (MRT) differential between air velocities of 0 to 45 fpm for both light- and medium-clothed occupants at sedentary to medium levels of activity. Thus, if a variable-volume thermostat is set for a wide-open condition of 76°F and is fully shut at 74°F, the 2°F MRT drop required with still air is automatically met for comfort equivalent to the 45-fpm air motion. Outlet selection to produce secondary air motion is desirable in VAV system design. Proper design may require a smaller Δt or higher aspirating-type outlets in order to improve air movement.

A typical pressure distribution throughout the supply and return air sections of a VAV system is shown in Figure 6-4. A basic design objective is to size the supply and return fans along with the supply and return air distribution systems to maintain a slight positive pressure in the space. But, since in a VAV system the positions of the zone dampers are constantly changing, one design combination cannot meet the pressure requirements at all operating conditions. The best compromise is automatically afforded by supply fan pressure control, which produces an energy conservation advantage as well. Reduction in capacity of the supply fan drops the pressure at points A and B in Figure 6-4 in comparison to full-capacity operation of the fan. The result is that the deviations from the desired zone pressures are minimized.

Air distribution is highly important in VAV applications. Give careful consideration to distribution and to sound levels at maximum and minimum flow. If the combined sound level of the terminal unit and diffuser at maximum flow is at least 3 decibels below the room ambient sound level, sound level variations will not be noticed. The subject of local velocities as they might affect variable-volume systems is covered in Chapter 31, Space Air Diffusion, in ASHRAE *Fundamentals* (1993).

6.7.2.4 Return versus Relief Fans

In VAV systems with an economizer cycle, performance can often be improved by replacing the return air fan with a relief air fan (Avery 1984). Figure 6-5 shows a typical economizer cycle. The return air fan handles all the return air and exhausts the relief air from the building. Typical pressures are shown for various locations in the system. Note that the return air damper has been sized for a 1.5-in. w.g. (water gage) pressure drop, three times the pressure drop across the outside air damper. If

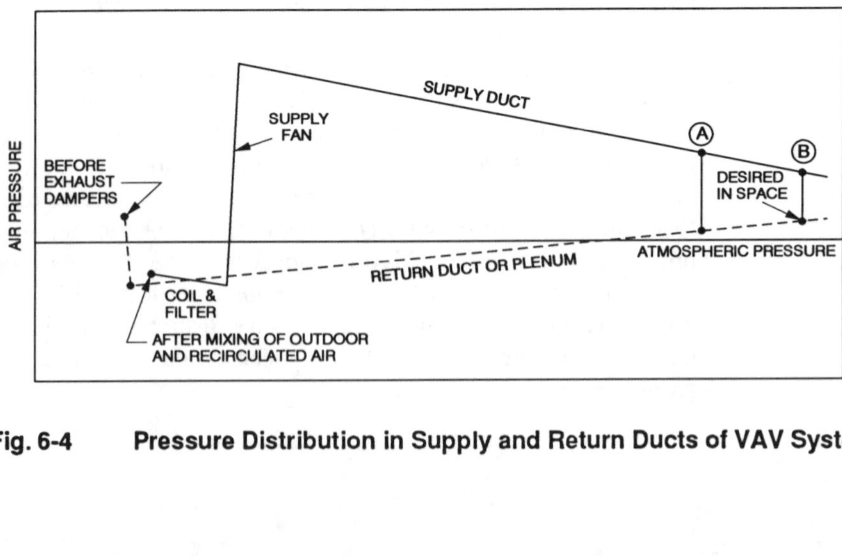

Fig. 6-4 Pressure Distribution in Supply and Return Ducts of VAV System

this is not done, damper instability can occur, and, since a 45-mph wind causes a pressure of 1 in. w.g., with some wind conditions and damper positions air may blow in through the exhaust opening and out at the air intake.

Fan power may be reduced by using the arrangement shown in Figure 6-6. Here the pressure drop across the return air damper has been reduced from 1.5 to 0.5 in. w.g. However, the work to move the supply air is unchanged. Only the relief air has to be pushed against the 1.5-in. w.g. relief air pressure drop, resulting in fan energy savings. The outputs of flow-measuring stations in the outside air duct and the relief air duct are fed into a comparator-controller. Its output controls the relief air fan so that the flow of relief air equals that of the outside air at all times.

The relief air damper closes whenever the economizer cycle is not in use. Minimum outside air volume is maintained because toilet and other exhausts provide sufficient relief to prevent overpressurizing the building. Advantages of the Figure 6-6 system are as follows:

- The return air damper is sized for the same pressure drop as other dampers.
- Total system pressure drop is reduced (20% in the example) whenever the system is off the economizer cycle.
- Wind effects are minimized.
- Heat generated by the return air fan no longer exists, thereby reducing the load on the cooling coils (Avery 1984).

6.7.2.5 Energy Savings

Variable-volume systems have the inherent potential to provide flexible cooling with minimum annual energy consumption compared to other systems. The design characteristics that make this possible are:

1. The reduction in system air volume coincides with the sum of the zone volume reductions. This permits direct savings in fan energy with indirect savings in refrigeration energy.
2. When the system layout includes all interzonal variations (*i.e.*, all exposures), the fan may be selected for the block load only (the coincident zone peaks) rather than for the sum of the zone peaks. However, as pointed out in Section 6.7.2.1, this could result in insufficient air capacity if the occupants choose to lower their thermostat settings below the design temperature.

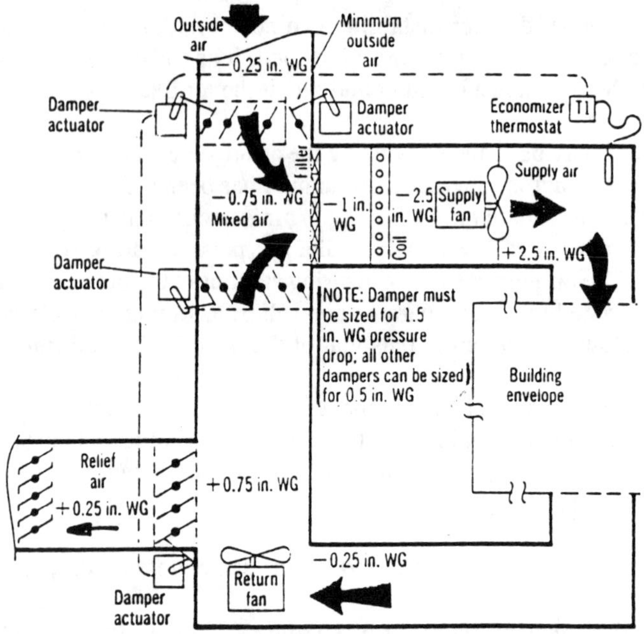

Fig. 6-5 Typical VAV Economizer Cycle with Return Air Fan

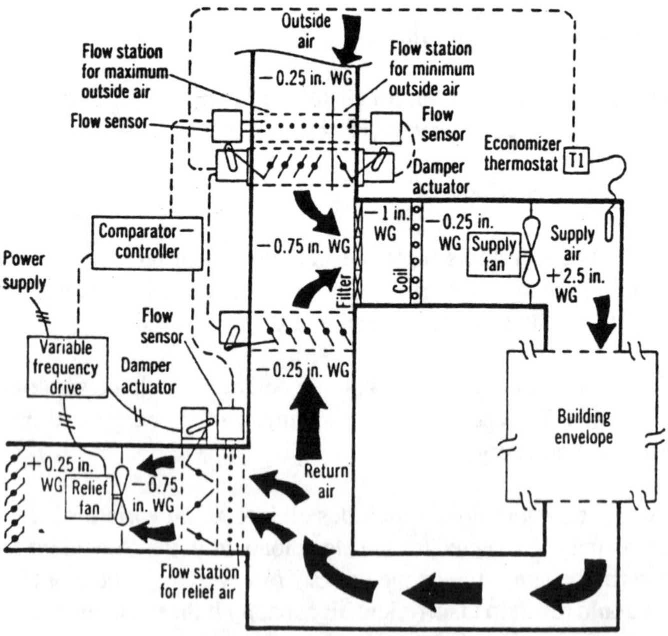

Fig. 6-6 Improved VAV Economizer Cycle with Relief Air Fan and No Return Air Fan

6.7.2.6 Controls

Proper design calls for a VAV terminal to be fully open when the zone under its control experiences maximum load. As the zone load decreases, the action of the controller reduces flow. This, in turn, increases duct pressure, which is transmitted to the central air handler and produces a reduction in system airflow. This airflow reduction is the major cause for energy savings of VAV systems compared to constant-volume systems. Figure 5-6, Chapter 5, illustrates the fan energy variations with airflow for different types of fan control. It is apparent that variable-speed fan control conserves the most energy, and you should use it where appropriate.

A means of ensuring minimum outdoor air as total flow is reduced may be installed at the central apparatus. A velocity controller located in the outdoor air intake will ensure that dampers are open sufficiently to provide the minimum flow desired. A less costly method is to schedule a variable minimum position of the outdoor air damper from the duct static pressure sensor, the fan control operator, or the velocity controller, any of which reacts directly to system volume reduction. Field tests best establish the two outdoor air damper positions to achieve substantially constant outdoor air intake during maximum and minimum system flow conditions. The outdoor air control damper can then be proportionally scheduled between these two points. If an economizer cycle is used, the constant airflow, outdoor air controller can be overridden by the mixing thermostat.

Preheating or tempering of the minimum outdoor air supply may be necessary during cold weather and low system volume. Resetting supply air temperature for economy or for minimum air motion may be accomplished by two means:

1. All zone thermostat signals are fed through a discriminating controller, which raises the supply air temperature as the last variable-volume controller starts to throttle, thus keeping the zone with the highest cooling load at optimum (low) temperature with its variable-volume damper wide open. If, under this mode of operation, any zone has a low enough load to be calling for less than its minimum ventilation, such a zone can be equipped to (a) sequence perimeter radiation valving or terminal reheat, (b) recirculate with a zone fan unit, or (c) use a suitable induction device when applicable. Methods (b) and (c) increase total air circulation only but do not increase the outside air supply.

2. The controller that senses a reduction in system volume may be used to raise the supply air temperature for maintaining maximum flow as long as all zones can be satisfied. When that temperature has been raised to its highest tolerable setting and excessive throttling still occurs, radiation temperature may be scheduled upward so that only enough auxiliary heat is imposed in the perimeter

areas to keep VAV dampers at a desired minimum open position. These procedures keep final zone control in the VAV damper.

For general information on controls, see Chapter 10. Appendix C contains detailed design calculations for a VAV system.

6.8 REHEAT SYSTEMS

6.8.1 System Concepts

The reheat system is a modification of the single-zone system. As the word *reheat* implies, heat is applied as a secondary process to either preconditioned primary air or cooled supply air. The medium for heating may be hot water, steam, or electricity.

The reheat system permits zone or space control for areas of unequal loading. It can also provide heating or cooling of perimeter areas with different exposures, maintain very close control of space humidity, and accomplish dehumidification independent of cooling, which other systems cannot do. It offers the designer infinite zoning capability in the initial design stages, which can be readily revised during construction phases as zoning changes are made. Field changes to accommodate later zoning revisions require only the addition of a heating coil or terminal unit. The major disadvantage of the reheat system is its high energy consumption and the need to provide piping and heating coils for every zone.

Reheat systems are applied to hospitals, laboratories, and office buildings, to spaces with wide load variations or high latent loads, and to process or comfort applications where close control of space conditions is desired.

Terminal units are designed to permit heating of primary air, or secondary air induced from the conditioned space, located either under the window or in the duct system overhead. Conditioned air is supplied from a central unit at a fixed cold-air temperature designed to offset the maximum cooling load in any zone. Each zone thermostat simply calls for heat as the cooling load in that zone drops below maximum.

Keep in mind that the application of heat to refrigerated air for space control is basically uneconomical. Reheat systems are inherently energy wasteful and have no self-correcting features. Many energy codes, including ASHRAE *Standard* 90.1, severely restrict their use; exceptions are detailed in Section 9.5.2 of that standard. Therefore, take special care in calculating design loads and system operational modes to ascertain whether reheat can be avoided. Where constant volume is maintained, energy use is maximum. Varying volume as a first step in control delays applying heat until a predetermined minimum volume is reached and thereby reduces operating costs considerably. Additional savings in operating costs may be accomplished by resetting the cold air supply temperature upward as outdoor temperature falls. *Discriminating controls* (see Section 9.3.3) accomplish this automatically. There is no energy penalty and, therefore, no objection to the use of reheat if it is accomplished through a heat recovery system using the reject heat of the refrigeration cycle or another internal heat source.

Figure 6-7a shows a schematic for a typical primary air reheat system; Figure 6-7b depicts a terminal unit. The corresponding psychrometric chart is also shown. Solid lines indicate average system conditions at full load; dashed lines pertain to individual zone conditions. Under full-load operation, space loads are picked up between *s* and *R*. Reheat is added to zones with lower loads from *prim* to *s*2 or *s*3. A zone with no latent load will maintain space condition *R*2, while zones with higher than average latent loads could produce conditions such as *R*3. Space humidities can be maintained at any level no lower than that indicated by *prim*, which is adjusted to correspond to the lowest desirable value of relative humidity in any zone.

6.8.2 Special Design Considerations

Procedures for determining air volume requirements for a single-duct reheat system are identical to those discussed in Section 6.6. The basic difference between constant-volume and variable-volume approaches is in the total air volume required for the system. The air volume for a constant-volume system is normally selected by the sum of the individual peak volumes required for each room. Ensure that all zones have sufficient reheat capacity at the block peak to maintain their desired humidity level. In constant-volume reheat systems, select the supply air temperature for the zone with the highest cooling load and require reheat for all other zones. Therefore, devote extra care to maximum cooling load zone calculations. In fact, at times disregard isolated high-load areas, assuming that a slight temperature rise in those areas may not be objectionable.

Low-velocity air systems require only the addition of a heating coil in the ductwork for zone control. For pressurized systems, a terminal reheat device containing a means of balancing and attenuating system noise is generally applied. The volume-regulating device is prebalanced for the desired air volume. As the room load falls, the room thermostat positions the heating valve to provide the energy to offset the load change. Reheat units may be located overhead in ductwork or under windows. Under windows is the preferred location in cold climates (design conditions below 20°F) to eliminate downdrafts. Primary air is fed through the floor or from column chases. Under-window types are usually provided with enclosures.

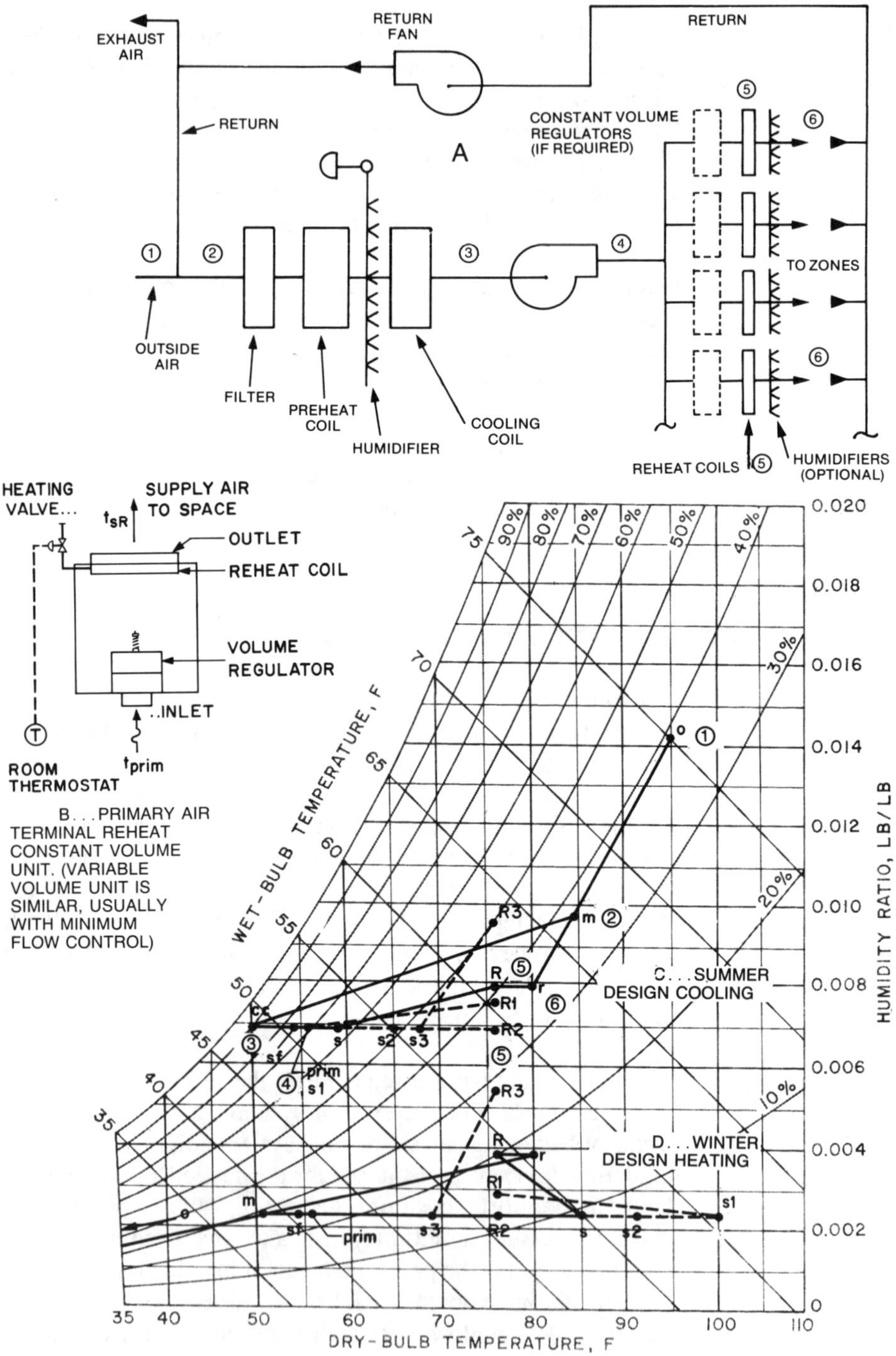

Fig. 6-7 Reheat System—Schematic Diagram and Psychrometric Chart

For interior areas, such as conference rooms (where a minimum flow setting is necessary) or top floors of buildings with roof loads, a heating means should be provided at the terminal.

Induction-type reheat units are usually installed at the building perimeter under the window. Routine maintenance is required to keep filters clean, as induction ratios are affected by increased pressure drop across the filter. Overhead installation is usual for all types. Access must be provided to service component parts. As coil surfaces are in the primary (not the secondary) recirculated airstream of low-temperature units, filters are not usually provided. Where units are installed in hospitals or hotels (high lint areas), routine cleaning is required.

The under-window induction type of reheat unit allows night heating with the air system shut down. Low-temperature types offer excellent control capability at reduced initial cost. For hospitals, return air systems are eliminated, reducing system complexity and air recirculation between patients' rooms.

6.9 DUAL-DUCT SYSTEMS

6.9.1 System Concepts

6.9.1.1 Basic Systems

Dual-duct systems distribute air from a central apparatus to the conditioned spaces through two parallel ducts. One duct carries cold air and the other warm air, thus providing air sources for both heating and cooling at all times. In each conditioned zone, a mixing damper controlled by a room thermostat mixes the warm and cold air in proper proportions to satisfy the heat load of the space. The return air is usually handled in a conventional manner through a single-duct system. Dual-duct systems designed to satisfy the conditions of no-load zones usually maintain acceptable room conditions at all times.

Among the advantages of dual-duct systems are good control of temperature and humidity, their ability to accommodate a variety of zone loads, ease of adding zones or subdividing existing zones, and adaptability to either constant- or variable-volume systems. Disadvantages are high energy consumption, high space requirements for two sets of ducts throughout the building, and the need for a large number of mixing boxes, which are expensive and may require maintenance. Moreover, economizer cycles are sometimes difficult to implement with dual-duct systems.

Dual-duct systems are used in hospitals, where VAV systems may not be appropriate, and they were extensively used in office buildings built during the 1950s and

1960s. Experience has shown that these systems produce good results when located in moderately humid climates, where outdoor design conditions do not exceed 78°F wet bulb and 95°F dry bulb, and when minimum outdoor air in the system does not exceed 35 to 40% of total air. Recently, dual-duct systems are less popular because of their high energy consumption. However, this can be mitigated by applying VAV control to reduce total supply air during periods of reduced cooling and heating loads, as explained in Section 6.9.1.4 and shown in Figure 6-11.

Figure 6-8 shows the simplest, least costly, and most compact apparatus arrangement for dual-duct air conditioning. The return fan shown may be eliminated on small installations if provisions are made to relieve the excess outdoor air from the conditioned spaces. They are generally required for economizer cooling (Section 9.3.7) in systems with substantial return air ductwork. The use of relief air fans instead of return air fans is gaining increased acceptance; they accomplish pressure relief with less energy use because they handle smaller air volumes—exhaust air only instead of all of the return air.

In summer, the cold air supply temperature should be kept just low enough to meet the space cooling and dew-point requirements. This is usually around 50 to 55°F with the air nearly saturated. In winter, it is sometimes reset 5 to 10°F higher for economy of operation; if internal loads are small, outdoor air may be used for the cold duct, permitting shutdown of the chiller. However, maintaining the cold air during the heating season at 55 to 60°F and raising the warm air temperature as the outdoor temperature decreases permit better humidity control and better air balancing between hot and cold ducts at the expense of increased energy use. In summer, the warm air temperature is governed by the return air from the conditioned spaces. It will always be above the average return air temperature, even though no heat is added by the heating coil, due to the contribution of outdoor air, fan energy, and heat from ceiling lights. In winter, the warm air temperature can be adjusted from the outdoor temperature. Bear in mind that an energy penalty results when the cold air temperature is lower than it needs to be to satisfy the cooling requirement of the "warmest" zone or, conversely, when the warm air temperature is higher than it needs to be to satisfy the heating requirement of the "coldest" zone.

The solid lines in Figure 6-8 represent average system air conditions, while specific zone conditions are shown by dashed lines. Thus, while the system as a whole is at peak cooling with an average supply air at state s yielding an average room condition R, a peak zone with a closed warm port would call for supply air at $s1$ yielding a room condition $R1$, and a conference room with full occupancy and no lights might call for $s2$ and yield $R2$. With all room thermostats set at 76°F, the supply air temperature to various zones can lie anywhere between $s1$ and $s2$, while room conditions are maintained at 76°F dry-bulb temperature but at a variety of humidity conditions between states $R1$ and $R2$. Figure 6-8c shows average winter conditions.

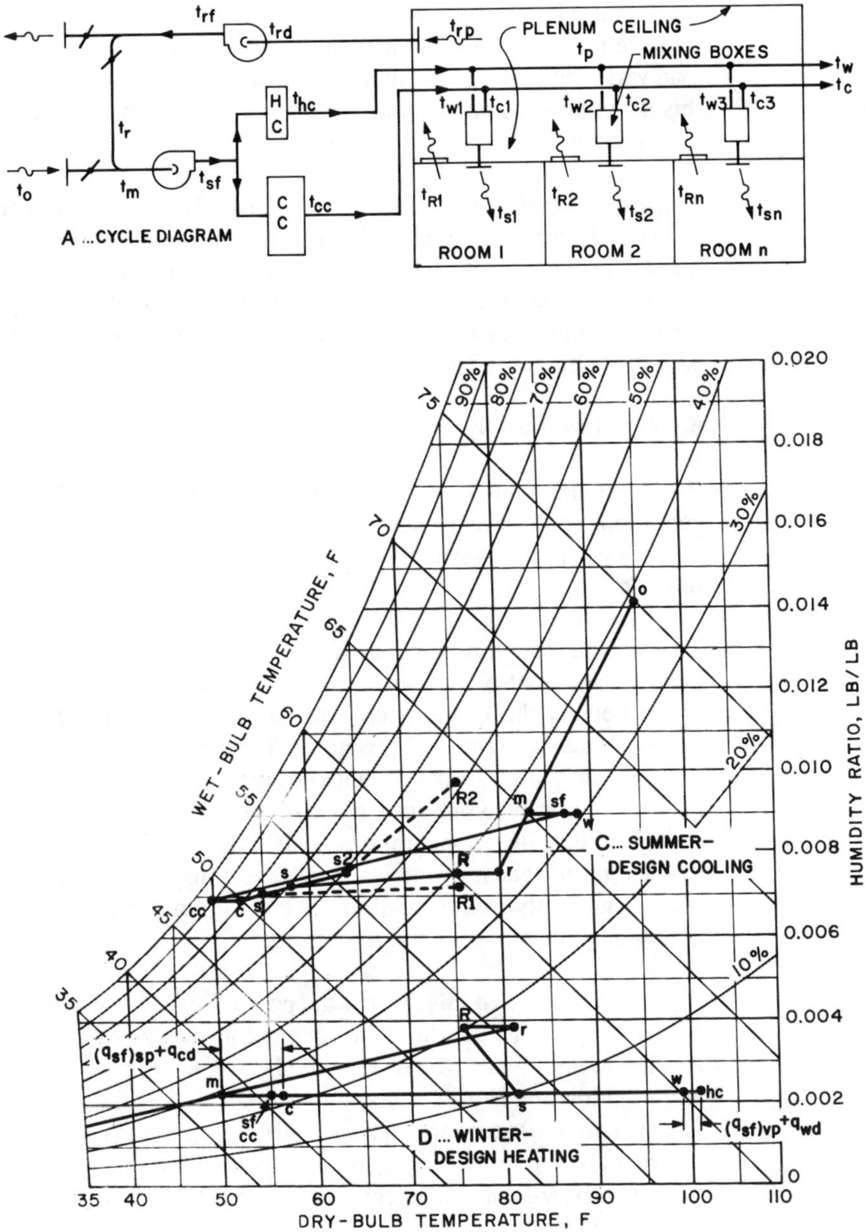

Fig. 6-8 **Single-Fan Dual-Duct System with Blow-Through Dehumidification—Schematic Diagram and Psychrometric Chart**

Figures 6-8b and 6-8c thus illustrate that a dual-duct system maintains accurate temperature in all zones year-round through the full range of each zone's internal sensible heat load. Although room humidities may rise moderately under normal load variations, only under unusual circumstances will they rise excessively. When this is not permissible, investigate dual-duct variations described in the following two sections.

Consider the fact that tail-end ducts with small quantities of airflow are much more influenced by duct transmission loads than large trunks and, particularly if uninsulated, may be substantially different in temperature from the supply air temperature leaving the central apparatus. While the mixing boxes automatically adjust for the changed duct temperatures, you must ascertain that these temperatures are sufficient to satisfy extreme zone heating and cooling demands.

6.9.1.2 Two-Fan System

When humidity must be kept below some specified value under all operating conditions, and psychrometric analysis shows that the Figure 6-8 cycle cannot accomplish that, investigate the two-fan blow-through cycle and the single-fan reheat cycle.

The two-fan blow-through system shown in Figure 6-9 is economical to operate and permits close control of room humidities in summer, when less than half the total air is handled by the warm duct. Each of the two supply fans is sized to handle approximately half the total system air. The bypass connections on the upstream and downstream sides of the supply fans do not require any flow control. Airflow through the bypass will always be determined by the thermal requirements of the conditioned spaces and by the temperatures maintained in the cold and warm ducts. The application of reheat to the warm duct coil may still be necessary to depress the relative humidity in the conditioned spaces during part-load operation coincident with high outdoor dew-point temperatures.

Another limitation of this cycle is the possibility of not meeting ventilating requirements in lightly loaded zones that are predominantly fed by the warm duct. Keep in mind, however, that with this type of all-air system, return air from full-load zones and from low-load zones is continuously mixed in the fan plenum. Since the same air never returns to the same zone and all the air to each zone is constantly mixed with air from all zones, no odor accumulation or environmental discomfort will occur, even if the low-load zone receives only used air from the rest of the building.

An alternative arrangement to the cycle shown in Figure 6-9 is obtained if the bypass connection on the discharge side of the supply fan is omitted and if both supply fans

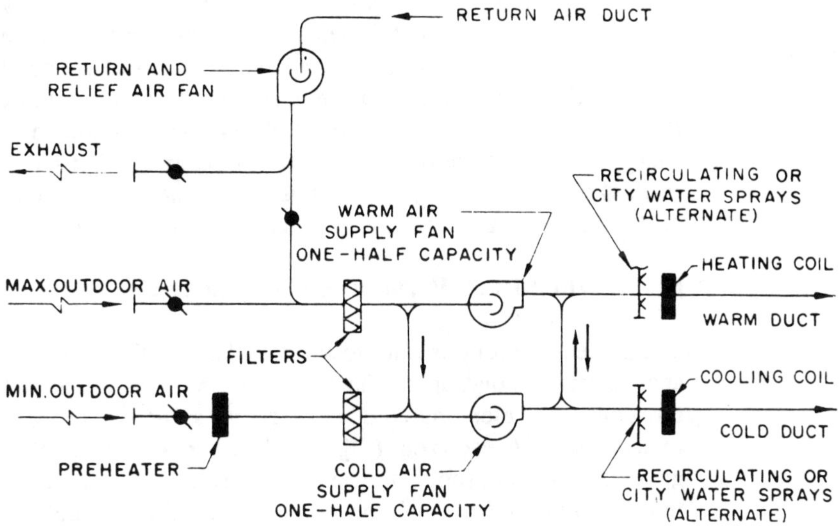

Fig. 6-9 Two-Fan Dual-Duct Blow-Through Cycle

are sized to handle approximately 100% of total system air. This arrangement eliminates the possibility of excessive air stratification in cold and warm chambers.

Its disadvantages are:

- Increased supply fan power requirements
- Increased complexity of controls due to the necessity of controlling static pressures in the duct systems
- Increased cost of apparatus rooms

6.9.1.3 Single-Fan, Dual-Duct Reheat System

Figure 6-10 shows an arrangement that permits high-limit control of relative humidities in the conditioned spaces through the entire range of operation. In this cycle, all the mixed air is cooled and dehumidified first, then divided into cold and hot decks, with reheat in the latter. Because of the continuous demand for reheat, the operating cost of this cycle is high. Also, additional refrigeration must be provided in most installations to treat the air bypassed to the warm-air duct under conditions of maximum loading.

Functionally, this cycle is equivalent to a conventional reheat system (Figure 6-7b). The only difference is that, instead of reheating air at a number of stations close to points of distribution, the reheat is applied centrally at one point. All operational and cost penalties of conventional reheat systems apply equally well to this cycle. According to ASHRAE *Standard* 90.1, this cycle should be used only when reheat energy is obtained from the refrigeration apparatus by hot-gas rejection or by another method of heat recovery. While this avoids most of the economic penalty due to heating, the refrigeration energy penalty remains the same.

6.9.1.4 Dual-Duct, Variable-Air-Volume Cycle

The preceding sections refer to systems that handle a constant volume in all zones at all operating conditions. The dual-duct VAV system accomplishes zone temperature control by blending cold and warm air subsequent to volume reduction of the total supply air to each zone (Figure 6-11). At maximum cooling (when zone cooling governs), the room supply temperature and air quantity requirements are identical to those of the constant-volume system; the volume regulators deliver maximum volume and the cold port is wide open. As the cooling load drops, the volume regulator reduces supply volume to the minimum acceptable value (see Section 6.7.2.1). On further decrease, the warm-air port begins to open as the cold air is decreased. Minimum volume is maintained below this cooling level and during the entire heating cycle. As an option, air volume from the warm-air duct may be increased as the heating requirement increases. Under this option, the warm duct temperature may be lowered for good mixing with room air and improved room air circulation.

This system enjoys all the advantages of the single-path VAV system and is superior to them at the expense of higher energy consumption. It consumes much less energy

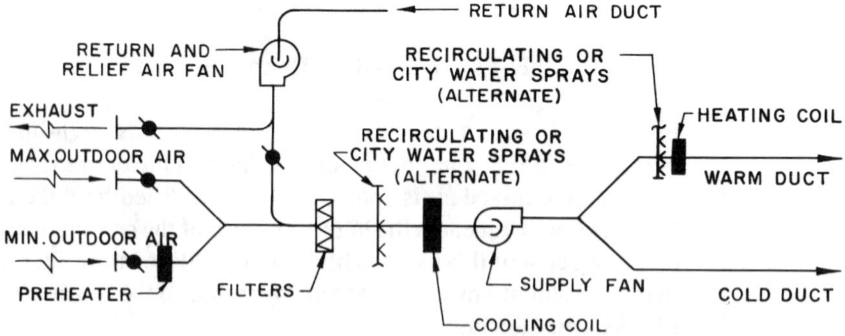

Fig. 6-10 Single-Fan Dual-Duct Draw-Through Cycle with Hot Deck Reheat

and is much lower in operating cost than constant-volume, single-path, terminal reheat systems.

The psychrometric chart of Figure 6-8 also applies to dual-duct VAV systems. However, VAV tends to shift the overall system process lines slightly because a fixed airflow raises the mixed-air temperature (state *m*) with reduced system air under any cooling load. Smaller air supply volumes in off-peak zones, whose loads are met with reduced cold air instead of a higher temperature blend of warm and cold air, tend to lower the supply temperature to such zones, thereby lowering the average system supply temperature and humidity.

6.9.2 Design Considerations

6.9.2.1 Overall Analysis

Equations (1) through (6) apply to all dual-path systems, which include dual-duct and multizone systems (see Section 6.10). Peak zone air supply requirements are complicated by the fact that mixing dampers of any kind always exhibit some leakage from the closed portion. The leakage ratio, expressed as a fraction of the total air handled by the damper, is a function of the system static pressure and the quality of the damper shutoff. For better-quality mixing boxes, it will vary from 0.03 to 0.07.

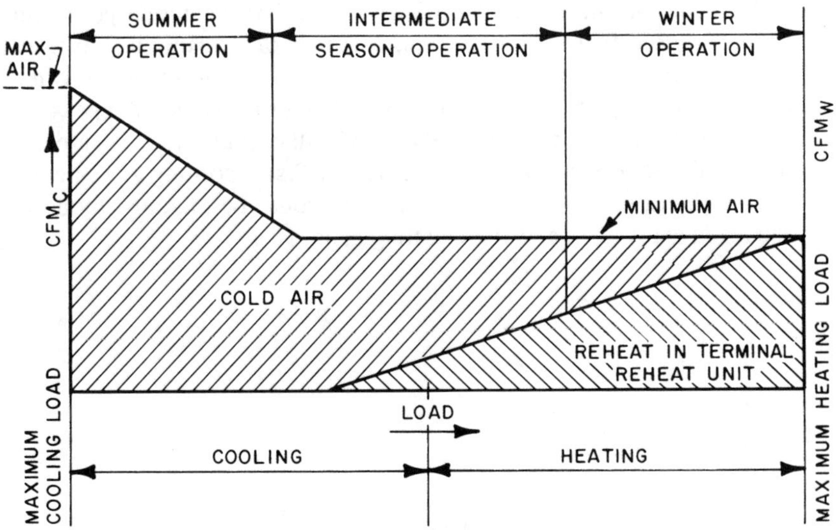

Fig. 6-11 VAV Dual-Duct or VAV Terminal Reheat System

For field-constructed damper assemblies (common in low-pressure systems) and multizone mixing plenum assemblies, it can vary from 0.10 to 0.20.

In any dual-path system, if the latent load in any zone peaks simultaneously with the system peak, some form of external or internal reheat may be supplied to that zone without affecting zone or system design air volume. However, if zone overcooling is avoided instead by admission of warm duct air, both zone and system volumes will be increased. If the SHF (sensible heat factor) slope is too steep, reheat applied at the zone is more desirable since it does not affect air volumes.

If a rational rather than an arbitrary solution for the summer warm duct temperature is desired, system supply volume must be obtained by trial and error. In this manner, that temperature may be optimized for minimum energy consumption by using only heat from fans, duct transmission, and ceiling plenums. Thus, heat need not be added from the heating coil at peak summer conditions unless some particular humidity control situation requires it.

6.9.2.2 Duct Sizing

Warm duct sizes are seldom derived from consideration of the warm air requirement under winter peak. Maximum flow in the warm duct will ordinarily occur during light summer loads or during intermediate season operation while warm duct temperatures are still kept at a low level. For this reason, warm duct sizes are usually fixed as a certain percentage of cold air duct area, as shown in Table 6-2.

Cold duct connections to each mixing box must be sized for the peak volume for each zone. Size cold duct branches serving a number of zone mixing boxes that have little or no diversity between them (*i.e.*, having the same exposure orientation with identical lighting loads and solar conditions) for the sum of the peaks of each zone. Size trunks, mains, or branches that serve either multiple exposures or zones with appreciable diversity (*i.e.*, noncoincident peaks) according to Table 6-2, column 2. This yields a somewhat larger area than would be obtained by a detailed study requiring analysis of several operating conditions.

TABLE 6-2 Duct Design Factors for Constant-Volume, Dual-Duct Design (from *Air Conditioning, Heating and Ventilating* 1964)

Ratio of Summer Cold Air to Total Air	Ratio of Zone Cold Air to Zone Total Air for Cold Duct Sizing	Ratio of Warm Duct Area to Cold Duct Area: All-Air Systems	Ratio of Warm Duct Area to Cold Duct Area: Systems with Supplementary Perimeter Heating System
1.0 to 0.9	1.0	0.80	0.70
0.89 to 0.85	0.95	0.80	0.70
0.84 to 0.80	0.9	0.85	0.75
0.79 to 0.75	0.85	0.85	0.75
≤ 0.74	0.80	0.90	0.80

In VAV dual-duct systems, warm air is not used to neutralize cold air at initial partial loads because cold air volume is reduced first. Therefore, oversizing of warm ducts is not necessary provided the terminal units accurately control minimum flow at varying system pressures. Table 6-2 does not apply to variable-volume systems.

In dual-duct systems where volume control is accomplished at terminal points, duct sizing becomes even less critical than for other systems and extreme precision is superfluous. Cold ducts may be sized by equal friction, static regain, or velocity methods or a combination of all three. Volume regulators in mixing units will absorb any pressure imbalance caused by the initial design, part-load operation, or future changes in load distribution, and they will produce a mechanically stable system under normal operating conditions regardless of the duct-sizing method used, provided the power needed in the longest duct is estimated correctly.

As pointed out previously, maximum flow through the warm duct usually occurs during intermediate season operation if low temperatures are maintained in a warm chamber. For this reason, warm duct sizes are usually determined as a ratio of cold duct sizes. The factors from Table 6-2 can be used for this purpose. However, some reduction in air distribution costs and space requirements can be achieved if the intermediate season warm duct temperature is elevated sufficiently to reduce the warm air demand at that time to be equal to, or less than, that required for winter heating.

6.9.2.3 Minimum Outdoor Air Control

In systems having constant-volume terminal units for all outlets, the fan capacity and static pressures on the fan itself remain fairly constant, and fixed minimum outdoor air dampers ensure adequate ventilation. Where the static pressure on the fan suction of VAV systems varies, however, it is important that the minimum outdoor air quantity be controlled to maintain ventilation needs. To do this, an instrument measuring the velocity pressure or the flow through the minimum outdoor air section can be installed. It controls the inlet vanes on a return air fan or overrides the thermostat operating the return air dampers to cause the proper amount of air to be drawn in through the minimum outdoor air connection.

6.9.3 Components

Two basic types of dual-duct mixing units are used—(1) under-window units, which are limited in size, capacity, architectural adaptability, and acoustical treatment, and (2) units designed for ceiling or remote installation, which are usually concealed or isolated from the conditioned space. Appearance, space, and acoustical treatment are, therefore, flexible. Capacities are generally from 100 to 4000 cfm, although, units up to 10,000 cfm may be considered. Larger capacities may justify separate air-conditioning systems.

Terminal units appropriate for dual-duct systems are shown in Chapter 2, All-Air Systems, in ASHRAE *Systems and Equipment* (1992). Terminal units, particularly those with capacities of more than 1000 cfm, frequently generate sufficient internal noise to require the installation of downstream attenuators or duct lining. Sound power spectra of noise generation for the range of unit sizes at various air deliveries and inlet static pressures are available from many manufacturers. The required attenuation can readily be determined by the procedures outlined in Chapter 42, Sound and Vibration Control, in ASHRAE *Applications* (1991). When units are installed within the occupied space with no intervening sound barrier, noise will be radiated directly from the terminal unit casing into the room. The amount of noise will approximate that emitted from the discharge. The sound from both sources in such situations should be added logarithmically to determine the resultant room sound levels.

6.10 MULTIZONE SYSTEM

6.10.1 System Description

In a multizone system, the requirements of the different zones of the building are met by mixing cold air and warm air through dampers at the central air handler in response to zone thermostats. The mixed conditioned air is distributed throughout the building by a system of single-zone ducts. Thus, the system is a multiplicity of single-zone systems. Return air is usually handled in a conventional manner by intermingling the return air from all zones. The two-fan cycle (Figure 6-9) has had little application in multizone system design, probably as a result of the higher first cost with smaller distribution systems, the lack of adaptability to packaged equipment design, and the availability of the cycle for more positive humidity control (Figure 6-10).

Advantages of the multizone system are good temperature control, the fact that ducts are the only system components outside the mechanical room, and ease of control. Disadvantages include the need for many individual ducts, which limits the number of zones; high energy consumption; inferior temperature control of the hot deck; and the difficulty of implementing an economizer cycle. Multizone systems are applied to small or medium-sized commercial buildings as well as to larger buildings with relatively few zones.

The multizone system is similar to the dual-duct system in all respects, except for the differences described in this section. It can provide a small building with some of the advantages of dual-duct systems at lower first cost with a wide variety of packaged equipment. However, most packaged air-handling apparatus lack the control sophistication for comfort and operating economy that can be built into dual-duct systems. The most common cycles used for multizone systems are shown in Figures 6-8 and 6-10.

VAV may be applied to multizone systems in a manner similar to that described in Section 6.7.1.2, with packaged or built-up systems having the necessary zone volume regulation and fan controls. However, it is seldom applied in this manner for entire distribution systems except for television studios and other critical noise level applications. More often, a few select rooms in a zone may incorporate VAV terminal devices when off-peak requirements permit this approach and air balance considerations indicate there will be no problems from omission of fan control or static pressure regulation.

Air volume and refrigeration capacities are treated the same as those described in Section 6.7. The air volume may be somewhat higher as a result of greater damper leakage.

6.10.2 Special Design Considerations

6.10.2.1 Single-Fan, Blow-Through Cycle

Zone mixing dampers are generally distributed horizontally along the hot and cold deck discharge plenum of the air handler. Each mixing box (Figure 6-8) may be considered analogous to a section of multizone damper allocated to each zone. The psychrometric analysis is identical to that shown in Figure 6-8b but for two exceptions:

1. The supply air blending occurs along line *cc-sf* for summer instead of *c-w*, and along line *sf-hc* for winter instead of *c-w*.

2. The duct temperature rise occurs along each zone duct instead of in the cold duct and warm duct.

6.10.2.2 Single-Fan, Draw-Through Cycle with Hot Deck Reheat

The cycle shown in Figure 6-10 is used more frequently in multizone than in dual-duct systems, mostly with packaged rooftop units (Section 6.11) employing hot-gas reheat for intermediate season fuel energy reduction, supplemented with steam, hydronic, direct electric, or direct fuel-fired heat exchangers. Supplementary heat is employed when the hot-gas reheat is inadequate (above 40 to 45°F ambient) and below these ambient conditions when the refrigeration apparatus is normally shut down in favor of economizer outdoor air cooling cycles.

6.10.3 Zone Dampers

The potential problems of zone volume variation increase with larger variance in zone sizes, complexity, and pressure of the distribution system, just as with dual-duct systems. Examine these aspects to determine the need for individual zone volume regulation and/or static pressure control, usually required when volume differences exceed a ratio of 2:1.

Normal quality manufactured dampers, whether field or factory manufactured, for built-up or packaged units have leakage factors as high as 0.10 to 0.20, especially if maintenance is poor. The effect of this leakage on supply air temperatures and volumes is appreciable and must be seriously considered from a practical standpoint. You can, however, specify dampers fabricated for 0.03 to 0.05 leakage; these are available for built-up jobs at an additional cost from some packaged unit manufacturers and from control damper manufacturers.

6.11 ROOFTOP SYSTEMS

Rooftop systems are small to medium-sized systems in which the bulk of the refrigeration machinery is located on the roof. Most have reciprocating or rotary compressors with air-cooled condensers. They are usually applied to low-rise buildings, such as small office buildings or shopping malls. *Unitary systems* are those in which all components of the refrigeration machine, including the air handlers, are located on the roof. *Split systems* are those in which compressor and condenser are located on the roof, but the evaporator(s) and air handler(s) are located inside the building. During the design, consider that unitary systems require relatively large roof penetrations for air ducts, while split systems require only small penetrations for the refrigerant piping, unless outside air intakes and exhausts are also ducted through the roof.

Rooftop systems may use single-zone, multizone, or VAV cycles. Their analysis is covered in earlier sections of this chapter. Small systems generally consist of only a single unit each of compressor, condenser, evaporator, and air handler, although more than one air handler is sometimes used for split systems. Larger single-zone, multizone, and VAV systems may have one or more of these components, depending on tonnage, space limitations, and other design criteria.

Among the reasons for the increasing popularity of rooftop systems are:

- They are easy to install and maintain; in low-rise buildings, roofs are easily accessible and relatively large compared to the size of the mechanical equipment.

- Machinery room(s) are eliminated, resulting in considerable first-cost savings.

- Placing the equipment on the roof does not reduce the rentable area, as does equipment located inside the building enclosure.

- Complete, prepackaged units with system performance guaranteed by the manufacturer are available. They relieve the designer of the need to match components in order to design a complete system, and contractors can even order them from a catalog. Some come skid-mounted for fast installation.

- A large number of small tenants can each own and operate a separate unit, relieving the landlord of the need for installing, servicing, and paying for the operation of the equipment. While the COP of small units is lower than that of large centrifugal units, each small unit is operated only as much as its owner requires. The total energy use of a number of rooftop units may, therefore, be smaller than that of a large central unit that must run whenever anybody in the building requires space conditioning.

Frequently, rooftop units are used for small jobs with tight budgets. Avoid pitfalls connected with these conditions, such as

- Rooftop units without return or relief air fans may encounter problems meeting exhaust air requirements specified in the applicable building code under fire conditions.

- Direct-expansion coils in rooftop units are usually provided in four rows. Their sensible heat factor (SHF) may be 0.6 to 0.75. What is needed in conventional comfort cooling (*not* in nightclubs, discos, or health clubs) is an SHF of 0.9 to 0.98. This means that you must select the next larger size unit if you wish to provide supply air no more than 20°F colder than room air. Thus, a 50,000-Btu/h load may require a 7.5-ton unit instead of a 5-ton unit, and a 100,000-Btu/h load may require a 15-ton instead of a 10-ton unit.

- Packaged equipment using direct-expansion cooling coils and hot gas, electric, or direct fuel-fired hot deck heat exchangers generally has step controls for hot and cold deck capacity. This substantially limits the quality of temperature and humidity control unless specific control provisions are made to avoid this.

- Air intakes should be placed as far as possible away from air exhaust stacks and plumbing vents, which may be difficult to do when a large number of rooftop units are installed on a relatively small roof.

- Failure to install access walkways to rooftop units may result in damage to the roof and cause water leaks into the space below.

Additional information on rooftop systems and equipment is presented in the *Systems and Equipment* volume of the ASHRAE *Handbook* (1992).

6.12 REFERENCES

Air Conditioning, Heating and Ventilating. 1964.

ASHRAE. 1989. Energy efficient design of new buildings except low-rise residential buildings. *Standard* 90.1-1989.

ASHRAE. 1991. *Handbook—Applications.*

ASHRAE. 1992. *Handbook—Systems and Equipment.*

ASHRAE. 1993. *Handbook—Fundamentals.*

Avery, G. 1984. VAV economizer cycle: Don't use a return fan. *Heating/Piping/Air Conditioning*, August.

Mutammara, A.W. and D.C. Hittle. 1990. Energy effects of various controls strategies for variable-air-volume systems. *ASHRAE Transactions* 96(1).

Chapter 7 AIR-AND-WATER SYSTEMS

7.1 INTRODUCTION

Air-and-water systems condition spaces by distributing air and water sources to terminal units installed in the spaces. The air and water are cooled or heated in a central mechanical equipment room. The air supplied is *primary air*; the water supplied is *secondary water*. Sometimes an electric heating coil is included in lieu of a hot water coil.

7.2 APPLICATIONS

7.2.1 Advantages and Disadvantages

Because of the greater specific heat and the much greater density of water compared to air, the cross-sectional area required for piping is much smaller than that for ducts to accomplish the same cooling load. Thus, the ducts in an air-and-water system are considerably smaller than those in an all-air system, which saves building space because a large part of the cooling load is removed from the building by water pipes. If the system is designed so that the air supply is equal to the ventilation requirement or to balance exhaust requirements or both, the return air system can be eliminated. The air-handling system is smaller than that for an all-air system, yet positive ventilation is ensured. All zones can be individually controlled and their cooling or heating demands can be satisfied independently and simultaneously. By operating the secondary water system, space heating can be accomplished without operating the air system. When all primary air is taken from outdoors, cross-contamination between rooms cannot occur.

Due to the low air quantities, high-quality filters are required, and design for intermediate season operation is critical. Changeover operation is difficult and requires a knowledgeable staff. These systems are generally limited to perimeter spaces. Controls are more complicated than for all-air systems. Induction and fan-coil units require frequent in-space maintenance, and humidity cannot be tightly controlled.

7.2.2 Suitability to Building Types

Air-and-water systems are applied primarily to the exterior spaces of buildings with high sensible loads and where close control of humidity is not a primary requirement. These systems work well in office buildings, hospitals, schools, apartment houses, research laboratories, and other buildings where their capabilities meet the performance criteria. In most climates, these systems are installed in exterior building spaces and are designed to provide (1) all the required space heating and cooling needs and (2) simultaneous heating and cooling in different parts of the building during intermediate seasons.

7.3 SYSTEM CONCEPTS

An air-and-water system includes central air-conditioning equipment, duct and water distribution systems, and room terminals. The latter can be induction units, fan-coil units, or conventional supply air outlets combined with radiant panels. Usually, the primary air system is constant volume. It provides all the ventilation, all the latent cooling, and some of the sensible cooling. The water system may be two-, three-, or four-pipe and provides the remainder of the sensible cooling. These water systems are discussed in detail in Chapter 8. Primary air is either all outdoor air or, if a return air system is provided, a mixture of outdoor and return air. The fraction of return air is always small. Thus, where freezing temperatures are encountered, a preheater is required at the central equipment room. Filters should be of higher efficiency than in an all-air system to compensate for the reduced dilution of space contaminants.

In a typical two-pipe system, primary air is cooled by a primary water chiller (although in smaller systems a direct expansion cooling coil may be used), and the room units, either induction or fan coil, are supplied by a secondary water system. The primary air system contains an optional reheat coil, while the secondary water system contains a chiller in addition to the heating coil.

In an alternative arrangement, horizontal fan coil units are suspended in each zone. Their coils are furnished with cold water in summer and warm water in winter. The ceiling of the corridor is furred down to form an insulated plenum in which conditioned (cooled and dehumidified) outside air is circulated. This air furnishes ventilation and dehumidification. The room units draw in some of the air from the corridor plenum, mix it with some return air from the room, condition it, and discharge the conditioned mixture into the room.

When individual zones are large, secondary water units are not located within the room but are some distance away, and their output air is ducted to the room terminals through a duct network.

Another type of air-and-water system is one in which individual complete units are used in each zone to perform the air conditioning. In this arrangement, units are installed around the outside wall of the building, usually one per room for small rooms and more when needed in the larger rooms. These units are complete with casing, fans, water coil, filter, and fan motor. In the back or bottom of the units is an opening that can be connected to an outside air intake. Another grilled opening in the front of the unit exists for drawing in return or room air. The coils in the units are supplied with chilled water from a refrigeration plant in the summer and with hot water in the winter from a hot water heater or heat exchanger. Each zone features individual control of the temperature. However, since only one coil is ordinarily used, the relative humidity depends on the room temperature, the length of time that the unit operates to maintain the temperature, and the chilled water temperature. To provide proper control of relative humidity, a separate heating coil is sometimes included so that dehumidification and reheat can be accomplished. This requires the use of a four-pipe water system.

Radiant panels for heating and cooling can also be applied to air-and-water systems. They are compatible with two-, three-, and four-pipe systems. These panels are discussed in Chapter 6, Panel Heating and Cooling, in ASHRAE *Systems and Applications* (1992).

The controls in air-and-water systems are usually arranged so that, at maximum required cooling, the primary air is supplied at lower than room temperature and cold water is circulated through the secondary coil. Between seasons, the primary air is reheated and cold water is circulated through the secondary coils. As the outdoor temperature decreases further, the warm primary air must be switched to cold primary air and the cold secondary water must be switched to warm secondary water. This is called the *changeover point* or *changeover temperature*. It is not the same for all sides of the building; therefore, the system should be zoned.

7.4 PRIMARY AIR SYSTEMS

Figure 7-1 shows the primary air system for a two-pipe system with induction terminal units. Such systems are comparable to the single-zone systems discussed in Section 6.6. If a given humidity ratio is to be maintained in winter, a humidifier must be installed. An air-water induction unit is shown in Figure 7-2 and a fan-coil unit in Figure 7-3.

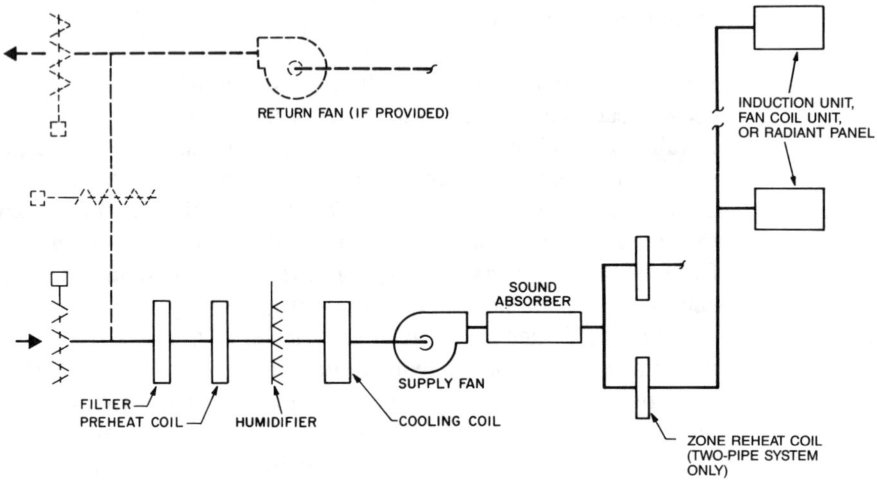

Fig. 7-1 Primary Air System

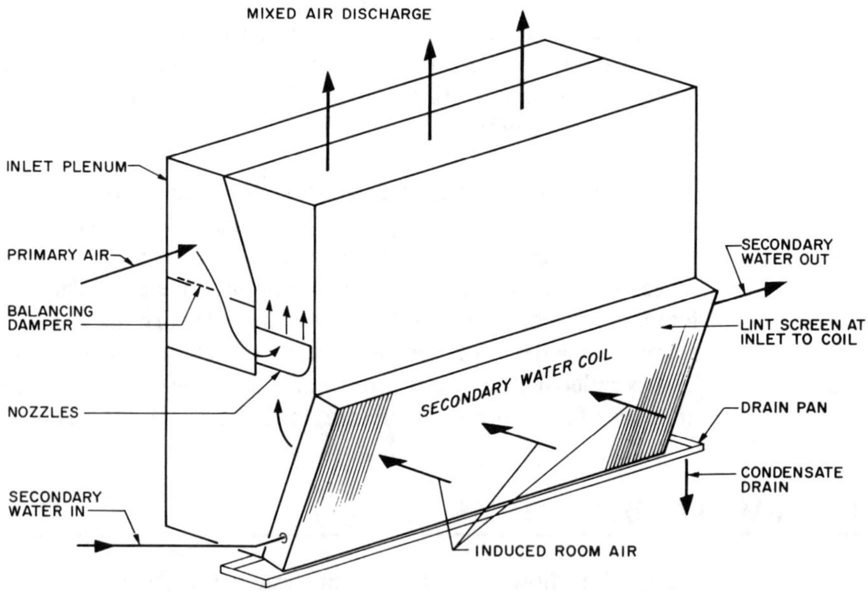

Fig. 7-2 Air-Water Induction Unit

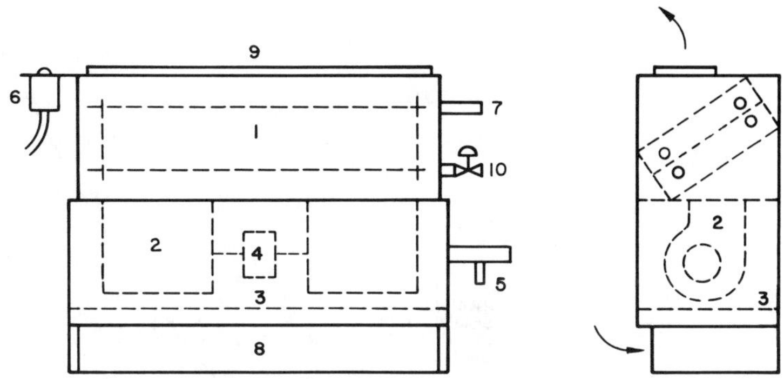

Fig. 7-3 Typical Fan Coil Unit

The quantity of primary air for each zone is determined by the

- Ventilation requirement
- Sensible cooling requirement at maximum cooling load [this equals the room cooling load less the sensible capacity of the space (water) cooling coil]
- Maximum sensible cooling requirement on changeover to the winter cycle when chilled water is no longer circulated through the coils

Since no drip pans are provided in the conditioned space, the primary air must be sufficiently dehumidified to prevent condensation at all times. The maximum permissible moisture content of the supply air can therefore be determined from

$$W_{pa} = W_r - q_l/(4840\ Q_p) \tag{1}$$

where

W_{pa} = humidity ratio of primary air, lb water/lb dry air
W_r = maximum desired humidity ratio in zone, lb water/lb dry air
q_l = room latent load, Btu/h
Q_p = primary air rate, cfm

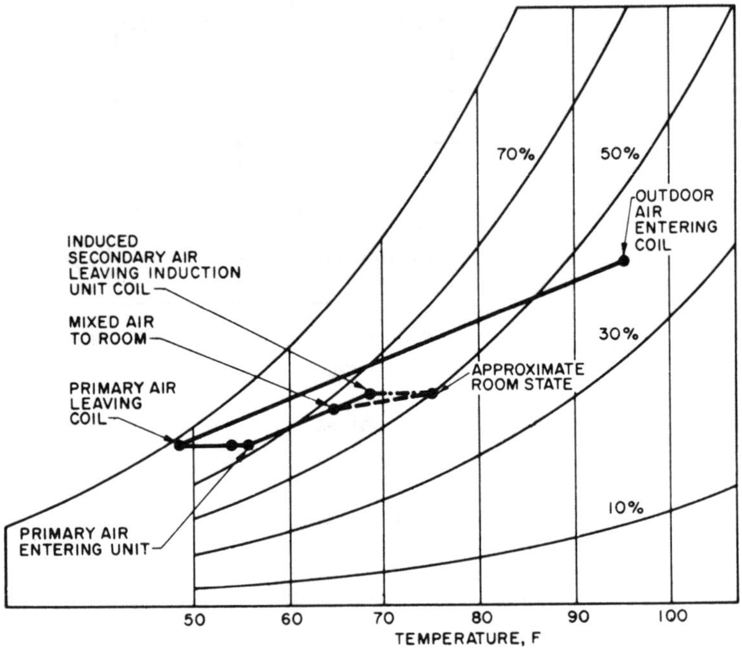

Fig. 7-4 Psychrometric Chart at Summer Peak

To achieve this, the air leaving the cooling coil must generally be 50°F or below and fully saturated, which requires deep coils and a chilled water temperature of about 42°F or lower. Reheat coils are required for two-pipe systems but not for four-pipe systems.

The psychrometric behavior of an induction air-and-water system is shown in Figure 7-4 at the summer peak and in Figure 7-5 at the winter peak. These figures are also valid for conditions in which the primary air is supplied by a fan coil unit with a directly connected air supply. The figures illustrate how, in the summer, all the latent cooling and part of the sensible cooling are performed by the primary air, while the secondary coil provides additional sensible cooling. During the winter, preheated primary air is shown to be humidified, while the secondary coil performs all heating.

7.5 SECONDARY WATER SYSTEMS

Water systems are discussed in detail in Chapter 8; refer to that chapter before attempting the design of an air-and-water system.

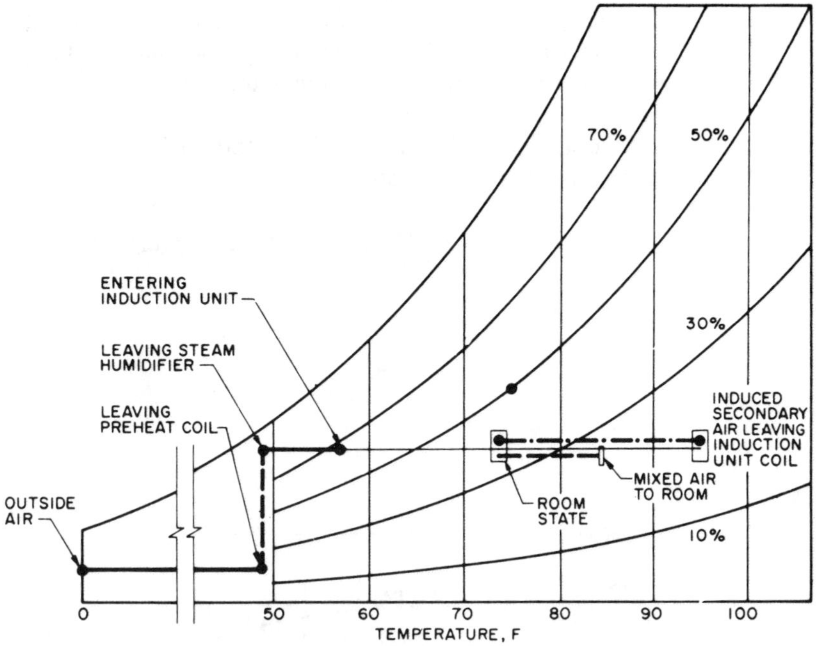

Fig. 7-5 Psychrometric Chart at Winter Peak

Individual room temperature is controlled by varying the capacity of the coil by regulating the water flowing through it or the air passing over it. During winter, the coil may be converted to a heating coil, or a separate heating coil may be provided. The cooling capacity of the coil is determined by the room sensible load reduced by the sensible cooling capacity of the primary air. Most air-and-water systems deliver cool primary air throughout the year. The heating capacity of the coil must therefore be designed for the room's heating load plus the primary air cooling effect. Design the coils conservatively to overcome any imbalances in the secondary water system.

7.5.1 Effect of Pump Curves on System Performance

The key variables defining the performance of a pump/piping system are the pressure rise in the pump (pressure drop in the piping) and the water flow rate. The relationship between the pump and piping determines the balance point where the system operation satisfies both. Typical performance curves of centrifugal pumps are shown in Figure 7-6 for a family of pumps operating at 1770 rpm. At low flow rates, the rise in pressure (head) is fairly constant; at high flow rates, it ultimately drops off to zero. The magnitude of the pressure rise is predominantly a function of the tip speed of the impeller, which is reflected in the higher heads with the larger diameter impellers. If the pressure rise provided by the pump of largest diameter available is too low, choose a pump that operates at a higher speed, *e.g.*, 3550 rpm.

For straight pipe, the head loss is proportional to the square of the velocity (and thus the square of the flow rate):

$$\text{Head loss (in feet)} = f(L/D)(V^2/2g)$$

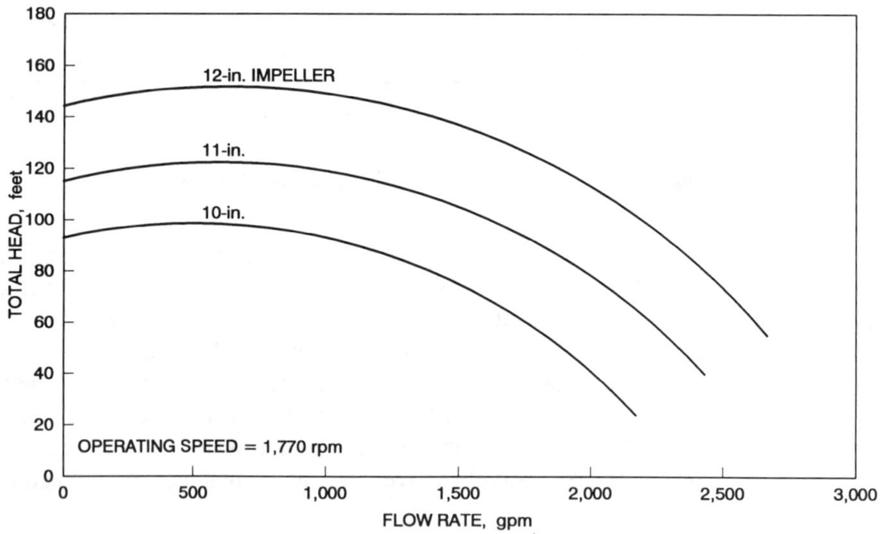

Fig. 7-6 Performance of Three Pumps with Different Impeller Diameters Operating at 1770 rpm

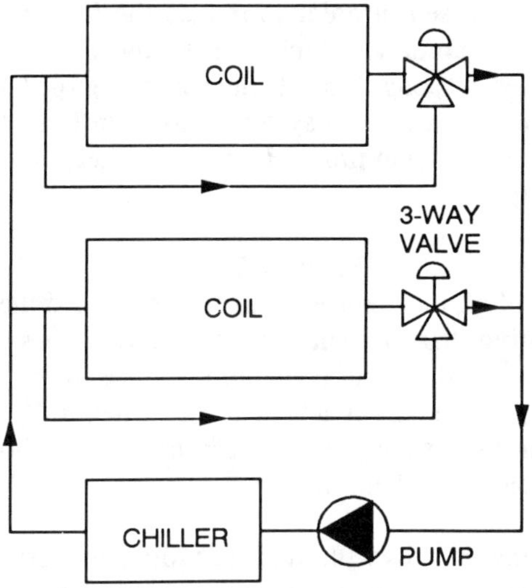

A Direct return with three-way valves

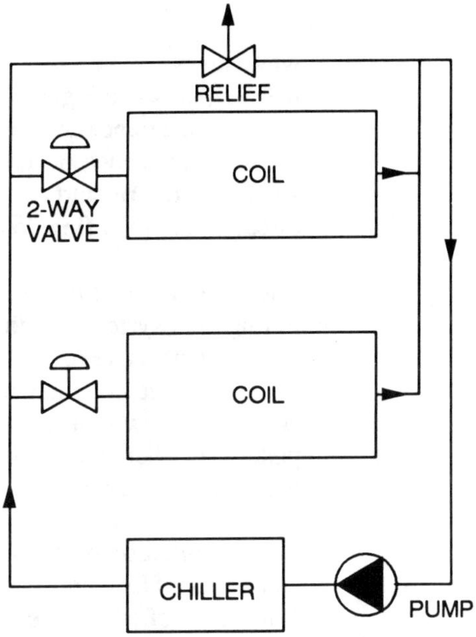

B Reverse return with two-way valves

Fig. 7-7 Flow Networks

where

f	=	dimensionless friction factor (see Chapter 2, Fluid Flow, in 1993 ASHRAE *Fundamentals*)
L	=	length of pipe, ft
D	=	internal pipe diameter, ft
V	=	flow velocity, ft/s
g	=	acceleration of gravity, 32.2 ft/s^2

For fittings, such as elbows and tees, the head is proportional to the square of the flow rate, as evidenced by the expression of head loss of a fitting as a certain number of velocity heads $V^2/2g$. In summary, the performance of a water distribution system is the combination of a pump characteristic such as the one in Figure 7-6 and a squared relation for the piping network.

A further consideration is how the pump/piping system behaves over a range of demands under different methods of control. In the two two-pipe networks shown in Figure 7-7, the coils in (A) are controlled by three-way valves while in (B) they are controlled by two-way valves. The intent of three-way valves is to maintain a somewhat constant pressure drop across the coil-valve combination, while with two-way valves the pressure drop is increased in order to reduce the flow rate. The network shown in Figure 7-7A is "direct return," which has the drawback of a higher pressure difference across the valves and coils closest to the pump. A more balanced pressure difference occurs in a "reversed return" system, shown in Figure 7-7B, where the return starts at the nearest coil and proceeds to the further ones before directing the total flow to the pump.

The relief valve at the end of the flow path in Figure 7-7B allows a low flow rate through the system and the pump if all the control valves close off and the pump discharge pressure rises. An operating characteristic of the two-way valve system is that, when the coils experience a low flow rate demand, the pressure difference across the flow control valve rises. The valves adjust to near their fully closed positions, where small changes in their stem positions can change the flow rate appreciably and bring with it the danger of hunting.

The combination of the pump characteristics with those of piping networks under some form of flow control illustrates the system behavior. Figure 7-8 shows the head flow curve of one of the pumps (the one with an 11-in. impeller from Figure 7-6). Pump efficiency and required pump power as functions of flow rate are also supplied. Superimposed on the pump characteristic are several piping network characteristics with their expected parabolic shape. The lowest piping curve might represent the three-way valve network of Figure 7-7A or the two-way valve network of Figure 7-7B with the valves fully open. The intersection of the pump and the piping curves is the "balance point," indicating the head and flow rate condition where that combination will operate. Computer simulation permits analysis of components so that a system of 20 coils, for example, can be optimized by using 4 three-way valves at the hydraulically most remote points and 15 two-way valves without the need for relief valves.

With the three-way valve control, the piping curve remains essentially constant regardless of the load demanded by the coils. The pump continues to deliver 2250 gpm against a head of 100 ft. With the two-way valve control, however, the piping network pressure drop increases as the flow rate is reduced, which produces a different piping network curve for each new adjustment. Two different curves with their different balance points are shown in Figure 7-8. Pump performance changes significantly over the range of operation with the changing piping network. As the flow rate is decreased and the operating condition moves upward and to the left along the pump characteristic, the pump efficiency increases from 75% to and through its maximum of 85%, then decreases again. The power requirement of the pump decreases as the water flow is throttled off.

One of the basic decisions pertaining to system cost is shown in Figures 7-6 and 7-8. Specify a system of large-diameter pipe with its resultant low pressure drop. While

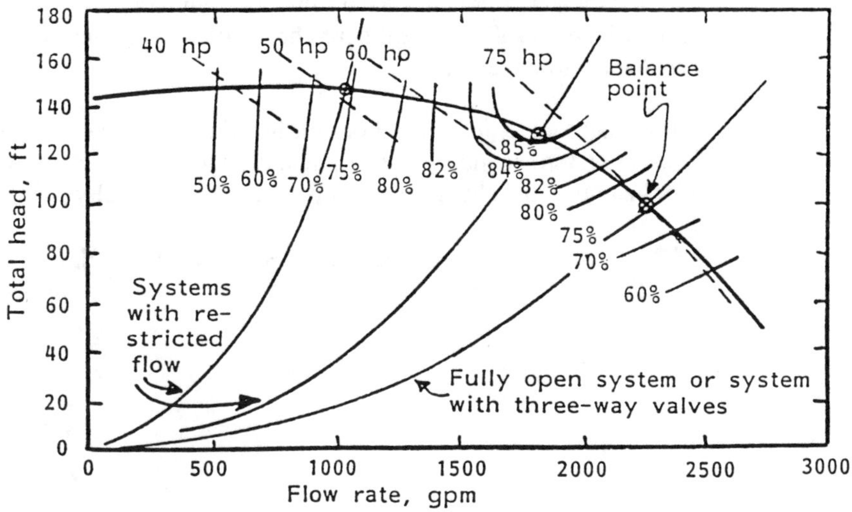

Fig. 7-8 Pump and Piping Network Curves

this may increase the piping cost, it may be possible to use a smaller pump, such as the one with the 11-in. diameter impeller shown in Figure 7-6. This pump might be slightly lower in first cost, but its big advantage lies in the lower energy costs for pumping. Both the system curve and the pump curve will lie lower on the graph in Figure 7-8, and the power requirements at the balance point will be reduced. Note how changes in one element of the system (the piping) impose changed operating conditions on another element of the system (the pump).

7.5.2 Two-Pipe System

7.5.2.1 Changeover

For a system using outdoor air, there is an outdoor temperature at which mechanical cooling is no longer required. At this point, the cooling requirement can be met by using outdoor air only. At lower temperatures, heating rather than cooling is needed. All-air systems capable of operating with 100% outdoor air rarely require mechanical cooling below 50 or 55°F. However, since air-to-water systems require considerably less air than all-air systems, they may require mechanical cooling well below an outdoor temperature of 50°F.

The transition from summer operation to intermediate season operation is gradual, consisting of raising the primary air temperature as the outdoor temperature decreases. This prevents zones with small cooling loads from becoming too cold. The secondary water remains cold during both summer and intermediate season operation. Figure 7-9 illustrates the psychrometrics of summer-cycle operation near the changeover temperature.

As the outdoor temperature drops further, the changeover temperature is reached. At this point, the secondary water system changes from cold water to provide hot water for heating. This changeover temperature is empirically given by (Carrier 1965):

$$t_{co} = t_r - [S + L + P - 1.1(t_r - t_p)]/T \qquad (2)$$

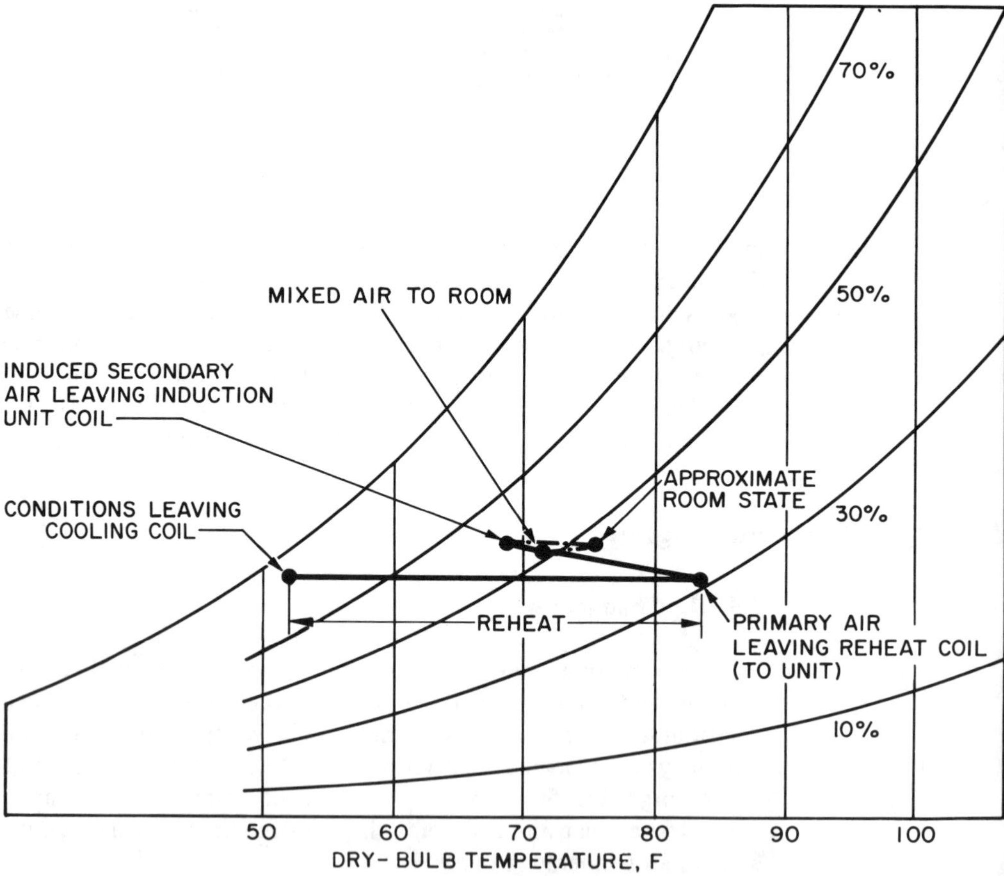

Fig. 7-9 Psychrometric Chart for Off-Season Cooling of Two-Pipe System Prior to Changeover

where

L = sensible heat gain from people, Btu/h
P = heat gain from lights, Btu/h
S = net solar heat gain at time of changeover, Btu/h
T = transmission per degree, defined as heat flow per °F difference between space and outdoor temperatures, which includes transmission through walls, windows, and roofs, if applicable
t_{co} = changeover temperature, °F
t_p = primary air temperature after changeover, normally taken as 48°F
t_r = room temperature at changeover, normally taken as 76°F

System changeover can take several hours depending on system size, and it usually results in a temporary upset of room temperatures. Good design, therefore, includes provisions for operating the system with either hot or cold secondary water over a range of 10 to 15°F below the changeover point. This minimizes the frequency of changeover by making it possible to operate with warm air and cold water on cold nights when the outdoor temperature rises above the changeover temperature during the day. This limits operation below the changeover point to periods of protracted cold weather.

Optional hot or cold water operation below the changeover point is provided by increasing the primary air reheat capacity so that it can provide adequate heat during colder outdoor temperatures. Figure 7-10 shows temperature variations of primary air and secondary water for a system with changeover.

7.5.2.2 Nonchangeover

Nonchangeover systems should be considered in order to simplify system operation for buildings with mild winter climates. Such a system operates on the intermediate season cycle throughout the heating season, as shown in Figure 7-11. Heating may be provided during unoccupied hours either by operating the primary system with 100% return air or by switching the system and room controls to hot water for gravity heating without air circulation.

7.5.2.3 A/T Ratio

A major design parameter in air-and-water systems is the A/T ratio, defined as

$$\text{A/T Ratio} = (\text{Primary air cfm})/(\text{Transmission per degree})$$

where the denominator is as defined earlier. All spaces on a common primary air zone should have approximately the same A/T ratio, which is used to establish the primary air reheat schedule during the intermediate season. Spaces with A/T ratios higher than the design base A/T will be overcooled during light cooling loads and spaces with ratios lower than the design ratio will lack sufficient heat during cool weather above the changeover point.

Calculate the minimum A/T ratio for each space by using the primary air quantity needed to satisfy the requirements for ventilation, dehumidification, and both summer and winter cooling as defined in Section 7.4. The design base A/T ratio is the highest ratio thus obtained, and the required airflow to each space is increased as required to obtain a uniform A/T ratio in all spaces.

For each A/T ratio there is a specific relation between the outdoor air temperature and the temperature of the primary air supplied to the unit that will maintain the room at 72°F or higher during conditions of minimum room cooling load. Figure 7-12

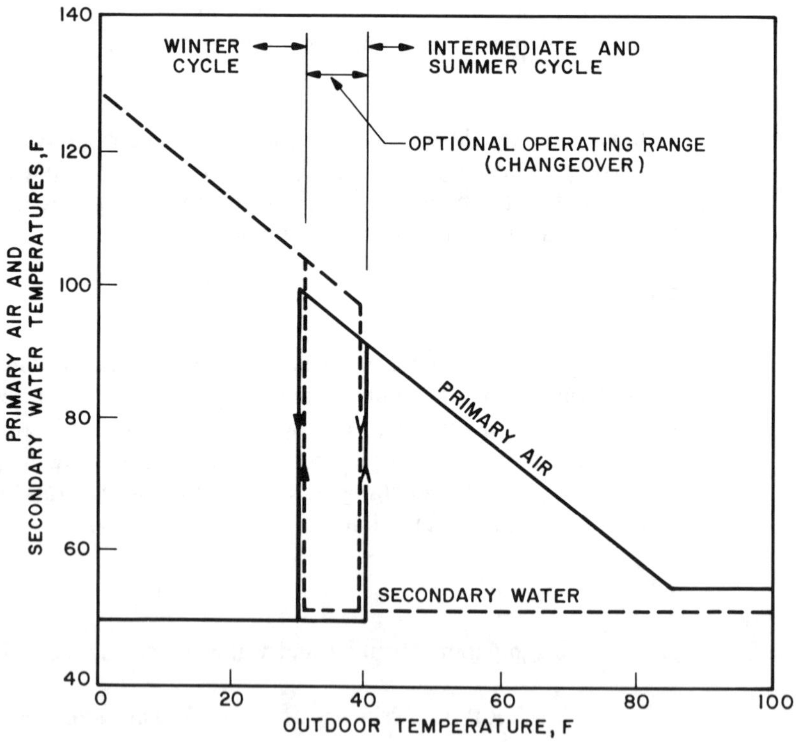

Fig. 7-10 Typical Temperatures for Changeover System

illustrates this relation based on an assumed minimum load of 10 times the T-value, *i.e.*, a difference of 10°F between outdoor and space temperatures.

Deviation from the A/T ratio up to 0.7 is permissible for massive buildings with high thermal inertia, but A/T ratios should be closely maintained for buildings with large glass areas, with curtain wall construction, or on systems with a low changeover temperature.

7.5.2.4 Zoning

A properly designed single-zone two-pipe system can provide good temperature control throughout the year. Its performance can be improved by zoning in several ways:

- *Primary air zoning to permit different A/T ratios on different exposures.* All spaces on the same primary air zone should have the same A/T ratio. If minimum A/T ratios for different spaces of one zone differ because of different solar exposures, increase air quantities for some of the spaces. It is often useful to zone primary air for northerly (north, northeast, northwest) exposures on a separate zone with reduced air quantities.

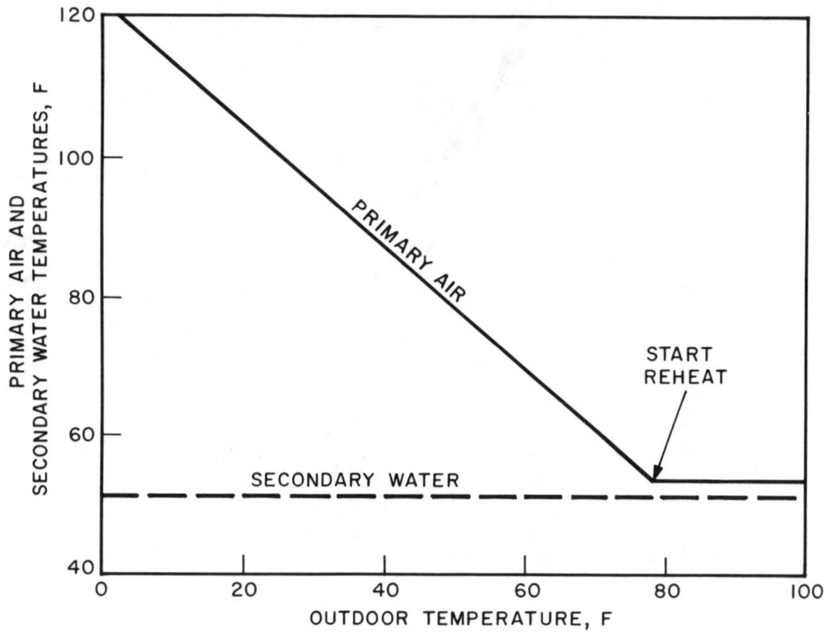

Fig. 7-11 Typical Temperatures for Nonchangeover System

- *Primary air zoning to permit compensation of primary air temperature.* Peak cooling loads for southern exposures usually occur during winter months because of high solar loads. Therefore, primary air zoning can be used to reduce air quantities and unit coil sizes for these exposures. Units can be selected for cold instead of reheated primary air, thus reducing operating costs on all solar exposures by reducing primary air reheat and secondary water refrigeration needs.

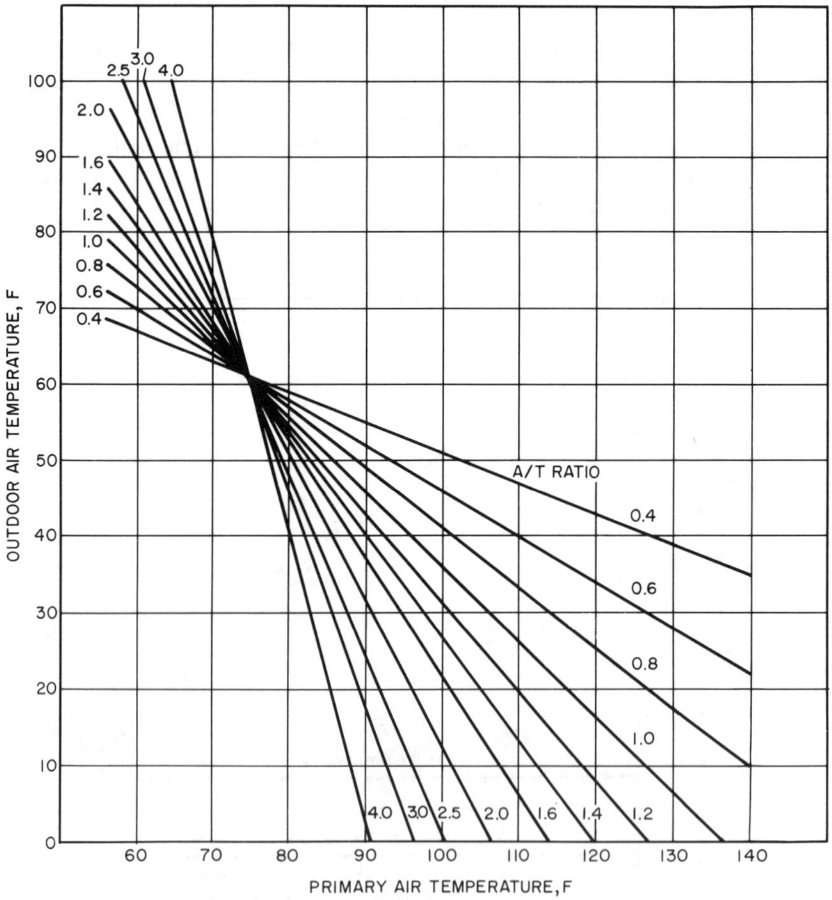

Fig. 7-12 Primary Air versus Outdoor Air Temperatures

- *Both air and water zoning to permit different changeover temperatures.* Separate air and water zoning permits northerly exposures to operate on a winter cycle with warm secondary water at outdoor temperatures as high as 60°F, whereas other exposures must operate at intermediate or summer cycles at those temperatures. Primary air quantity can be lower due to separate A/T ratio, and operating costs due to reheat and refrigeration are reduced.

7.5.2.5 Evaluation

The two-pipe air-and-water system was the first to be developed and is frequently installed. Least expensive of these systems to install, it is less capable of handling wider variations in loads than three- or four-pipe systems. Changeover is cumbersome, and operating costs are higher than for four-pipe systems.

7.5.3 Three-Pipe Systems

Because of their wasteful energy performance (see Section 8.6), these systems are rarely installed. Detailed discussions can be found in Chapter 4 of the 1973 and 1976 ASHRAE *Systems* volumes.

7.5.4 Four-Pipe Systems

In four-pipe systems (see Section 8.5), the terminal unit is provided with two independent secondary water coils—one served by hot water, the other by cold water. As an alternative, a common coil may be used that is served by two supply and two return pipes from the chilled and hot water loops, respectively (Figure 7-13). The primary air is cold and remains at the same temperature year-round. During peak cooling and heating, the four-pipe system performs in a manner similar to the two-pipe system with essentially the same operating characteristics (Figures 7-4 and 7-5). During intermediate seasons, any unit can be operated at any capacity from maximum cooling to maximum heating without regard to the operation of any other unit. Thus, zoning is not required for either the primary air or the secondary water system.

Standby secondary water pumps and heat exchangers should be considered for critical applications, such as hospitals. However, on two-pipe systems, a substantial degree of standby protection for heating is obtained by operating the primary air system on its summer cycle with warm air temperatures when the secondary system fails. No such capability exists for a four-pipe system, whose only source of heat is the secondary water system. Therefore, the installation of standby equipment or the

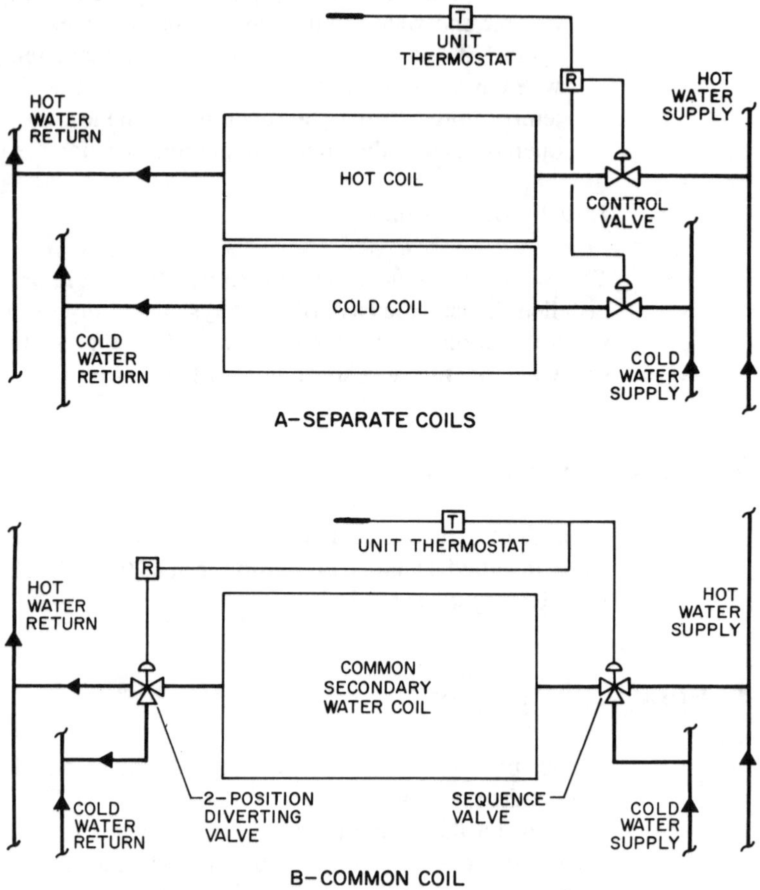

Fig. 7-13 Room Unit Control for Four-Pipe System

stocking of critical parts for the secondary water system should be considered for four-pipe systems with critical applications.

This system is more flexible than two- or three-pipe systems. While its installation cost is higher, its efficiency is higher, and its operating cost is lower than that of the other two systems. Operation is simpler than for the two- or three-pipe system. No changeover is required.

7.6 REFRIGERATION LOAD

The refrigeration load equals the total building block load to be removed by the combination of air and water systems. It equals:

$$R = 4.5\, Q_p\, \Delta h + q_s - 1.1\, Q_p\, (t_r - t_s) \tag{3}$$

where

q_s = room sensible heat for all spaces at peak load, Btu/h
Q_p = primary airflow rate, cfm
R = refrigeration load, Btu/h
t_r = average room temperature for all rooms at time of peak, °F
t_s = average primary air temperature at point of delivery to rooms, °F
Δh = enthalpy difference of primary airstream across cooling coil, Btu/lb

The first term on the right side of Equation (3) is the load on the primary air cooling coil, while the remaining two terms represent the load on the secondary water system.

7.7 ELECTRIC HEAT OPTION

Electric heat may be used either by installing a central electric boiler and hot water terminal coils or individual electric resistance heat coils in the terminal units.

One design approach is to size the electric terminal resistance heaters for the peak winter heating load and to operate the chilled water system as a nonchangeover cooling-only system, thus avoiding the problems of hot water/chilled water changeover. One disadvantage of this system is the wide swing in supply air temperature if the electric coil is cycled instead of modulated.

Another approach is to use a small electric resistance terminal heater for the intermediate season heating requirements in conjunction with a two-pipe changeover-type chilled water/hot water system. The electric heater is used for outdoor temperatures of more than 30 or 40°F. System or zone reheating of the primary air, as required for a two-pipe system, is greatly reduced or eliminated. Changeover to hot water occurs for lower outdoor temperatures and is thus limited to a few times per year. Simultaneous heating/cooling capacity is available, except in extremely cold weather when little cooling is required.

7.8 CLOSED LOOP WATER-SOURCE HEAT PUMPS

Many large buildings have substantial core areas that experience no transmission losses but do have internal heat gains. Therefore, these internal areas require cooling all year during occupied hours. Rather than reject the heat extracted from these core areas to the atmosphere, this reject heat is used as a heat source for a heat pump system serving the perimeter of the building. The closed-loop heat pump system (Figure 7-14) ideally serves this purpose.

The closed water loop serves as a source of condenser water for the core air conditioning; it also serves as a heat source to one or more water-to-water or water-to-air heat pumps used to heat the perimeter, thereby increasing the overall system coefficient of performance (COP). The temperature of the closed water loop is

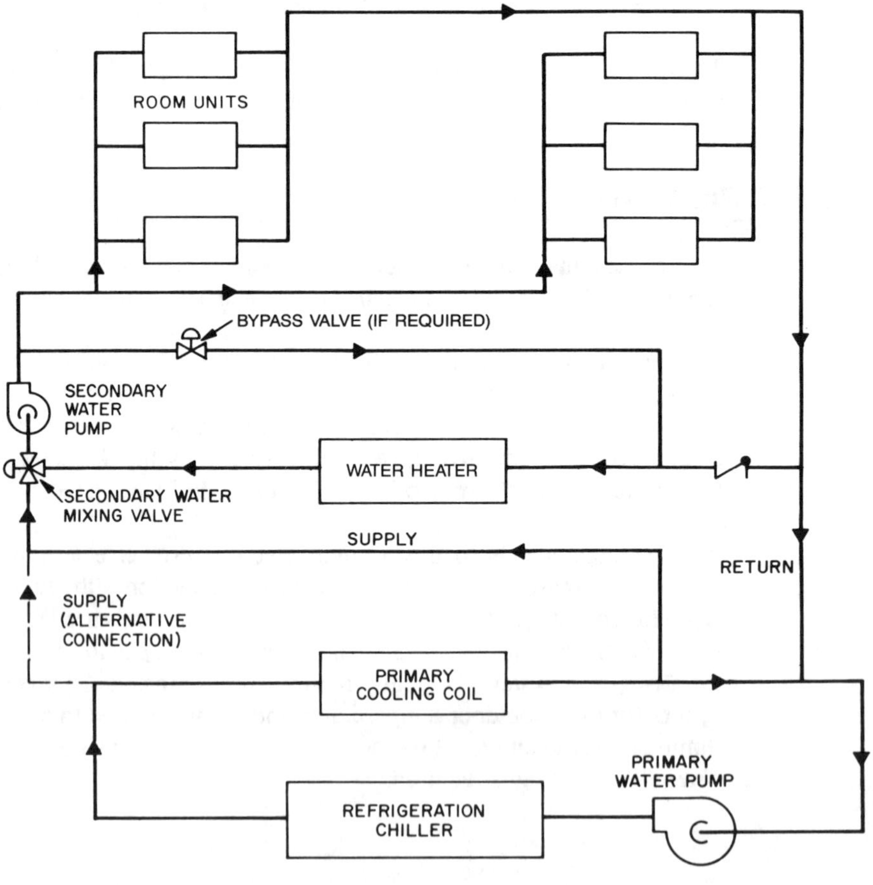

Fig. 7-14 Closed-Loop Water-to-Air Heat Pump System Schematic

usually kept between 65 and 90°F, but the unit manufacturer should be consulted as to the range the unit can withstand. When the loop temperature drops below the low limit, a boiler is used to raise it. When the high limit is exceeded, heat is rejected by way of a cooling tower and, if necessary, a chiller. No pipe insulation is required for the water loop as long as the ambient temperature surrounding the piping is in the 60 to 80°F range.

Water-source heat pumps require 2 to 3 gpm in circulation per ton of refrigeration based on the sum of the nominal tonnage of the units (usually much higher than the block load). This flow rate is generally constant during both cooling and heating modes. Since water must circulate continuously through the entire closed loop as long as a single user requests heating or cooling, the annual pumping costs for these systems are high. When the core areas need only minimal or no cooling during a significant portion of the day (*e.g.*, apartment houses or hotels with infrequently used public areas), excessive use of the supplementary boiler can wipe out the benefits of the heat pumps' high COP.

The closed loop heat pump system permits the landlord to wire the multiple terminal heat pump units to the respective tenant's electric meter, thereby effectively charging each tenant in proportion to the individual's use. This may extend beyond general use when the tenant operates around-the-clock computer rooms or performs extensive night and weekend work. Performance comparisons of water-source heat pump systems with fan-coil all-water systems are given in Section 8.11. Frequently, the two are closely compared for system selection.

For further information, refer to Chapter 5, Unitary Refrigerant-Based Systems for Air Conditioning, and Chapter 8, Applied Heat Pump and Heat Recovery Systems, in ASHRAE *Systems and Equipment* (1992).

7.9 REFERENCES

ASHRAE. 1973, 1976. *Handbook—Systems.*

ASHRAE. 1992. *Handbook—Systems and Applications.*

ASHRAE. 1993. *Handbook—Fundamentals.*

Carrier. 1965. *Handbook of air conditioning system design.* New York: Carrier Air-Conditioning Company.

Chapter 8 ALL-WATER SYSTEMS

8.1 INTRODUCTION

An all-water system is one in which the space-cooling function is performed by chilled water circulated from a central refrigeration plant to heat exchangers (terminal units) located in, or immediately adjacent to, the conditioned space. Heat transfer to the room air occurs essentially by forced or natural convection. Radiant heat transfer is either secondary or virtually nonexistent. Radiant panel cooling systems, other than those used for industrial process cooling, are invariably linked with a dehumidified air supply. Therefore, for purposes of this manual, they are considered to be air-and-water systems. All-water systems can be employed for both heating and cooling. Heating water is supplied either through the same piping network used for the chilled water in summer or by an independent piping system. These options are discussed in Sections 8.4 and 8.5.

The attributes and weaknesses of water as a medium for distributing cooling and heating are discussed in Chapter 4, All-Water Systems, in ASHRAE *Systems and Equipment* (1992). All-water systems are often used based on three characteristics:

1. With regard to both space and energy use, water is a more efficient medium for energy transport than air.

2. Recirculation of air through a central system is unnecessary; therefore, comingling of odors and other contaminants from various tenants and concerns over fire and smoke spreading from one space through a central air system are eliminated.

3. First cost is often less than for other systems, depending on the desired number of control zones and terminal locations. If a terminal is wanted at every window, an all-water system could be advantageous.

To prove the first statement, consider the size of a duct with air flowing at a velocity of 2000 fpm, supplied at 55°F to maintain a space at 75°F, and compare the duct area required with that for two pipes circulating water at 6 fps (360 fpm) operating at a 10°F rise to produce the same rate of cooling.

$$\frac{\text{Duct area}}{\text{Pipe area}} = \frac{\text{Specific volume of air}}{\text{Specific volume of water}} \times \frac{\text{Specific heat of water}}{\text{Specific heat of air}} \times \frac{\Delta t\,(\text{water})}{\Delta t\,(\text{air})} \times \frac{\text{Velocity of water}}{\text{Velocity of air}} \times \frac{\text{No. of ducts}}{\text{No. of pipes}}$$

$$= \frac{13.33\,\text{ft}^3/\text{lb}}{0.016} \times \frac{1.0\,\text{Btu/lb}}{0.24} \times \frac{10^\circ\text{F}}{20} \times \frac{360\,\text{fpm}}{2000} \times \frac{1}{2} = 156$$

While this may be an oversimplification, it indicates the relative merits of the two media for transporting heat. Space must also be included for ducts to exhaust or return the air, for insulation on ducts and pipes, and for drains to remove condensate from pans under condensing coils in terminals. Rectangular duct shapes are more space-efficient than circular pipes, but pumps require far less power than do fans.

8.2 APPLICATIONS

8.2.1 Advantages and Disadvantages

Among the major advantages of all-water systems are:

- Less building space is required.
- They are well suited for retrofit applications.
- Little or no space is needed for central fan room or duct clearances.
- There is individual room control with little or no cross-contamination of air between rooms.
- Because of the low-temperature water used for heating, they are well suited for solar heating and heat recovery applications.

Among the disadvantages are:

- Maintenance is high and it must be performed in occupied areas.
- Unless dehumidification is accomplished by a central ventilating system, condensate drain pans and a drain system are required; they must be equipped with filters and cleaned periodically.
- Ventilation is not easily controlled and is often accomplished by opening windows.

- Relative humidity levels may be high in summer, particularly if modulating chilled water valves are used to control room temperature.

More detailed comparisons between all-water systems and other systems are presented in Section 8.11.

8.2.2 Suitability to Building Types

In the past, all-water systems have been widely applied to office buildings, multifamily residences, hotels, and schools. They are well suited to perimeter spaces with seasonal cooling and heating requirements. When selected for spaces with nearly constant or year-round cooling demands, chilled water must be supplied throughout the year or alternative means of cooling by outdoor air must be provided. Consequently, all-water systems are not well suited for interior spaces. Also, they are not good choices where close control of ventilation and relative humidity is important. On the other hand, they offer individual room temperature control capability. The fact that chilled water requires significantly less space and energy than air circulation accounts for their popularity for apartment houses and hotels. Even more important for these applications is the separation and independence that can be achieved between tenancies.

For better hotel and motel guest rooms, all-water fan coil systems compete with water-source heat pumps. Either system should be supported with a ventilation air supply for quality air conditioning. These two systems also compete for selection in large apartment houses, where a ducted supply of ventilation air is also desirable but is less frequently installed. For perimeter classrooms in schools, unit ventilator all-water systems are frequently considered along with unitary-type unit ventilators or variable-air-volume (VAV) systems with supplemental perimeter heat. The trend for office building applications tends to favor VAV systems and, other than for larger structures, water-source heat pump systems.

All-water systems can be far easier to adapt to existing structures than any other type because of their limited space requirements. All-water fan-coil systems are common in churches and other monumental or historic buildings because of this attribute, even where humidity control is important for artifact preservation (in spite of the fact that such systems are inferior in that respect to all-air constant-volume systems).

8.3 SYSTEM CONCEPTS

All-water systems, as currently installed, can be classified into three basic types:

- Two-pipe systems, in which the terminal units are served by two water pipes—a supply and a return. Chilled or hot water is circulated alternately through a single closed circuit system to the terminals.

- Three-pipe systems, in which each terminal has separate chilled and hot water supplies but only a single, common return.

- Four-pipe systems, in which two separate piping systems supply and return either hot water or chilled water to separate coils within the terminal units.

8.4 TWO-PIPE SYSTEMS

8.4.1 Basic Two-Pipe System

This system is composed of central water cooling and heating equipment, terminal units, pumps, distribution piping, and system and terminal controls. Each terminal is connected to a single supply pipe and a single return pipe. During cooling, chilled water circulates to the terminals to accomplish both sensible and latent cooling; hot water circulates through the system for space heating. Room temperature is controlled at the terminal by automatically or manually regulating air or water flow or both. In its simplest form, the entire system must be either on cooling or on heating. Some spaces served by the system may require cooling while others need heating, and unsatisfactory performance will result. It may be possible to group spaces that have a common heating or cooling requirement together in a single zone. By design of the circulating system and its control, some zones can be operated on cooling while others are on heating. The zoning, piping arrangement, and controls to achieve this flexibility demand careful design.

Zones can frequently be divided according to exposure. Zoning is practical for spaces where solar heat gains during spring and fall are the predominant factor in calling for heating or cooling. However, shadows from other buildings or wings that vary throughout the day (not just at peak time) can defeat this approach. Occasionally each major branch, perhaps a single set of risers, is designed to serve one zone. The mode (heating or cooling) for terminals connected to that branch may be indexed by a key space thermostat or discriminator control circuitry. This arrangement is actually a cross between the two-pipe concept and the four-pipe concept described later.

The piping arrangement illustrated in Figure 8-1 contains both a heating and a cooling pump. When zones require both modes, both hot and chilled water are available. The chilled water pump must have adequate capacity to maintain the minimum required flow through the chiller. If cooling only is needed, the hot water pump can be turned off. If heating only is needed, the chilled water pump can be turned off. It is feasible to use the two pumps operating in parallel to circulate chilled water. Operating in this mode, hot water is not available to the system, but total pumping capacity can be saved. During intermediate seasons, and possibly throughout the heating season as well, partial system water flow is usually adequate to meet the space thermal needs. It is important to lay out the return piping in a manner that will avoid the blending of hot and chilled water upstream of the chiller or hot water generator.

Control valves on both supply and return are necessary to switch to the appropriate source for the zone. The basis for changeover from one mode to the other can be a difficult choice, since the required mode depends on several factors, such as outdoor temperature and solar load. Solar-compensated outdoor temperature sensors lagged to simulate the building thermal storage or "flywheel" characteristics have been employed. Key room thermostats and manual switching are other techniques of zone

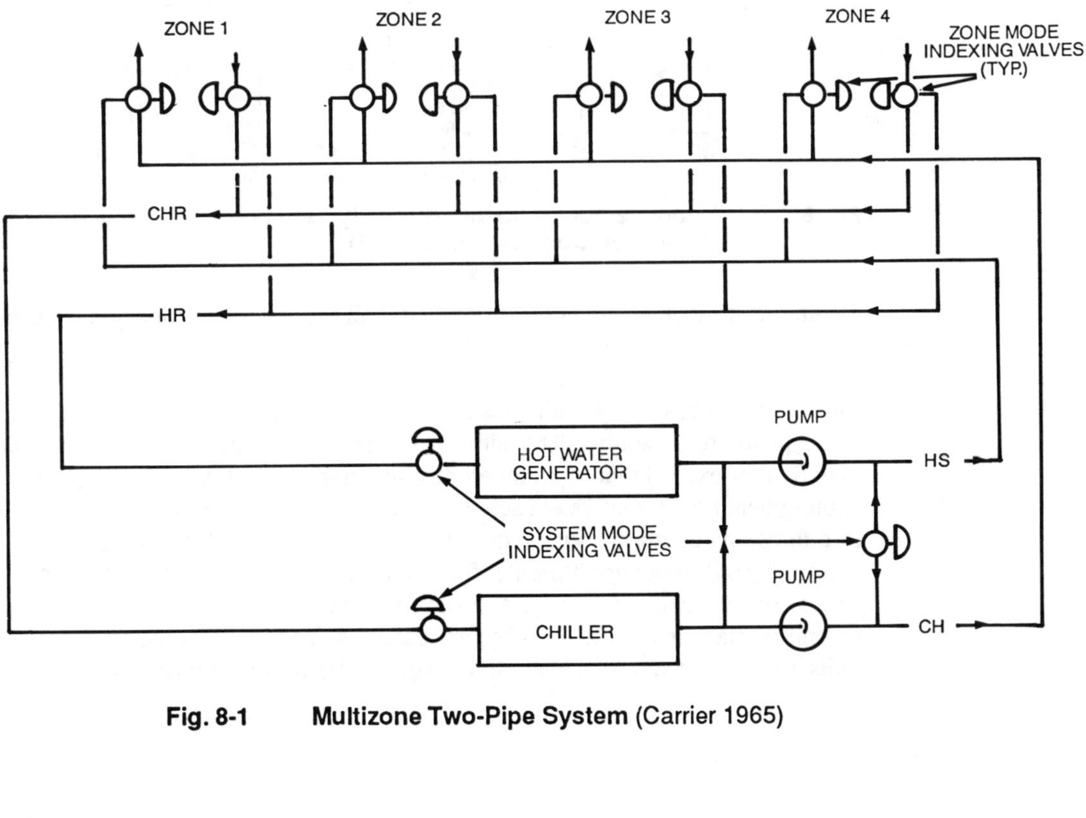

Fig. 8-1 **Multizone Two-Pipe System** (Carrier 1965)

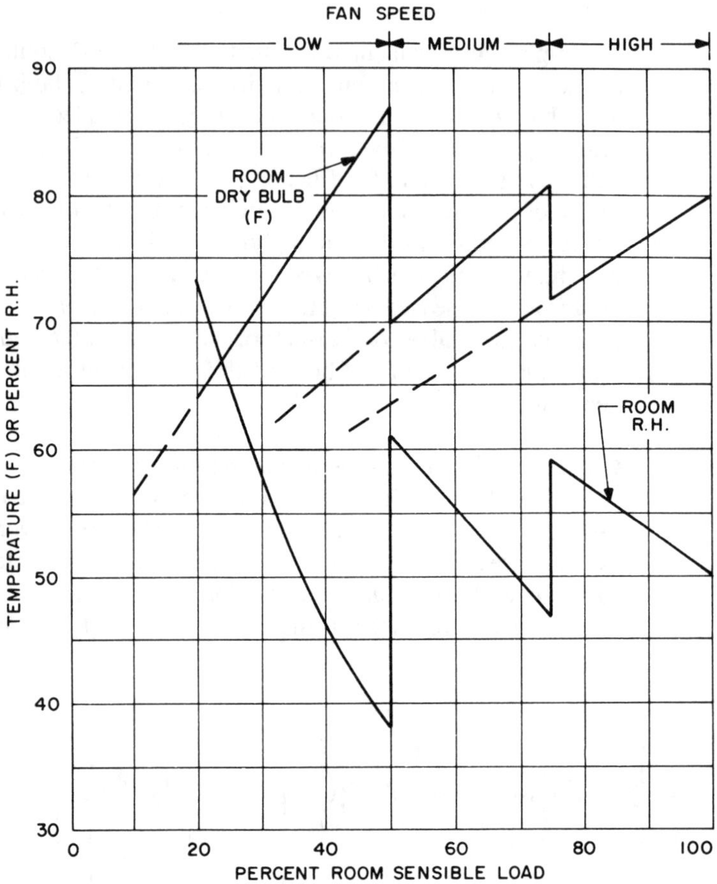

Fig. 8-2 Room Temperature and Humidity Variation under Manual Three-Speed Fan Control (Carrier 1965)

cooling/heating changeover. Unfortunately, all these control concepts have shortcomings.

The foregoing discussion indicates that system design may become complicated to accomplish proper zoning. The additional expense for piping, pumps, and controls may rival or exceed that of an inherently superior concept, which is to provide the cooling/heating alternative at each terminal with a four-pipe system. The temperature in rooms equipped with fan-coil units is maintained by fan on/off or speed control, by on/off or variable water flow control, or by both. Vertical riser units with series-connected heat exchangers can be controlled by bypassing air around the heat transfer surface using face and bypass dampers. Varying unit water flow through units arranged in series as a method of room temperature control is not feasible.

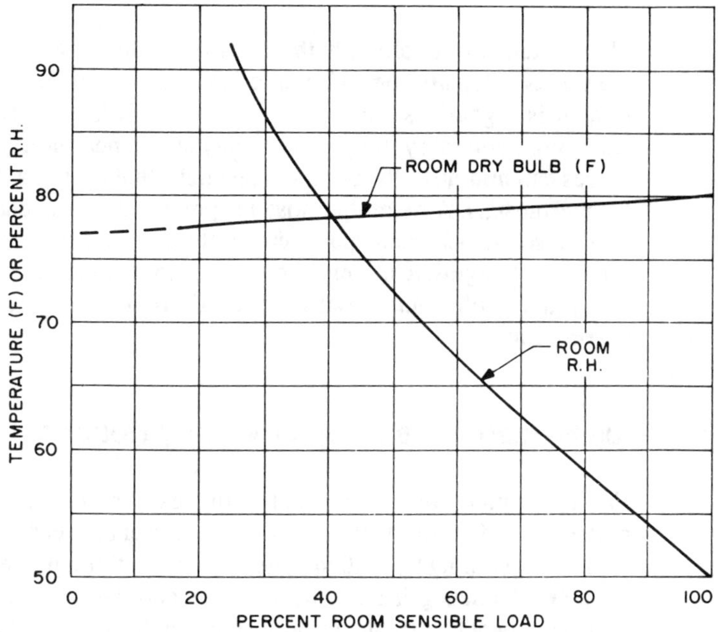

Fig. 8-3 Room Temperature and Humidity Variation under Automatically Modulated Water Control with Continuous Fan Operation (*Carrier* 1965)

Unless the conditioned space is dehumidified by a supplemental air-conditioning system, both sensible and latent cooling are performed by the terminal unit coil. The sensible heat ratio of the cooling process or a given space over time will vary widely. This variation will be most extreme if ventilation air is introduced through windows or unit intakes in the exterior wall. Since room temperature is the basis for unit control, humidity control is incidental. Relative humidity can vary widely, as depicted in Figures 8-2 and 8-3. Control solely by manually controlling the fan is unsatisfactory. Automatic control of fan operation with manual selection of fan speed results in intermittent airflow through the occupied space, temperature swings near the outside wall, and potentially annoying sound level variations. However, these characteristics are similar to those of residential forced air heating systems and, in this application as well as in many multiresidence applications of fan coil units, they have been accepted as tolerable. As Figures 8-2 and 8-3 indicate, superior control of humidity is more likely with fan control than with water flow control because humid outside air may continue to be brought in even though the thermostat is satisfied and the chilled water is at a minimum while the fan continues to run. Tightly sealing dampers on wall openings should be automatically closed when the unit fan is off. Use of water flow control in combination with ventilation by wall apertures incurs a real risk of coil freeze-up in cold weather.

Manual fan control along with on/off or modulating control of water flow can be viewed as a step up from automatic fan control in other respects. This is true of cost (which is higher) as well as performance. On/off water flow control can cause annoying frequent cycling of the supply air temperature. Quietly operating solenoid valves can minimize the bothersome clicks that occur as the valves are actuated by the thermostats. A room thermostat is preferred over a return air thermostat, since it will sense the occupied space temperature even if the fan is off. When the system mode is changed, a second thermostat that senses system water temperature, an aquastat, obtainable with the terminal, must reverse the action of the room thermostat.

8.4.2 Two-Pipe System plus Supplementary Heating

Fan coil units can be obtained with auxiliary electric heating coils. This can materially reduce the difficulties of changeover between cooling and heating modes. As long as cooling by chilled water circulated through the terminals is necessary to counter solar heat gain and other transient cooling loads, overcooling of other spaces with a net heat loss on the same system can be prevented by energizing the electric coil. This will probably eliminate the need for the complex zoning previously described. Even if total heating by electric resistance heat is not cost-effective, electric heating during the intermediate seasons may be justified by reduced piping and control costs as well as by significant performance improvement balanced against increased unit and electrical distribution costs.

Control of cooling must be by chilled water flow valves or by face and bypass dampers. Such control is then sequenced with on/off control of the electric heater energized through a magnetic contactor. Both cooling and heating are under control of a common thermostat. When hot water is supplied for heating, the unit aquastat reverses the control action of the room thermostat on the water valve and deenergizes the electric heater circuit. Only heating by hot water is then available. The electric heater is deenergized whenever the fan is not operating. The system mode changeover decision is less critical but still involves the numerous variables previously mentioned. More often than not, the pivotal rationale for the decision is economic. Owners are reluctant to incur monthly electrical demand charges for short-period operation of their refrigeration plants. Therefore, the changeover decision is governed by the calendar—so many months on cooling and so many on heating. Not infrequently, the months when cooling will be available become a lease provision, but many tenants will not accept such a lease restriction.

8.4.3 Two-Pipe Systems with Total Electric Heat

If electric heating is economically justifiable throughout the year, the need to change from cooling to heating and vice versa is eliminated. Unit control is simplified; reversal of the room thermostat action is unnecessary. Chilled water circulation and refrigeration can be discontinued when there is little or no need, and zoning is unnecessary. A significant advantage is that the electric power source for heating and air circulation can be the tenant connection and can thus be separately metered. For some occupancies in some jurisdictions, this is a legal requirement as an energy conservation measure.

8.5 FOUR-PIPE SYSTEMS

Figure 8-4 shows a schematic layout of a typical four-pipe system including basic system controls. Providing two independent water distribution systems—one dedicated to chilled water and the other to hot water—permits either cooling or heating by the room terminals throughout the year. Each unit is equipped with a heating and a cooling coil and two water-flow control valves (or equivalent coil bypass control) and is connected to the distribution system by four pipes—supply and return for chilled and hot water. Both valves are controlled in sequence by the room thermostat. A dead band of several degrees is normally incorporated into the control. In this band, both valves are closed. This feature prevents valve cycling with frequent shifts between heating and cooling and reduces energy use. The four-pipe concept and the two-pipe concept with total electric heating offer significant advantages over the simple two-pipe concept, such as

- All-year availability of heating and cooling for individual space temperature control
- Elimination of zoning cost and complexity
- Simpler changeover decisions

In practice, the lowest and quietest unit fan speed is adequate most of the time. In actuality, the usual operating practice is to discontinue the circulation of chilled water in winter and that of hot water in summer. Based on this premise, some designs have incorporated system changeover modes to employ both distribution systems for cooling in summer and for heating in winter. Both chilled and hot water remain available during the intermediate seasons. The impetus for this variation is reduced pumping capacity and smaller pipe sizes, offset to some degree by the loss of

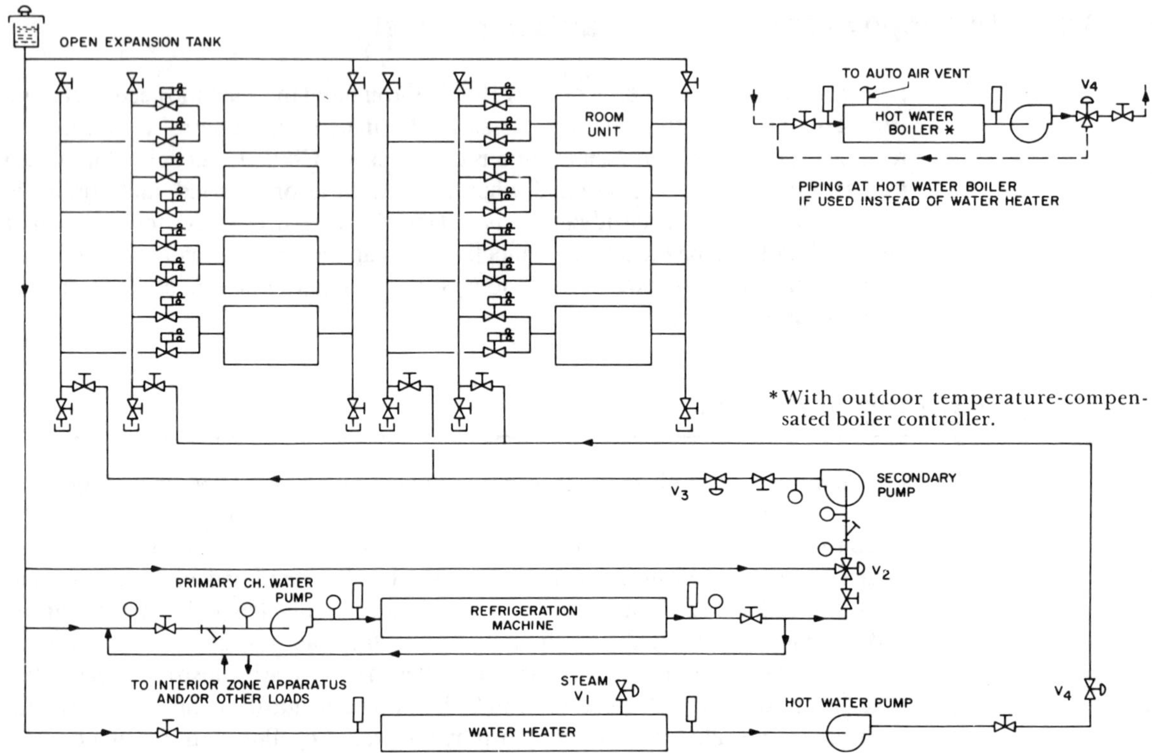

Fig. 8-4 Piping Schematic for Four-Pipe Fan-Coil System (*Carrier* 1965)

versatility, the increased burden of changeover decisions, the complexity of terminal valve control, and the added provisions needed to switch the pumps to different duties.

8.6 THREE-PIPE SYSTEMS

Three-pipe systems have separate chilled and hot water supply piping, a single coil, and common return piping. The terminal control is a three-way valve that admits water from either the hot or the cold water supply as required but does not mix the supply streams. The valve is specially designed so that the hot port gradually moves from open to fully closed and the cold port gradually moves from fully closed to open. At mid-range, there is an interval in which both ports are completely closed. The water returns through common piping to either the chiller or a hot water heater.

Once popular for perimeter spaces of large buildings because they avoided many of the changeover problems associated with two-pipe systems, three-pipe systems have been virtually replaced by other options, especially four-pipe systems. While some versions proved successful, many were plagued with excessive mixing of hot and chilled water in the common return and consequent energy waste.

8.7 PIPING DESIGN CONSIDERATIONS

Several considerations are of special importance for the design of water distribution systems for all-water systems.

8.7.1 Flow Diversity

The system (block) peak cooling load is likely to be less than the sum of the space peak loads. If the cooling capacity of the terminal units is automatically controlled using two-way valves, it is improbable that all valves will be fully opened concurrently. Consequently, some reduction from the sum of the individual terminal flows is possible in selecting pumps and sizing pipe mains. In theory, the diversity factor may be viewed as the ratio of the block cooling load to the sum of the peak loads for each terminal. Such a theory is based on the supposition that all space thermostats are set at the design space temperature and that flow is reduced in direct proportion to terminal load reduction. Neither is correct because other operating conditions also affect the degree of diversity. A rule of thumb is to reduce the aggregate flow of all terminals by a diversity factor equal to the square root of the ratio of the block load to the sum of the peaks. Diversity should not be considered for risers or horizontal mains serving only a single exposure of a building.

8.7.2 Water Pressure and Flow Balance

Balancing the water flow through a large number of parallel connected coils is at best difficult and laborious. However, with reasonable care in design, the water distribution network can be sufficiently self-balancing to make field adjustment of terminal coil flow unnecessary. The effect of chilled water flow variations within 25% of the design rate on a fan coil unit's cooling capacity is usually on the order of 10%. This provides considerable tolerance. Consequently, some difference in the flow resistance of the piping circuits conveying water to the many terminals is tolerable without significantly affecting the performance of the terminals situated on the longer runs. Two conditions must be met:

1. The flow resistance in the parallel terminal subcircuits is at least equal to 40% of the variation of the pressure drops in the distribution piping circuits, *i.e.*, the difference between the greatest and the least pressure drop in the circuits.

2. The pressure drops of the parallel subcircuits have reasonable consistency; *i.e.*, where possible, avoid mixing high and low loss terminals on common circuits. The subcircuit includes the control and manual valves, the connecting piping, and the unit coil.

Any closed piping loop may be either direct return or reverse return (see Figures 8-5 and 8-6). The direct return system is popular because it requires less piping; however, balancing valves may be required on subcircuits. Since water flow distances are virtually the same in a reverse return system, balancing valves require less adjusting or can be omitted.

Quantitative evaluation of the flow variations that circuit imbalance creates is beyond the normal system design protocol. As a general guideline, an imbalance of 10 to 15 ft of head is safe. This latitude allows the use of direct return risers. The imbalance can be controlled by judicious pipe sizing. Reverse return design of the horizontal mains looping within the building perimeter feeding the risers usually

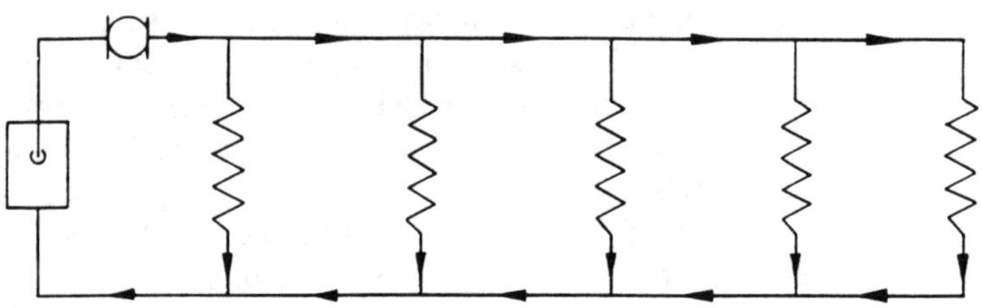

Fig. 8-5 Direct Return Piping System

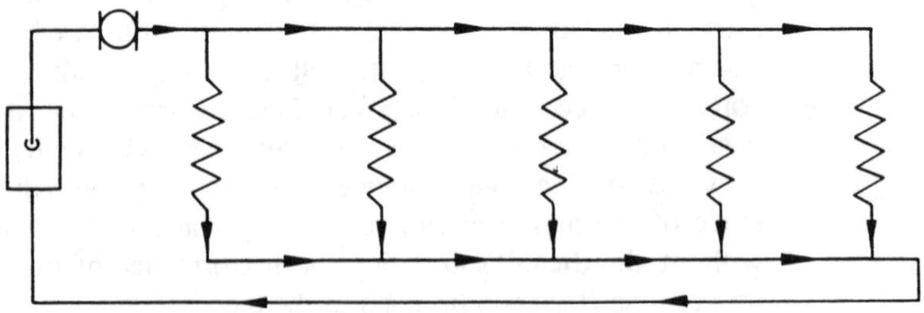

Fig. 8-6 Reverse Return Piping System

does not incur significant additional piping expense over direct return and is good practice. Mains that serve major sections of the building and are parallel connected with each other should be provided with the means for balancing and flow measuring.

8.7.3 Control of System Pressure Differential

Where two-way valve control of terminal coil water flow is employed, the system flow will be reduced at partial load. The valves reduce flow by adding flow resistance to the piping subcircuit. As flow is reduced, pipe friction also decreases and pump head increases, requiring even further closure of the terminal valves. If the differential pressure imposed across these valves becomes great enough, the maximum shutoff pressure rating of the valves could be exceeded. The valves would not close, defeating the space temperature control. Even barring that, wire drawing at the valve seat is a real possibility, which could lead to costly replacement of the terminal valve seats. Several preventive measures are available to the designer:

1. Control the system pressure differential with taps in the supply and return piping, usually near the midpoint of the piping network, to vary the speed of the pump, reducing its capacity and developed head at part load, as described in Section 8.8.

2. Regulate a throttling valve in the supply mains near the pump discharge from a system pressure differential controller (in lieu of speed control). This shifts part of the task of restricting flow from the terminal valves (see Figure 8-4).

3. Install three-way valves at the terminals in place of two-way valves. At reduced load, the valves bypass the coils. This incurs some disadvantage since these bypasses are, in effect, short circuits unless a resistance equal to the terminal water coil is built into each bypass. Without this, at partial system load, system flow will actually increase, as will pumping horsepower. Also, some terminals may be starved of water, since the pressure differential at those terminals may fall below that needed to deliver the water flow desired by the space thermostat. The installed cost of three-way valves is somewhat higher than for two-way valves. Some designers favor a single bypass valve located near the extremities of the mains controlled by system pressure differential. The effect is similar to a main throttling valve but can have the effect of increasing system flow and horsepower. A spring-loaded relief valve is unsuitable for this purpose, and an industrial-quality control valve is required.

4. Install a combination of two- and three-way terminal valves. This practice limits the reduction of flow and the buildup of pressure differential across the terminal valves without needing to control flow velocity, main throttling valves, or a bypass valve. A little study will indicate the percentage mix, but 30 to 50% three-way valves generally suffice. Locating the three-way valves in the greater resistance paths ensures good flow throughout the extremities of a direct return system.

If a system is solely equipped with two-way valves, a no-flow condition can occur at the pump. Serious damage to the pump could result. Prevent this either by adopting a pressure control option that will ensure flow or equip the pump with a bypass. Another potential hazard relating to the use of two-way valve control is the freezing of sections of the distribution piping if exposed to low outdoor temperatures, possibly in exterior wall cavities under no-flow conditions.

8.7.4 System Cleanliness

The performance of an all-water system can be seriously impaired by dirt and corrosion products in the piping system. The following safeguards avoid this impairment:

- Protect piping material on the job site prior to installation
- Protect openings in the piping during installation
- Flush and chemically clean the piping system prior to operation
- Locate strainers upstream of pumps, chillers, and heat exchangers
- Install scavenging filters in pump bypasses selected for 5 to 10% of system flow
- Provide isolation valves and accessible drain connections for risers

8.8 VARIABLE-SPEED PUMP OPERATION

The characteristics of centrifugal pumps and piping networks follow some special interrelations when the pumps are equipped to operate at variable speed. Pump theory states that the shutoff head (the head at zero flow rate) is proportional to the square of the rotative speed, as illustrated in Figure 8-7. This figure also illustrates the relationships between head, flow rate, pump speed, and efficiency.

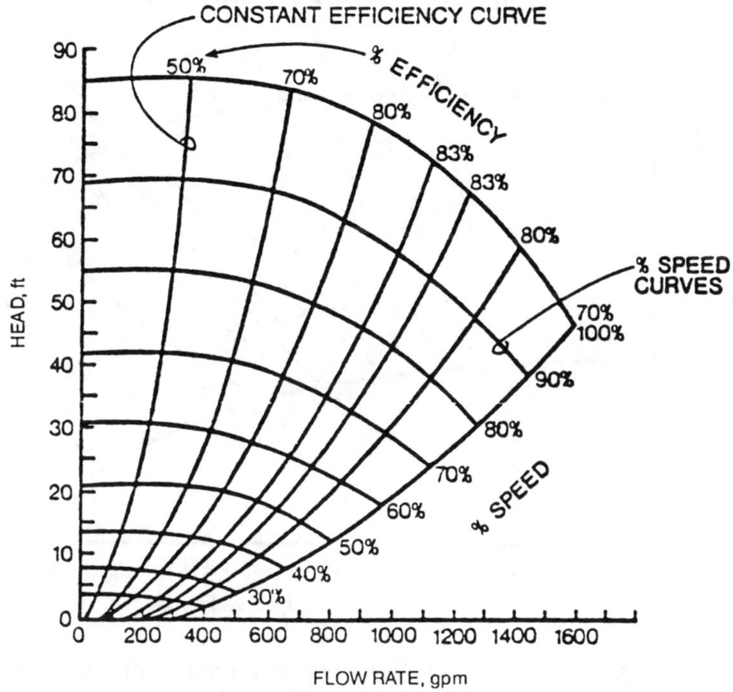

Fig. 8-7 **Pump Head, Flow Rate, and Efficiency as Function of Pump Speed** (ITT 1985)

Analogous to the fan laws (see Section 6.5.2), the pump laws for centrifugal pumps are:

- Flow rate varies with speed (pump rpm)
- Head varies with square of speed
- Head varies with square of flow rate

The last relation is precisely the characteristic of a piping network. The combination of the pump and piping relationships is illustrated in Figure 8-8. If a pump operates at maximum efficiency at the design point, then that pump serving a constant system at various flow rates operates at maximum efficiency throughout. This relation must be tempered with several qualifications. The piping network might serve just a single thermal load and the load requirements would control the speed of the pump. If the piping system is a two-way valve network, as shown in Figure 5-12, no single head-flow curve represents the complete performance map. The head-flow curve of the

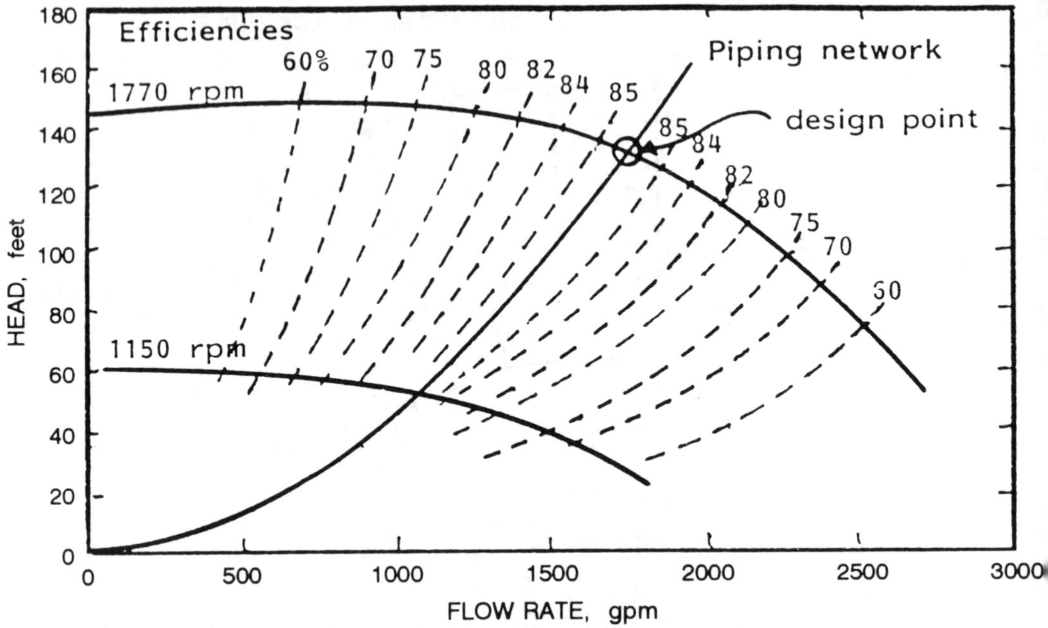

Fig. 8-8 Combination of Characteristics of Piping Network with Those of Variable-Speed Pump for Maximum Efficiency

three-way valve network in Figure 5-13 suggests a single head-flow curve, but actually it is only one point—the design point—on that curve.

The opportunity to lower the pump delivery pressure must be communicated to the speed control of the pump. One possibility of sensing the adequacy of the pressure supplied by the pump is to monitor the position of the two-way valves controlling the rate of water flow to the coils. Ideally, only a single valve in the system should be completely open. If the valves are pneumatically actuated, for example, the highest control pressure (or lowest in the case of normally open valves) should be used as the controlled variable to regulate pump speed.

Less pumping power is required when the flow rate is adjusted downward by reducing the pump speed instead of by throttling. Many successful installations have been reported (ASHRAE 1988). Other designers prefer to achieve efficient variable-volume pumping by installing multiple pumps and shutting down one or more pumps to approximately match the required flow rate. System curves to illustrate these conditions are shown in Figure 8-9 for two pumps in series and in Figure 8-10 for two pumps arranged in parallel.

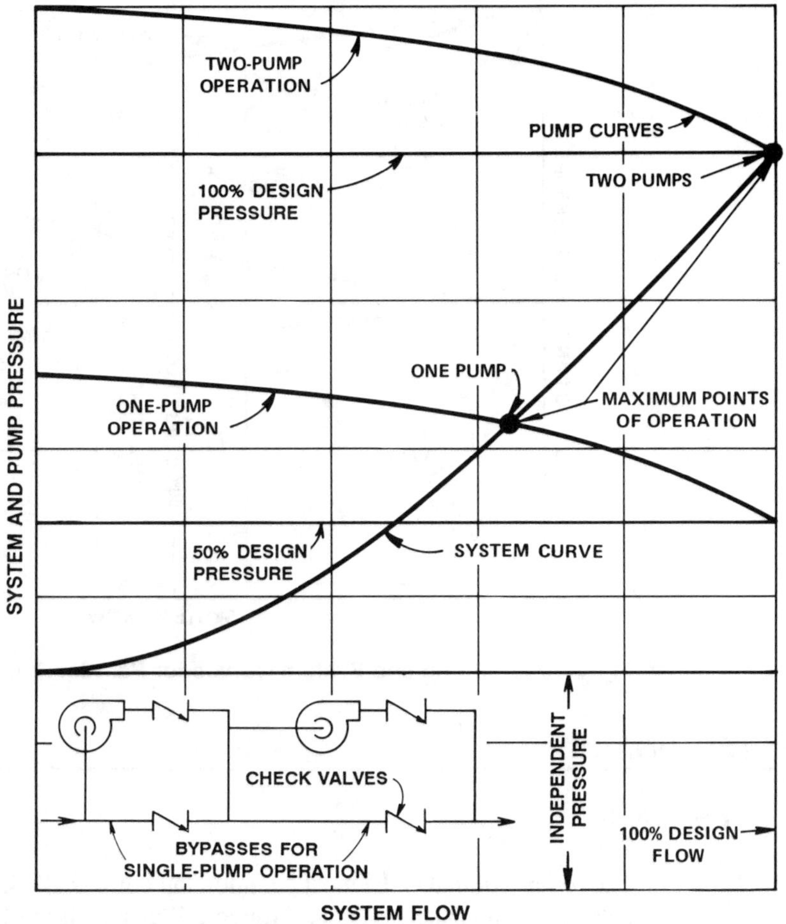

Fig. 8-9 Pump and System Curves for Series Pumping

Variable-speed pumping should not be used with reset of the chilled water leaving temperature, which is based on constant volume pumping. Temperature reset decreases chiller efficiency but keeps pumping power high. Compare the reduction in chiller power from reset against the reduction in pumping power from variable-speed pumping. Variable-speed pumping is best suited to piping networks having a high fraction of head loss that varies with the flow rate (long piping runs) and a small fraction of constant head loss (terminal units, valves, differences in elevation). Thus, in an equipment room serving air handlers located close to the chiller, there is little variable head loss, and a variable-speed pump would act very much like the constant speed pump whose characteristics are shown in Figure 8-7 (ITT 1985).

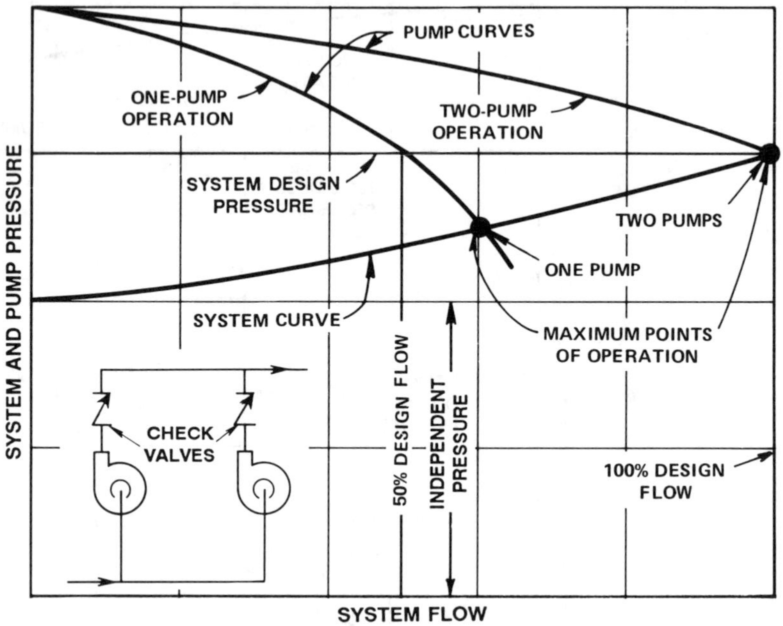

Fig. 8-10 Pump and System Curves for Parallel Pumping

8.9 TERMINALS

8.9.1 Fan Coil

The room terminal is the most common fan coil installed. It consists of a direct- or belt-driven fan that induces air through a filter and discharges it through an extended surface coil for delivery into the room. The air source is room air or a mixture of room air and outdoor air for ventilation. An insulated drain pan beneath the coil collects moisture condensed from the air as it passes over the cooling coil. The unit may have an additional downstream coil for heating by electricity, steam, or hot water.

One type of vertical fan coil unit integrates the extended heat transfer surface with the supply and return chilled water (or dual-temperature) riser section. When stacked in a multistory building, these *column units* virtually eliminate the need for separate water distribution risers in a two-pipe system. Water flows through the unit heat exchangers in series. With a direct return riser, the average water temperature is approximately the same in all units on a riser. On an upfeed system, the lowest unit has the coolest supply water and the warmest return water. Thus, unit capacities are independent of their vertical location on the riser. For a four-pipe system, additional heating coils are parallel connected to dedicated hot water heating risers. Some column units are designed with parallel-connected coils, factory connected to internal riser sections, and therefore are closely akin to the conventional unit type. The unit enclosure is also the riser enclosure with air supplied from a grille (or grilles) near the ceiling and returned near the floor. With units located at the exterior wall, the direction of air discharge is commonly parallel to that wall to offset the exterior heat loss or gain. Various fan-coil unit arrangements are shown and described in Chapter 4, All-Water Systems, in ASHRAE *Systems and Equipment* (1992).

8.9.2 Unit Ventilators

Unit ventilators are similar to fan-coil units but have a more elaborate provision for the direct introduction of outdoor air for ventilation and, as an option, free cooling by outdoor air usually with economizer cycle control (see Section 9.3.7). Unit ventilators are most often applied in assembly areas, such as classrooms and meeting rooms. They were the terminal of choice for heating and ventilating classrooms prior to the increased application of air conditioning for such spaces, and they continue to be installed either as components of an all-water system or as unitary equipment.

8.9.3 Valance Units

The valance terminal offered by one manufacturer depends on natural convection for heat transfer from a coil located in a valance enclosure near the ceiling line, usually along the building perimeter wall. Filters are not employed, and ventilation is not a feature of this delivery system. Space temperature is controlled by coil water flow. Consult with the supplier to evaluate appearance, maintenance, and cost considerations.

8.10 VENTILATION

If ventilation air is introduced by the unit, it is most frequently drawn in through a dampered opening in the building wall. The damper is motorized to close when the fan is off. Far less frequently, a preconditioned air supply will be ducted to the unit and blended with recirculated air. The rate at which ventilation air is introduced directly through an aperture in the building wall is almost impossible to regulate because of wind pressure and stack effect. Coils have been frozen and cold drafts across the floors have been the result of excessive flow. On the lee side of a building, airflow may be reversed through the intake even with the fan operating. In a high-rise building during cold weather, excessive outdoor air can enter through the unit openings on lower floors and exfiltrate through the unit apertures on the upper floors. The list of shortcomings of this approach to ventilation also includes poor humidity control, wasted energy, insect problems, and smoke entry from fires on lower floors. Yet those tell-tale grilles beneath apartment house windows continue to appear.

Ventilation must be provided, however, if the all-water concept is to fulfill the function of an adequate means of air conditioning. The most common approach is to assign the function to exhaust systems, for example, kitchen and bathroom exhaust in multifamily residences. An ample source of makeup air is required for exhaust systems to function properly. Windows, if operable, are usually permitted by building codes. This source is effective only when windows are open or if they are sufficiently leaky. The shortcomings to natural ventilation and air makeup through windows include poor humidity control, excessive heating and cooling loads, and entry of dirt and noise from outdoors. In practice, many occupants would rather put up with very limited ventilation than incur the discomfort of open windows, especially in extreme weather. If the windows are tightly closed and there is a significant exhaust air requirement (as in the case of residential buildings), the conditioned space will be subject to substantial negative pressure. Some makeup air may enter from common corridors or hallways, although most fire codes prohibit undercutting of entranceways or the use of exits as makeup air plenums. Fireplaces installed in some apartments have become unanticipated makeup air passageways.

This discussion highlights a common design pitfall inherent in all-water systems—lack of ventilation. The proper solution is a ducted, preconditioned (at least by filtration and tempering), fan-powered ventilation air supply. This adds significantly to project cost and is at risk in any cost-trimming exercise, sometimes by an unwary owner who may later hold the design engineer responsible for not pointing out the consequences.

8.11 SYSTEM EVALUATION AND COMPARISON

Each air-conditioning system type offers advantages and disadvantages that are of varying importance to different applications. Generalizations should therefore be used with care. Nonetheless, in the following tabulations, all-water fan-coil systems are compared with other system types with which they frequently compete. You may find these comparisons helpful as a checklist to ensure that pertinent traits of the systems are considered in the selection process.

Comparison of All-Water Fan-Coil System to All-Air, Variable-Air-Volume (Non-Fan-Powered) System with Perimeter Heating (Section 6.7.1.3)

All-Water Fan-Coil System Advantages

- Air motion in space not dependent on temperature control (with water flow or coil bypass control)
- No duct space required
- More efficient energy transport
- No central air recirculation
- No cross-contamination of different tenancies
- Lower first cost for some applications
- No duct penetrations requiring fire barriers
- Better containment of smoke from fires or tobacco
- Air conditioning can be readily discontinued in unoccupied areas

All-Air Variable-Air-Volume System Advantages

- Integrated and positive ventilation supply
- Improved air cleanliness
- Better flushing of space contaminants
- Cooling capability whenever free cooling by outdoor air is available
- Superior humidity control
- No moisture removal at terminals (no drip pans)
- No odor accumulation on wet coils
- No motors, fans, and/or coils in occupied space
- Fewer potential harbors for allergens and pathogens in room terminals
- No water leak potential
- Inherently quieter
- Ceiling or remote location of terminals more feasible
- Suitable for perimeter and interior spaces
- Greater flexibility for partition changes

Comparison of All-Water Fan-Coil System to Air-and-Water Induction System (Section 7)

All-Water Fan-Coil System Advantages

- Versatility of location and style
- No air duct connection required
- Frequently less costly
- Facilitates electric energy metering
- Less frequent maintenance of terminal filter/lint screen
- Less thermal transport energy
- Off-hour conditioning does not require central air system operation
- Air conditioning can be readily discontinued in unoccupied areas

Air-and-Water Induction System Advantages

- Integrated and positive ventilation supply
- Superior humidity control
- Superior air cleanliness
- Condensation on cooling coil normally eliminated
- Two-pipe concept offers intermediate season heating and cooling (over basic two-pipe fan-coil system)
- Reduced depth of coil, easier to clean
- No motors, drives, and/or fans in occupied spaces
- Inherently quieter
- Underwindow units not as deep

Comparison of All-Water Fan-Coil System to Unitary Water-Source Heat Pump (Section 7)

All-Water Fan-Coil System Advantages

- Inherently quieter
- Potentially superior cooling and heating efficiency
- Less water in circulation
- Smaller pipes and pumps
- Lower pumping costs
- Terminals less costly
- Lower capacity terminals available
- Less equipment to be maintained in occupied space
- Life-cycle cost of equipment likely to be lower
- First cost of two-pipe system lower
- Lower winter electric power demand and consumption (if other than electric heat)

Water-Source Heat Pump System Advantages

- Immediate all-year availability of heating and cooling with only two water pipes
- No seasonal changeover required
- Range of circulated water temperature requires no pipe insulation
- Cooling available without operation of central refrigeration plant, resulting in odd-hour flexibility
- Can transfer energy from spaces being cooled to spaces being heated
- Waste heat recovery potential
- No demand charge penalty for operating central refrigeration plant equipment
- Reduced dependence on reliability of central plant for heating and cooling
- Suitable for both perimeter and interior space conditioning
- Lower operating staff requirements

8.12 DESIGN SEQUENCE

8.12.1 Concept Selection—Getting Started

Several concepts will usually be compared prior to determining the one most appropriate for a specific project. Common starting points for the preliminary modeling of systems under consideration are (1) establish design criteria and (2) estimate cooling and heating for the portions of the building to which the system is to be applied. This can be done by load calculations or by estimating based on experience. Engineers commonly use a parameter of tons of refrigeration or Btu per hour per square foot of gross or net building area for a quick estimate or check on more detailed calculations (see Table 2-1, Chapter 2). Such rules depend on specific application and location. Cooling load figures based on several parameters may be found in the ASHRAE *Pocket Guide* (1993). However, many engineers prefer to develop their own check figure files based on their previous project experience. Design criteria include building envelope characteristics, such as type and amount of fenestration, lighting levels, occupancy, ventilation, and space temperature objectives.

8.12.2 All-Water Concept Modeling

In the preliminary design of an all-water system, establish the method of ventilation 1 first. Consider then whether the cooling and heating of this ventilation air will be imposed on system terminal conditioners. Following this, determine the system to be evaluated along with the types of terminals. In deciding whether to use fan coil units, or two- or four-pipe systems, consider physical arrangement, performance, and economic objectives.

The system control strategy should also be developed to a point where the building's system capabilities are well understood. Try to answer such questions as: How is space humidity to be controlled? What is the likely changeover point? Is exposure zoning necessary? Will such zoning be workable and cost-effective? Once the concept to be evaluated is fairly well set, the system model can be further developed by "first cut" component selection.

8.12.3 Changeover Point

Before proceeding to the next step in preliminary design, an explanation of changeover point is in order. In simple terms, the changeover point for cooling is the outdoor temperature below which chilled water is no longer required to offset heat gain in any zone. The changeover point for heating is the outdoor temperature above which hot water is no longer required to offset heat loss. A given space no longer needs cooling by chilled water when

Solar and internal heat gain = Space cooling by other systems + Heat transmission loss

Changeover temperature can be calculated as

$$t_{co} = t_r - (\text{Solar and internal heat gain} - \text{Other cooling})/VA$$

where

t_{co} = changeover temperature
t_r = governing room temperature
VA = heat transmission per degree of temperature difference through exterior envelope

The calculation is approximate because the terms in the equation vary with time. If solar and internal heat gain is taken as the maximum for the time of year when t_{co} is likely to occur, the *lowest* outdoor temperature at which cooling is required will be obtained; if solar and internal heat gain is taken as the minimum anticipated, the *highest* outdoor temperature at which heating is required will be obtained. The spread between the two values represents the range within which either heating or cooling may be required. If the range is wide, as it might be for an office with southern exposure and substantial but variable internal heat gain, a two-pipe (either/or) system is going to be a poor choice and a two-pipe plus supplementary heat or a four-pipe concept should be considered. Note that northern or shaded exposures will have decidedly different t_{co} spread characteristics than those exposed to the sun.

Zoning must be considered for the two-pipe concept to group spaces with similar changeover characteristics to allow some zones to be cooled while others are being heated. Thereafter, the means of zone mode control must be devised as discussed in Section 8.3. Realize that changing over a system from hot water to chilled water (and vice versa) takes time. Introducing warm water into a chiller could blow the refrigerant charge. Once a decision (automatic or manual) is made to shift modes, time must be allowed for the water temperature to neutralize before the shift can be completed. Occupants may be uncomfortable during this waiting period. By understanding changeover, the limitations of the two-pipe concept should become clear.

8.12.4 Preliminary Component Selection and Design

A schematic piping layout can now be prepared with terminal unit and central cooling, heating, and pumping locations. Some study is required to decide whether horizontal mains feeding vertical risers, vertical mains with horizontal piping loops or branches for each floor, or a hybrid is most efficient. Reverse return piping arrangements should also be evaluated. If exposure zoning is to be used, the design must allow independent operation of each zone on heating or cooling without mixing of water from zones on different modes in a common return. With the number of units now established, the total amount of water for fan-coil units may be roughly estimated at 3 gpm per ton of block load. If the maximum temperature rise

$$\Delta t_{max} = 24(\text{Block cooling load, tons})/(\text{Terminal water flow, gpm})$$

is lower than the desired temperature drop in the chiller, or if the flow through the terminal pumping system is not constant during the cooling mode (which occurs in two-way valve terminal control), primary/secondary pumping circuits will be needed. This is commonly the case for all but smaller systems. Figure 8-1 shows a single chilled water pump, and Figure 8-4 indicates both primary and secondary pumps.

Preliminary modeling for construction cost and energy cost estimating and space considerations may be further developed by selecting a typical terminal unit, pump, chiller, and heat exchanger. From this system model, the all-water concept can be compared to other concepts considering initial cost, energy cost, performance capabilities, maintenance requirements, acoustics, longevity, and space needs. This preliminary design and evaluation, concluding with a recommendation (usually during the concepts phase of building design, as explained in Section 2.2), will facilitate the final system selection by the owner.

8.12.5 Final Design

The final design process is as follows:

- Calculate room cooling and heating load.

- Select terminal units based on established criteria for space configuration, supply water temperature, sound levels, cooling and heating loads, and air distribution

- Select terminal control

- Design system control and instrumentation
- Finalize block cooling and heating loads; then incorporate them with other system loads to determine total building refrigeration and heating capacity requirements
- Determine system water flow(s)
- Finalize flow diagram and piping distribution layout, delineating service valving, and provide for water expansion, air venting, and makeup and pipe drains
- Size piping, including condensate piping
- Analyze and provide for controlled pipe expansion, including anchor points, guides, and expansion bends or joints
- Determine water system pressure loss and select pumps, heat exchangers, and other accessory equipment, such as strainers and filters
- Define scope, thicknesses, and type of pipe insulation
- Specify equipment, components, materials, and system commissioning

8.13 REFERENCES

ASHRAE. 1988. Variable volume water systems, design, application and trouble-shooting, a symposium. ASHRAE *Transactions* 94(2):1427-1466.

ASHRAE. 1992. *Handbook—Systems and Equipment.*

ASHRAE. 1993. *Pocket Guide for Air Conditioning, Heating, Ventilation, Refrigeration.*

Carrier. 1965. *Handbook of Air Conditioning System Design.* New York: McGraw-Hill Book Co.

ITT. 1985. Bulletin No. TEH-685. ITT Fluid Handling Division.

Chapter 9 SPECIAL SYSTEMS

9.1 DESICCANT PREDRYING

9.1.1 System Description

Conventional air-conditioning systems accomplish sensible cooling (lowering the air dry-bulb temperature) and latent cooling (dehumidification) simultaneously. Air must be chilled deeply below its dew point to remove moisture by condensation, frequently to a temperature far below what is needed for sensible cooling. In desiccant predrying, the two air-conditioning functions are separated. A liquid or solid desiccant removes moisture from the air by absorption or adsorption, and then conventional refrigeration equipment performs sensible cooling at a higher temperature (55 to 58°F), which results in a higher COP. The principal benefit of desiccant drying is that it reduces the refrigeration load and the electrical demand of the refrigeration equipment by eliminating the need to chill the air below its dew point. In addition, handling sensible and latent cooling separately enables the designer to manipulate HVAC design to a greater extent and to select from a broader range of energy sources and temperature levels to accomplish the goals. As a result, desiccant drying makes possible a significant reduction in energy costs for air conditioning.

Desiccant dehumidification requires thermal energy to regenerate (dry) the desiccant and subsequent cooling to remove the heat released during sorption. Regeneration can occur at a relatively low temperature (as low as 120°F) and may therefore use relatively inexpensive thermal energy sources, such as reject heat from building equipment, cogenerated heat, solar heat, and industrial waste heat. Cooling to remove the heat of sorption can be done at a relatively high temperature (up to 85°F). This cooling can be provided by nonrefrigerated water from a cooling tower or from evaporatively cooled building exhaust air, by well water, or by chilled water.

Figure 9-1 shows an air-and-water system with desiccant predrying. Outside air is dehumidified at the central plant (A to B) and distributed to terminals (B to E), where it is mixed with recirculated air and supplied to the occupied space (E to G). Outside air (A) is heated (not shown) and used to regenerate the desiccant. Exhaust air (I to J) cools the reconstituted desiccant (J to K) and is then discarded.

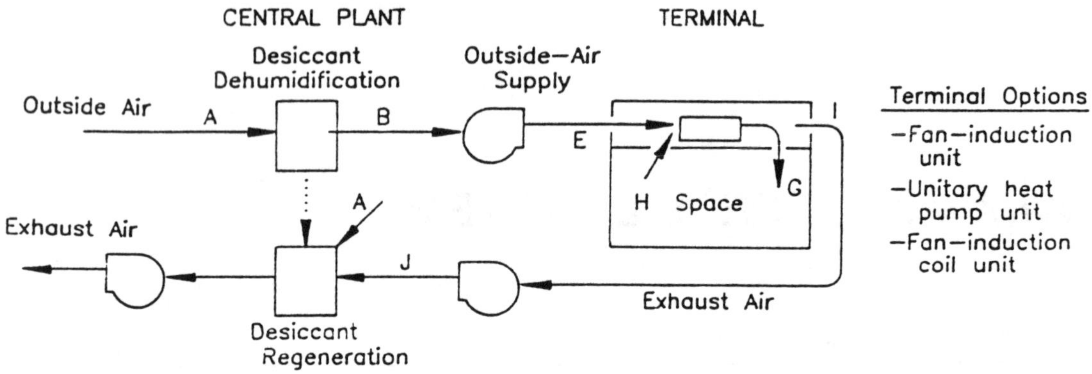

Fig. 9-1 Terminal Recirculation Cooling with Central Desiccant Dehumidification in Air-Water System

9.1.2 Applications

Construction of desiccant predrying systems is most economical when the HVAC system is configured as an air-and-water system in which the quantity of dry primary air is limited to the outside air required for ventilation, *e.g.*, 0.1 to 0.2 cfm/ft^2. This air should be dried to about 30 to 50 gr/lb (grains of moisture per pound of dry air), which is approximately 0.0045 to 0.0070 lb of water per lb of dry air. This low level of humidity provides all the latent cooling required. Fan-coil units, unitary heat pumps, or fan-induction coil units can then provide sensible cooling without the need for condensate drains at the terminals. In large buildings with long distribution runs, the construction costs for the smaller duct system can be substantially less than those for an all-air system. When combined with the savings for a smaller cooling unit, they can, to a large extent, offset the cost of the desiccant equipment.

Desiccant predrying can also be cost-effective in an all-air system where (1) 100% outside air is required, as in restaurants and certain laboratories, or (2) a special humidity control problem exists, as in supermarkets and pharmaceutical and electronic manufacturing areas.

9.1.3 Desiccants

Desiccants used in HVAC systems are of two types—solid or liquid. Solids are lithium chloride, molecular sieve, silica gel, and alumina gel. Commonly used liquids are aqueous solutions of either lithium chloride or triethylene glycol. In general, solid desiccants *adsorb* water on the surface of the desiccant, while liquid desiccants *absorb* water by chemically combining with it. In solid desiccant systems, the temperature rises in the dehumidification chamber as the sorption process releases heat (adiabatic dehumidification); the air being conditioned is precooled, aftercooled, or both. In liquid desiccant systems, cooling to remove the heat of sorption occurs as the heat is released in the dehumidification chamber (isothermal dehumidification).

Three dehumidification processes are compared schematically and psychrometrically in Figure 9-2. In the condensation system (Figure 9-2a), a conventional refrigeration system provides simultaneous dehumidification and sensible cooling (A to B in the psychrometric sketch on the right). The air temperature must then be raised to achieve the required supply air temperature (B to C) by reheat or by mixing with warmer room air.

In the desiccant predrying systems (Figures 9-2b and 9-2c), nonrefrigerated water from a cooling tower removes the heat of sorption, and subsequent sensible cooling is to the supply air temperature only. In the solid desiccant system (Figure 9-2b), incoming air is dehumidified by passing through a wheel that contains the solid desiccant (A to B) and is then sensibly cooled by nonrefrigerated cooling to remove the heat of sorption (B to C), followed by refrigeration cooling (C to D). The exhaust air is heated and used to regenerate the desiccant embedded in the slowly rotating wheel before being dumped. At any time, one-half or more of the desiccant wheel dehumidifies the incoming air, while the other half or less is being regenerated by the heated exhaust air.

In the liquid desiccant system (Figure 9-2c), incoming air is simultaneously precooled and dehumidified in the dehumidification chamber (A to B), followed by sensible cooling (B to C). This system contains two chambers—a conditioner (dehumidifier) and a regenerator (concentrator). In the dehumidifier, moisture is absorbed from the incoming air by a spray of cool desiccant. Heat is removed by water from a cooling tower. The moisture-laden desiccant circulates to the regenerator, where it is heated and sprayed into a scavenging exhaust airstream, thus removing its moisture. The scavenger air is then dumped to the outside.

The exact conditions causing absorption, adsorption, and desorption of water by desiccants are explained in Chapter 19, Sorbents and Desiccants, of ASHRAE *Fundamentals* (1993) and in Chapter 22, Desiccant Dehumidification and Pressure Drying Equipment, in ASHRAE *Systems and Equipment* (1992). For an example of a liquid desiccant drying system with engine chiller and fan-coil terminals, refer to Appendix E.

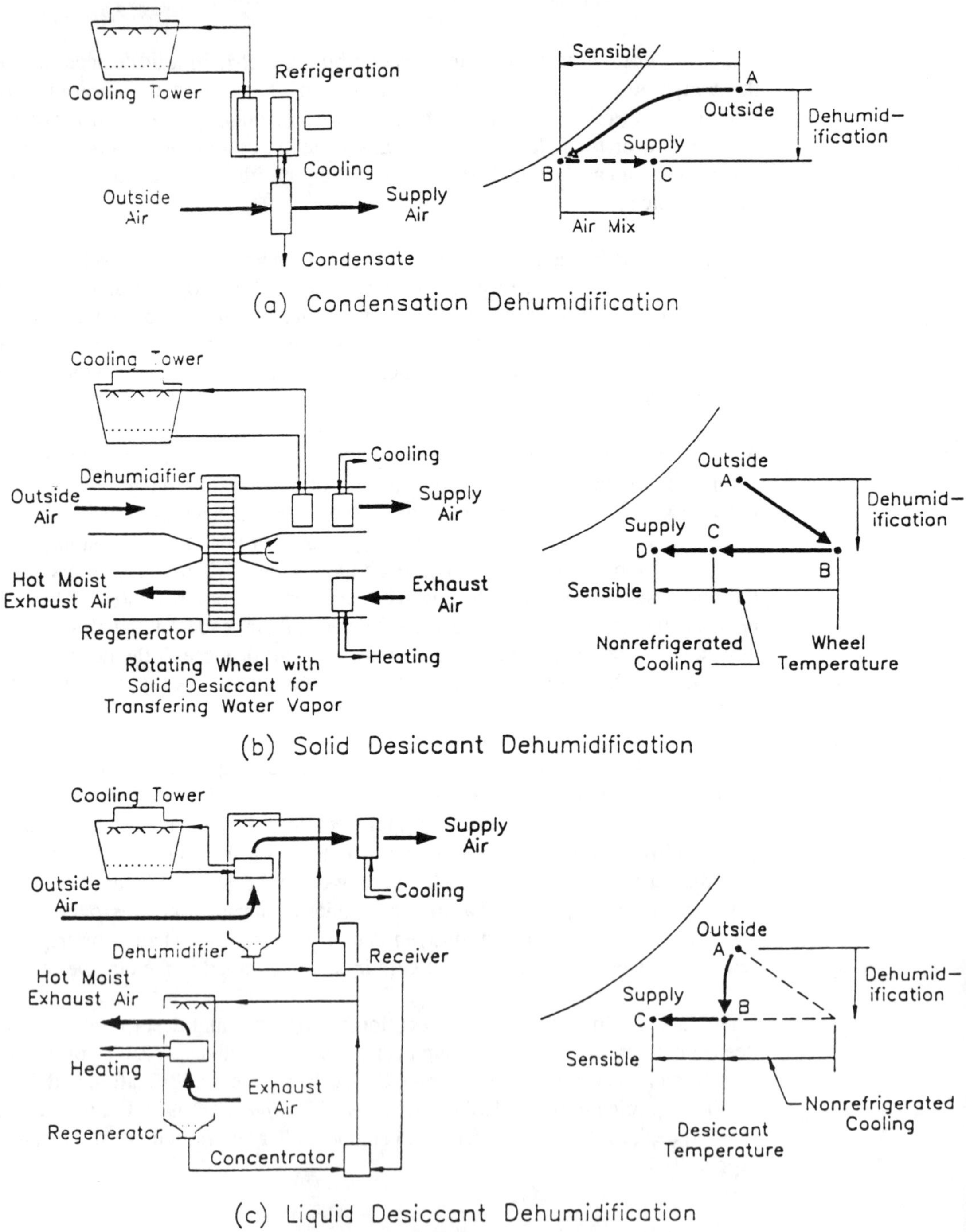

Fig. 9-2 Dehumidification Options

9.2 THERMAL STORAGE

9.2.1 General

Energy conservation usually involves using less energy, but it can also mean using energy more efficiently. Thermal storage allows cooling to be done at night for use the following day. This improves efficiency for utilities because it allows them to provide more energy with the same generating capacity. The resulting lower cost of utility operation is passed on to the customers in the form of lower rates. For an explanation of utility rates, see Section 2.7.5. Thermal storage, as such, is not an energy conservation measure but a cost-saving technique. It can save the building operator money in several ways:

- By decreasing or eliminating chiller operation during utility peak periods, demand charges are reduced
- By displacing energy use from peak to off-peak periods, a lower energy charge may be incurred
- Many electric utilities promote thermal storage by offering an incentive for displaced power, which can amount to several hundred dollars for each kilowatt displaced from peak to off-peak demand

Carefully investigate whether thermal storage can be advantageous for a particular project. Among the applications most favorable to thermal storage are buildings with high cooling demands of short duration, extensions of or additions to existing buildings, and projects in the service area of utilities with generation capacity problems and, consequently, high demand charges. For a more detailed discussion of thermal storage, see Chapter 39, Thermal Storage, in ASHRAE *Applications* (1991) or a number of publications by the Electric Power Research Institute (EPRI) referenced there.

9.2.2 Design-Day Cooling Load

When storage is involved, the cooling load for air-conditioning systems requires a different method of analysis. In systems without thermal storage, only the maximum instantaneous cooling load needs to be calculated. Capacity-reduction devices then allowed the refrigeration plant to follow the load profile during periods of smaller demand. No analysis of daily cooling loads was necessary. For storage systems, however, the cooling requirement for every hour of the design day must be calculated. In Figure 9-3, the base load is seen to vary from 13 tons at 8 A.M. to 20 tons between noon and 6 P.M. After some lights are extinguished and some air

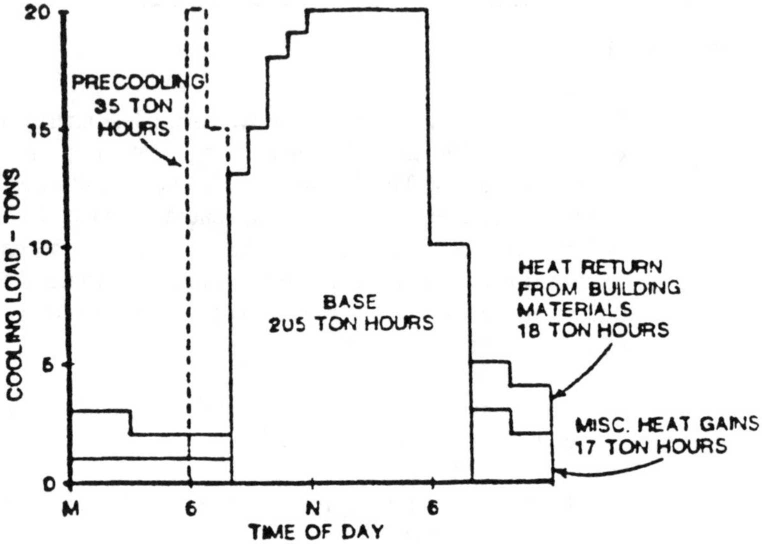

Fig. 9-3 Variation of Cooling Load During the Day

handlers are turned off, the load drops to 10 tons until the cooling systems are totally stopped at 8 P.M. At this point, the daily load would appear to integrate to 205 ton-hours.

It would be a mistake, however, to assume that this represents the entire cooling load. In reality, there are continuing heat gains of two types. One originates from sources such as emergency lights and video displays. The other derives from concrete floors, walls, and furniture, which absorb radiant heat from lights and solar sources and release it during unoccupied periods. These nighttime heat gains must be eliminated prior to occupancy. They are usually handled with a precooling cycle by starting the cooling plant before occupants arrive. In this example, the aggregate of these heat gains is calculated to be 35 ton-hours. The total daily cooling load is therefore 205 + 35 or 240 ton-hours. Only now can you begin to select a storage solution.

9.2.3 Design of Storage and Cooling Plant

Conventional refrigeration requires a unit with sufficient capacity to handle the largest rate of cooling and a capacity control to handle all lesser rates. From Figure 9-4, a 20-ton unit starting at 6 A.M. and following the cooling load profile would

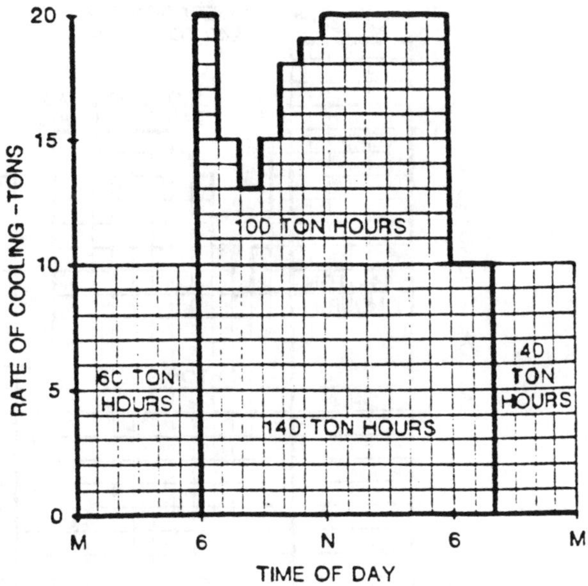

Fig. 9-4 Cooling Equipment Operation Options to Match Instantaneous Load of Figure 9-3

manage the load until it is shut off at 8 P.M. Each square under the cooling load profile represents one ton-hour. By counting squares, it can be seen that the cooling system provided 240 ton-hours during the day. While it required a unit with a 20-ton capacity to accomplish this, it is obvious that a 10-ton unit operating through 24 h could also have done the job.

Trying to cover the daily cooling load with this 10-ton unit requires 100 ton-hours of stored cooling between 6 A.M. and 6 P.M. to help the undersized cooling unit manage the load. This new aggregate of cooling capacity may not be the least expensive, but it accomplishes a useful function by reducing electric demand, since a 10-ton unit will use less power during the time of peak demand than a 20-ton unit. This system is referred to as *partial storage* because it is only large enough to help the undersized cooling unit get through the day.

If the refrigeration plant needs to be shut down entirely during the daytime period of high demand, the storage must be enlarged to handle the entire cooling load, as shown in Figure 9-5. Now the system is referred to as *full storage* because it can store the entire daily cooling demand of 240 ton-hours. If the electric utility peak period lasts from 6 A.M. to 8 P.M., 10 h are left for the cooling unit to regenerate the full storage. In this case, the unit would have to be as large as 240 ton-hours/10 h, which

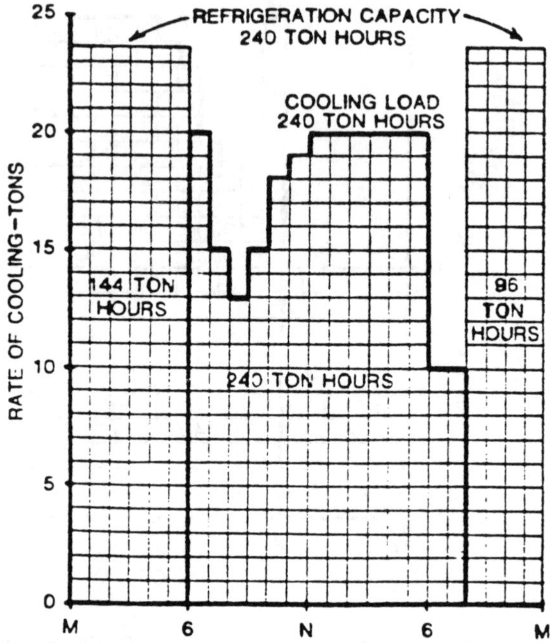

Fig. 9-5 **Cooling Equipment Off-Peak Operation to Satisfy Load of Figure 9-3**

equals 24 tons. In this case, which is typical, full storage is costlier to install than partial storage because both the refrigeration plant and the storage will have to be larger.

Storage is particularly viable when serving cooling loads of short duration. Figure 9-6 describes a church that requires 300 ton-hours of cooling for the entire week. Conventionally, this would require a 75-ton unit operating for 4 h on Sunday morning. If storage were provided, a two-ton ice builder operating continuously could serve the load with 36 ton-hours remaining for storage losses from the ice/chilled water tank (Figure 9-7). This soluiton not only reduces electric demand but may save on initial cost as well; also, consider that the three-phase service to feed the 75-ton unit may not be available in some residential neighborhoods.

Another situation might involve a coliseum that can be maintained in the comfort zone without people or lighting for 200 tons but requires 2000 tons for 4 h to serve 25,000 people during nightly rock concerts (Figure 9-8). The conventional approach would involve the purchase of 2000 tons of refrigeration to operate 4 h per day at full load and 20 h per day at 10% of capacity. The storage concept would manage the

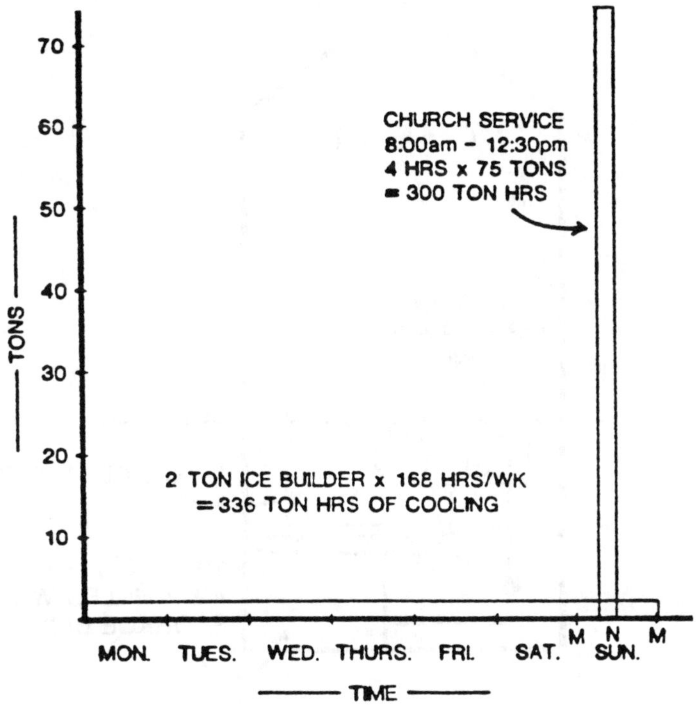

Fig. 9-6 Typical Cooling Load for Church

same 12,000-ton-hour daily requirement less expensively with a 500-ton cooling plant operating around the clock.

9.2.4 Sizing, Cost, and Location of Storage Tanks

Storage tanks of 50,000 gal or less are usually constructed of steel or reinforced plastic. Larger tanks are constructed more inexpensively of concrete. The cost of the tank is an important element for water storage because of the large size required—it varies from \$0.25 to \$1.00 per gallon. Storage volume is a function of the amount of cooling to be stored. This often works out to be between 0.5 and 1.0 gal/ft^2 of conditioned space for water storage and about 25% as much for ice storage. Because of its bulk and weight, storage is frequently located in or under the basement or next to the building. On the other hand, "topside" storage eliminates the energy required to transfer chilled water from open storage containers into basement piping circuits

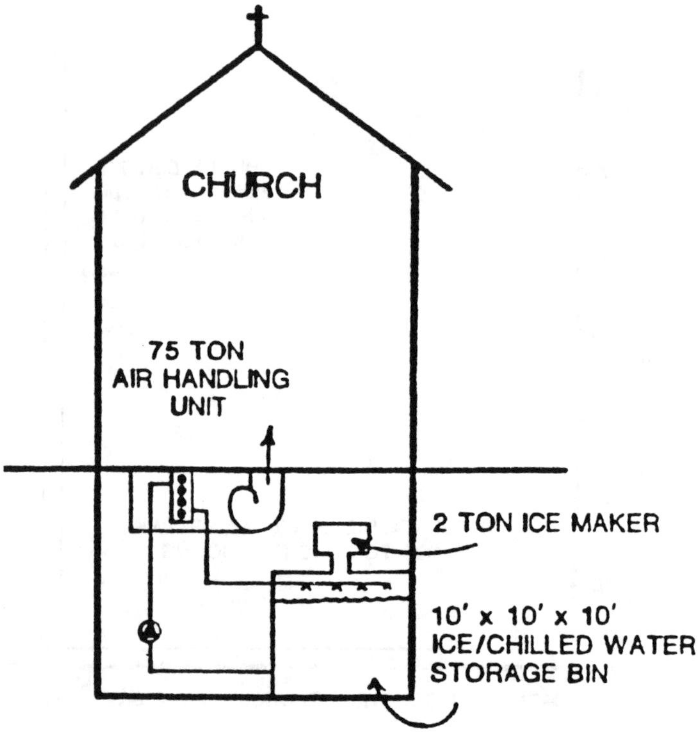

Fig. 9-7 Thermal Storage Installed in Church

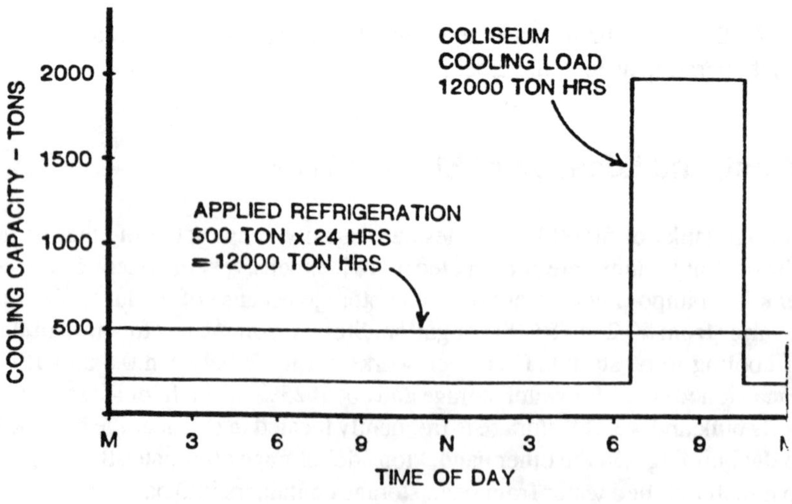

Fig. 9-8 Typical Cooling Load for Sports Arena

pressurized by the static head of the building [see Chapter 39, Thermal Storage, in ASHRAE *Applications* (1991)].

9.2.5 Chilled Water Storage Circuitry and Control

Figure 9-9 shows the elements of a good control system for chilled-water storage. Once the demand limiter has been set for the chiller, it provides chilled water at a constant temperature but at some fraction of full-capacity flow. The demand limiting device forces the control valve to open more flow to bypass. This reduces the inlet temperature to the chiller and causes it to operate using less energy. The balance of the chilled water required by the load is drawn from storage. System demand for chilled water is based on the position of throttling valves at the coils. These, in turn, may be controlled by thermostats in the air leaving the coils. The total flow is regulated by a pressurestat, which controls the pump with a variable-speed motor or

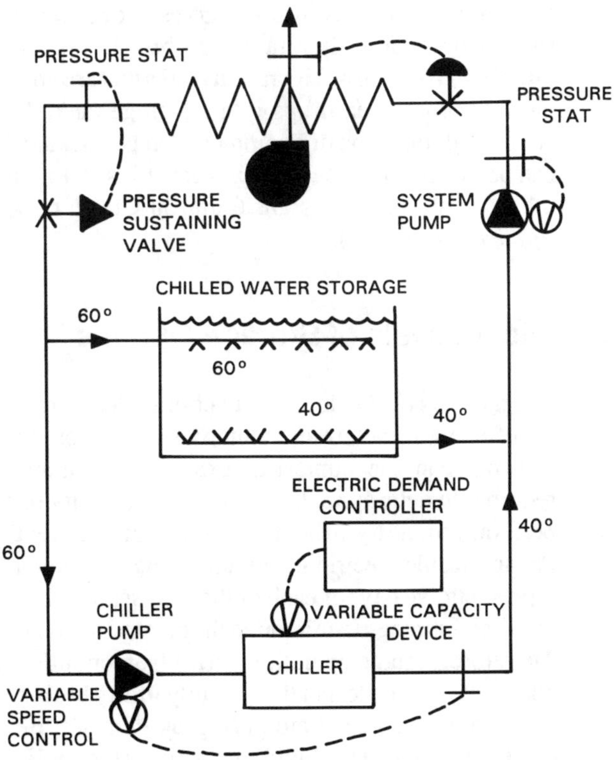

Fig. 9-9 Typical Chilled Water Storage System Schematic

bypass control. In some cases, there may be an override control to maintain the temperature of water returning to storage. As shown, the controls maintain the design supply and return water temperature to ensure that storage retains its rated capacity. The pressure-sustaining valve should be controlled from a pressurestat located at the high point of the system and set for some minimum positive pressure, such as 5 psi.

9.2.6 Ice Storage Concepts

The concept of chilling water and storing it in tanks is straightforward. By comparison, the techniques for ice storage are more varied and complex. One widely applied ice storage system consists of refrigerated pipe coils submerged in a water tank. During charging, ice builds on the coil, and during discharge, the chilled water circulating through the tank melts the ice. Another type of ice storage module is made so that it can be frozen and thawed by circulating brine through coils in cylindrical water tanks. Advantages claimed for this technique include the fact that melting and refreezing always take place adjacent to the piping instead of remote from it. Other types of ice storage consist of harvester or shucker ice packages, which generate ice to be stored in an insulated ice bin. Ice is frozen to a thickness of approximately 3/16 in. on refrigeration plates that are up to 7 ft long. A defrost cycle releases the ice, which then drops into a storage tank. As far as the refrigeration cycle is concerned, this type of equipment can be packaged in sizes up to 400 tons. The tank can be of any size to suit the process. Illustrations and operating cycles for these and other ice storage types are found in Chapter 39, Thermal Storage, in ASHRAE *Applications* (1991).

9.2.7 Economic Analysis of Storage Options

The analysis of the design-day cooling load profile shown in the previous sections is sufficient to determine tentative values for storage and cooling plant sizing, as illustrated in the numerical examples. A more thorough analysis, however, is required to determine the economic feasibility of a thermal storage project. This is preferably done by an hour-by-hour analysis over the entire year, taking into account the applicable electric utility tariff. As a minimum, it must cover the analysis of a representative day for each of the four seasons because optimum use of cool storage requires a strategy to minimize the electric demand of the chiller during peak periods. Therefore, storage operation varies from month to month as a function of the daily cooling load profile and the monthly utility billing demand, which is not necessarily identical to the actual monthly peak demand. Be aware of the fact that the entire electric demand of the building (not only the cooling plant demand) must be included in the calculations to obtain the optimum tradeoff point between chiller and storage size.

For example, Figure 9-10 illustrates a 500-ton chiller managing a 12,000-ton-hour cooling load on the hottest day of the year. The chiller and 7000 ton-hours of storage are both being used to full capacity. In October, the maximum cooling day requires only 9000 ton-hours (Figure 9-11). Thus, if the storage is used to its full extent, the chiller may be limited to only 200 tons of input. It would be normal to leave the chiller at the 200-ton setting for other October days of lesser cooling requirement because the chiller demand has already been created, and the more cooling is generated and directed to load, the less loss through storage. For these lesser days, the operator would precharge the storage only as much as needed to get through the day, since

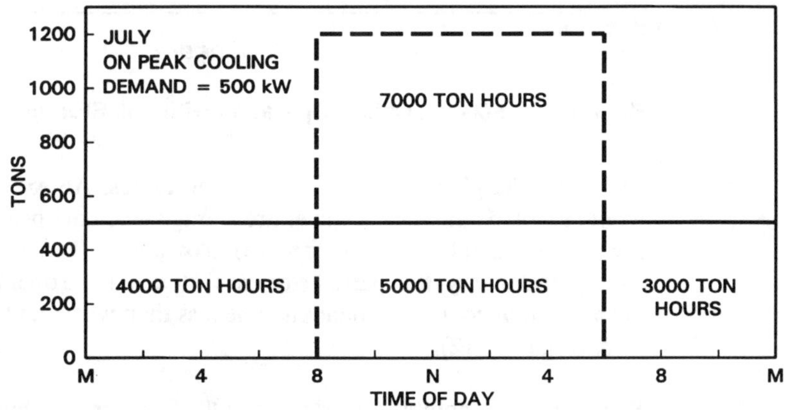

Fig. 9-10 **500-ton Chiller and Partial Storage Satisfying 1200-ton Demand on Summer Design Day**

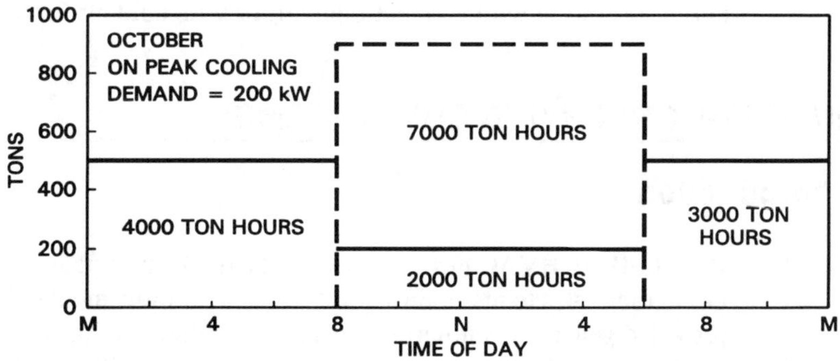

Fig. 9-11 **500-ton Chiller and Partial Storage Operating on Autumn Day**

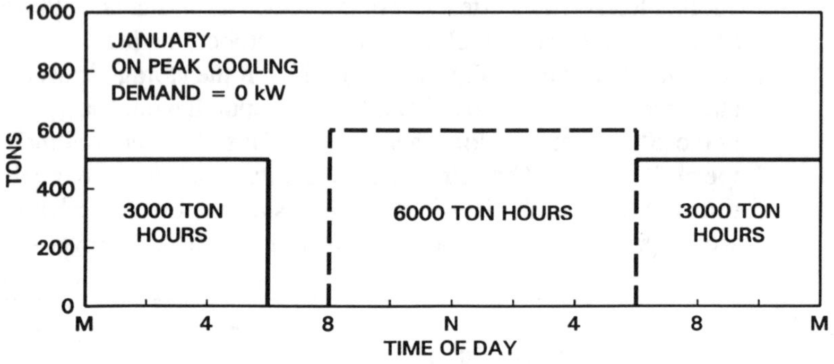

Fig. 9-12 500-ton Chiller Operating with Full Storage on Winter Day

partial precharging helps to limit storage losses. An exception to this procedure would involve daily use of the entire storage when off-peak rates were sufficiently lower (for example, 2¢/kWh or more) to overcome the penalty of transfer pumping energy. In January, this partial storage system may be operated as full storage since the maximum cooling demand may be less than what can be generated entirely off peak (Figure 9-12).

Some designers combine ice storage with a cold air distribution system (see Section 9.3.8) to take advantage of the lower air temperature available from ice storage. Bhansali and Hittle (1990) compared a variety of VAV systems with economizer cycles in five climates under four different utility tariffs. In spite of the additional energy use (20 to 45%), the combination of ice storage and cold air resulted in lower annual utility costs than ice storage with conventional temperature air or regular chiller operation with conventional air, or with cold air for all cases investigated.

9.3 ENERGY-CONSERVING SUBSYSTEMS

9.3.1 Introduction

Prior to 1970, HVAC designers found it expedient to plan systems for generous air circulation rates to maintain a good sense of air movement for occupants. They also provided generous ventilation rates for the dilution of airborne contaminants. Less thought was given to achieving these fundamentals with the least possible energy use because energy cost constituted a relatively small part of operating a building. The oil embargo of the 1970s led to a fundamental reappraisal of how to optimize the energy use of existing systems and how to design for minimum energy use.

Commercial HVAC systems were found to be consuming from 60,000 to 300,000 Btu/ft^2·yr with an average of, perhaps, 160,000 Btu/ft^2·yr. Where modifications were pursued vigorously, this range has been reduced to 40,000 to 150,000 Btu/ft^2·yr. New systems are typically targeted within the range of 30,000 to 85,000 Btu/ft^2·yr. To understand the design philosophy of energy conservation, it is important to examine what was wasteful about conventional design, what has been done to optimize existing systems, and what is now considered appropriate for new systems.

It is energy-efficient to allow the space temperature to rise above design conditions in the summer and float downward in the winter during unoccupied hours. This may be done simply by resetting the thermostat or temperature sensor automatically or manually. However, when there are a number of zones in a building, some of which are occupied while others are unoccupied, a more complex control system is required. Modular equipment, variable-speed fans and pumps, reset controls for hot and chilled water, zone control valves or mixing boxes, and unitary equipment are options to accommodate varying occupancy profiles or varying loads due to climate or building use.

9.3.2 Simple Systems

Many constant-volume systems were set up to heat and cool a zone (Figure 9-13). Cooling could be provided fully by outside air when it was below 55°F, with help

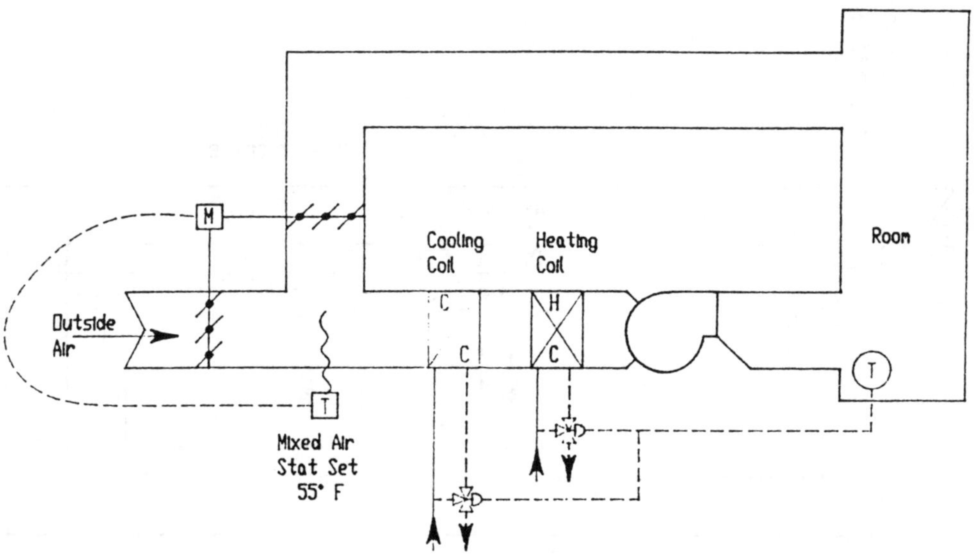

Fig. 9-13 Conventional System

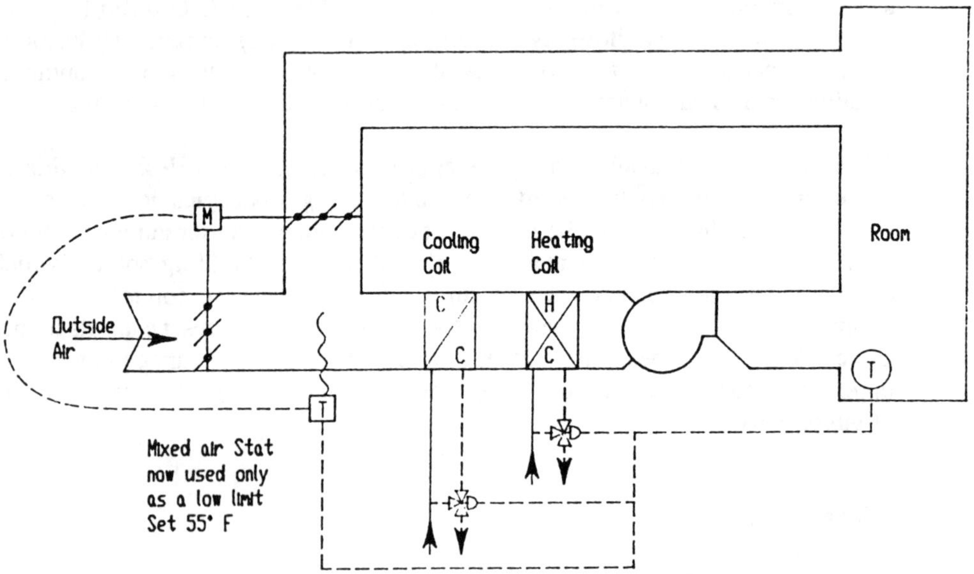

Fig. 9-14 System Modified to Eliminate Parasitic Heating

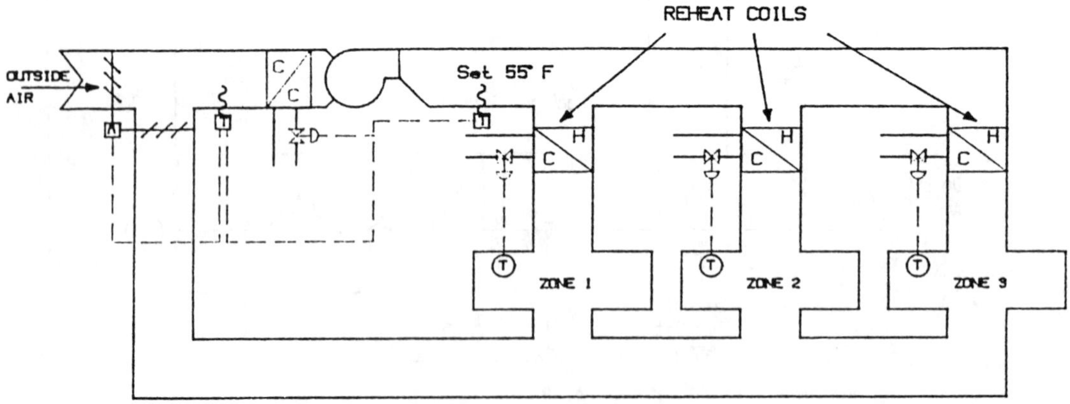

Fig. 9-15 Conventional Control

from mechanical refrigeration between 55 and 70°F and by refrigeration alone above 70°F. While this system offered comfort for occupants, it wasted heating energy when the room thermostat found 55°F air too cold and turned on the heating coil to raise the supply air temperature. Figure 9-14 illustrates how the room thermostat has been modified to control the mixed air directly. This avoids overcooling the zone and eliminates the parasitic heating in the system depicted in Figure 9-13.

9.3.3 Discriminating Temperature Controls

In Figure 9-15, a ductstat (a thermostat located inside a duct) uses outside air at first and then applies mechanical cooling to establish a base supply air temperature to more than one zone. Reheat coils respond to zone thermostats to raise the supply air temperature, if required, to match the actual cooling requirement in each zone. The problem with this scheme is similar to the previous one, since most of the time, none of the zones will require air as cold as the base supply air temperature, which was initially set to suit some maximum design condition. The solution is to "discriminate" the supply air temperature to respond to feedback from the zone thermostats. The supply air temperature will now be only as low as that required for the zone with the greatest cooling load. Parasitic reheat may now be a small fraction of what it was before, as is apparent from Figure 9-16.

This concept of discrimination applies equally to the control of most basic HVAC systems. For example, multizone deck temperatures need be only as high or as low as demanded by the greatest zone requirement for heating or cooling. The hot and cold supply temperatures of a double-duct system need be only as deviant from room temperature as required by the strongest call for heating and cooling—a call usually

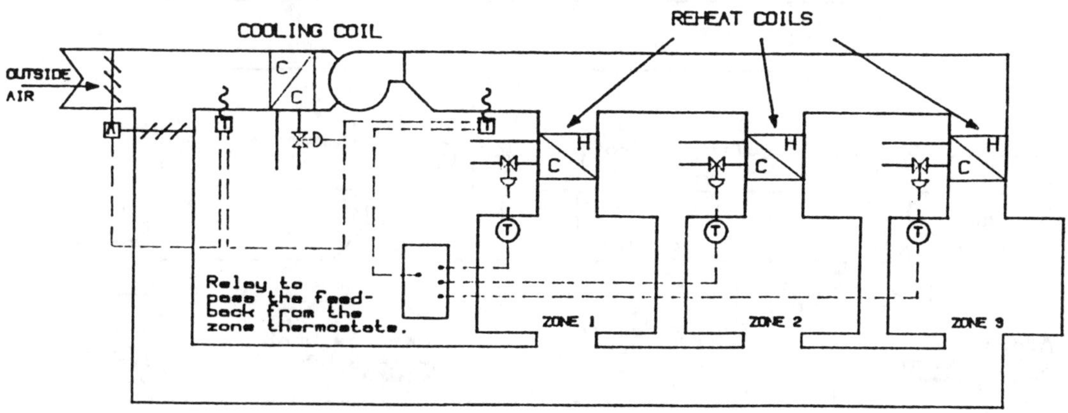

Fig. 9-16 Discriminating Control to Minimize Reheat

indicated by an extreme valve or damper position in one of the subzones. The air and water temperatures for all systems should be responsive to feedback from zone controls so that the least possible energy is required to provide them. In this way, the flexibility for individual zone temperature control can be provided with a minimum of parasitic heating and cooling.

9.3.4 Eliminating Energy Waste in Simple Systems

An ideal HVAC system would use the least possible energy for air and water circulation and generate only the amount of heating and cooling necessary to offset space heat losses and gains. Reheat energy for temperature control, as used in the past, would be eliminated because it adds to both heating and cooling requirements. How can such an ideal system be constituted?

First, the amount of air and water circulated should only be sufficient, at any given instant, to convey the necessary heating or cooling to the space. In this respect, it is prudent to circulate a minimum of four air changes per hour to provide occupants with the sensation of adequate air movement and to provide adequate indoor air quality. Above this value, circulation rates will vary to match fluctuating loads. Second, the generated heat must be only enough to serve the aggregate zone heating loads, and the generated cooling must be only enough to serve the aggregate zone cooling loads. If these concepts are applied to the simple system shown in Figure 9-13, the zone thermostat would reduce the volume of air being circulated as the zone temperature was being satisfied. This would optimize fan energy. When

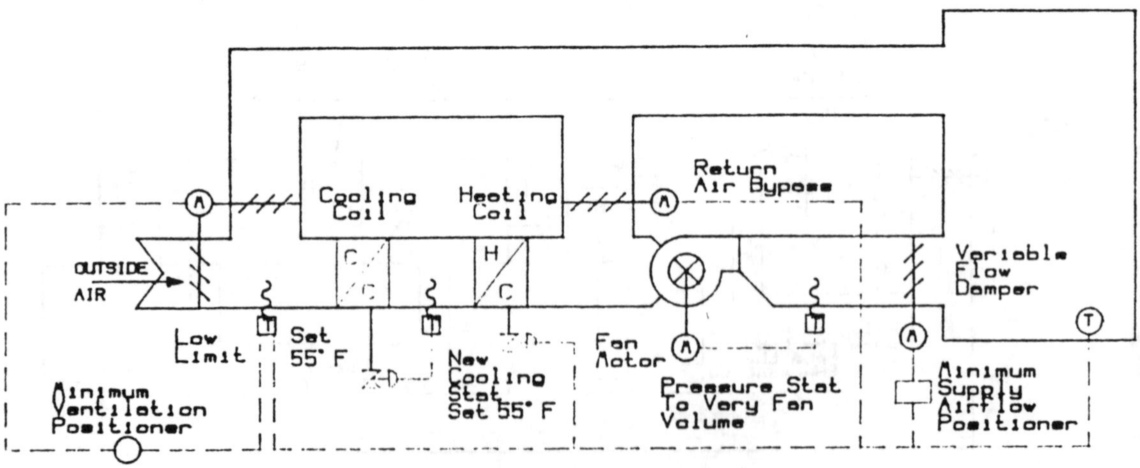

Fig. 9-17 Modifications for Energy Conservation

the minimum of four air changes per hour is reached, the thermostat should have the authority to raise the supply air temperature so that the space is not overcooled.

Designers will recognize a flaw in this step because air that is not cooled to 55°F may carry too much outside moisture to the space to maintain the desired relative humidity of 50% or less. Figure 9-17 illustrates control modifications that act to conserve fan energy while preserving minimum air circulation and a satisfactory relative humidity in the space. The return air bypass ensures that no moisture-laden outside air circumvents the cooling coil. Variable supply airflow allows reductions in fan energy down to a minimum rate of air circulation. In this scenario, mechanical cooling would be initiated when the outside air temperature rose to more than 55°F. As space temperature rises, the thermostat would first throttle the volume of 55°F air to the minimum setting and then raise the supply air temperature by opening the return air bypass. The air mixing and bypass air dampers must be subjected to a minimum outside air volume for adequate ventilation.

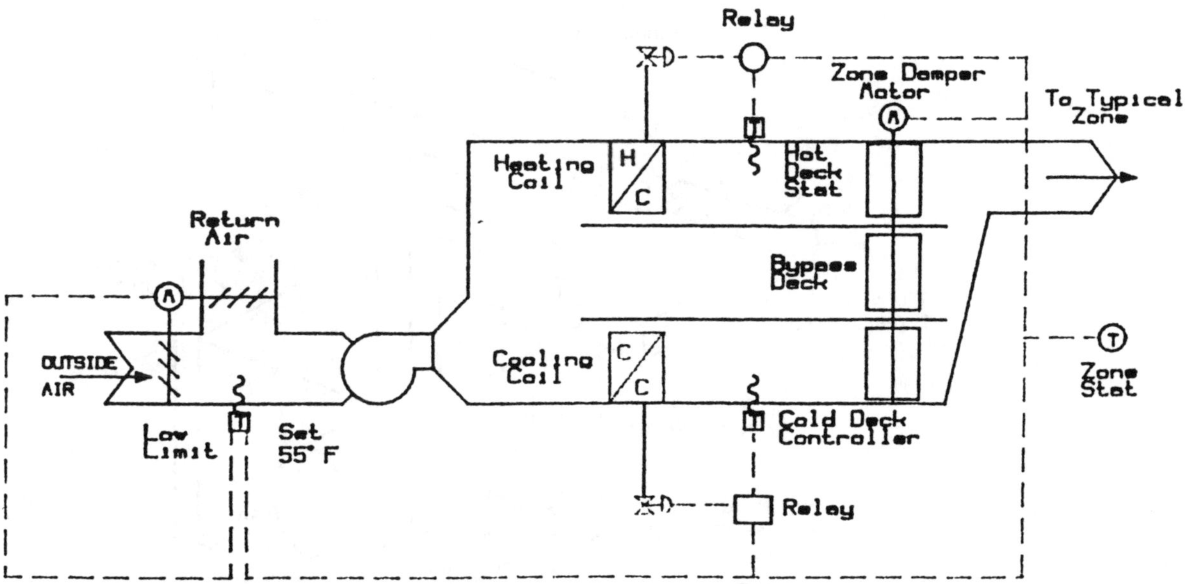

Fig. 9-18 Modified Multizone Unit

9.3.5 Eliminating Reheat in Multizone Systems

Historically, the multizone air unit has wasted energy substantially. In many cases, the deck temperatures were permanently set to satisfy the greatest winter heating requirement and the greatest summer cooling requirement. Automatic discrimination of deck temperatures to suit real-time zone demands can do much to reduce the parasitic air-mixing losses, but total elimination of cooling and heating waste can only be managed by adding a bypass deck to each zone, as shown in Figure 9-18. The zones are now able to modify the deck temperature to suit the requirement of the highest zone demand. In addition, the bypass deck eliminates the mixing of hot and cold air. The zone thermostat can call, in sequence, for full hot air, hot air and bypass mix, full bypass, cold air and bypass mix, and full cold air. Manufacturers now offer this type of multizone unit.

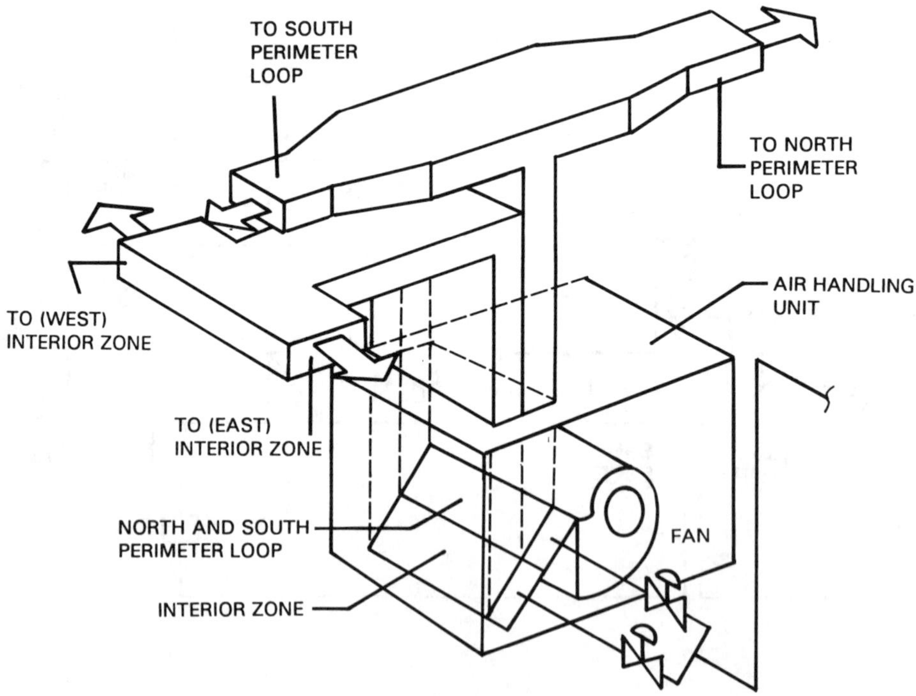

Fig. 9-19 Two-Zone Air-Handling Unit

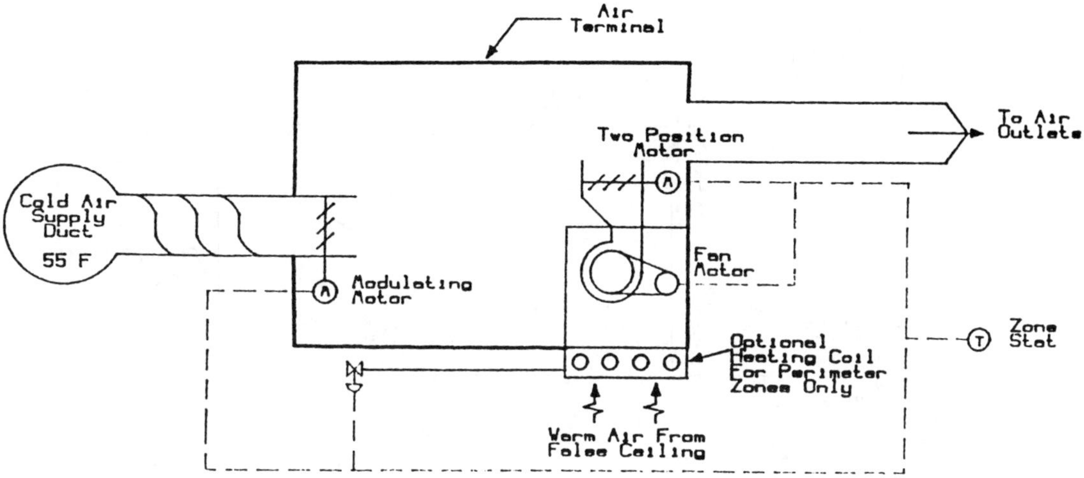

Fig. 9-20 Air Terminal with Side Pocket Fan, Sized for at Least Four Air Changes per Hour

9.3.6 Eliminating Parasitic Heating in VAV Systems

In the past, VAV systems have saved on fan energy but not without countervailing side effects. When 55°F supply air to interior zones is reduced to match internal heat gains, the volume may be reduced below the minimum air change rate, leading to air quality complaints. The antidote involving reheat of the minimum supply air volume is also counterproductive for energy reasons. When the same 55°F air is supplied to perimeter zones, it must be reheated to room temperature in winter before it is heated further to overcome envelope heat loss. The heat input from 55°F to room temperature is parasitic. One way to overcome this is to supply air that is thermostatically controlled for each solar zone. At the least, this implies separate air supply zoning for interior, south-facing perimeter, and other perimeter zones. For compartmented air-handling systems, the air unit can be configured as shown in Figure 9-19.

Another solution that is gaining in popularity is to add a fan to each air terminal, as shown in Figure 9-20. The side-pocket fan in the air terminal would be sized to provide the minimum rate of air circulation. It would be started only when the cold air supply fell below this critical value.

Reheating the 55°F air supply to perimeter zones would be avoided by ensuring that perimeter heat would not be activated until the cold air damper was completely shut. Since the addition of fans in the false ceiling adds unpopular maintenance for bearings and filters in occupied spaces, the same effect can be achieved with one

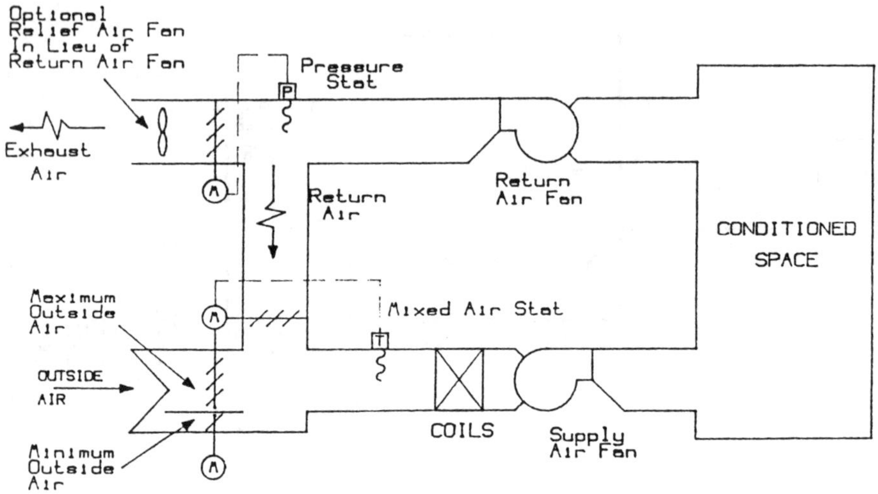

Fig. 9-21 Air-Side Economizer Cycle

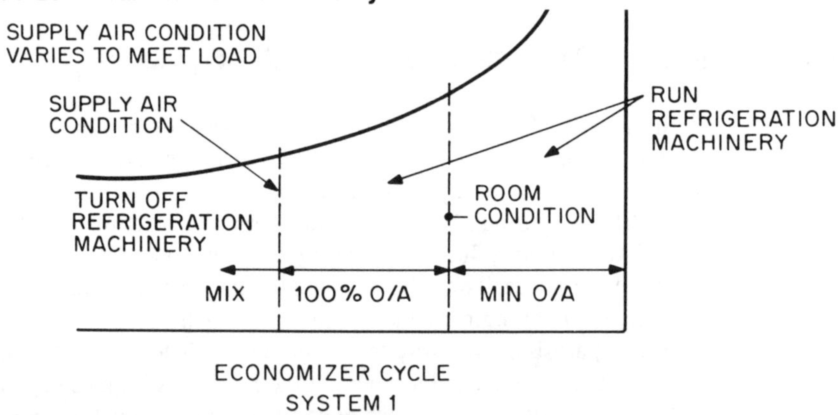

Fig. 9-22 Dry-Bulb Temperature Economizer Cycle

central fan for each floor or major zone. The central fan would then recirculate air at a neutral temperature to double-duct-type terminals connected jointly to the cold air supply and neutral air supply. Perimeter air terminals would have the option of adding heating coils, which would be activated only after the cold air damper was totally closed.

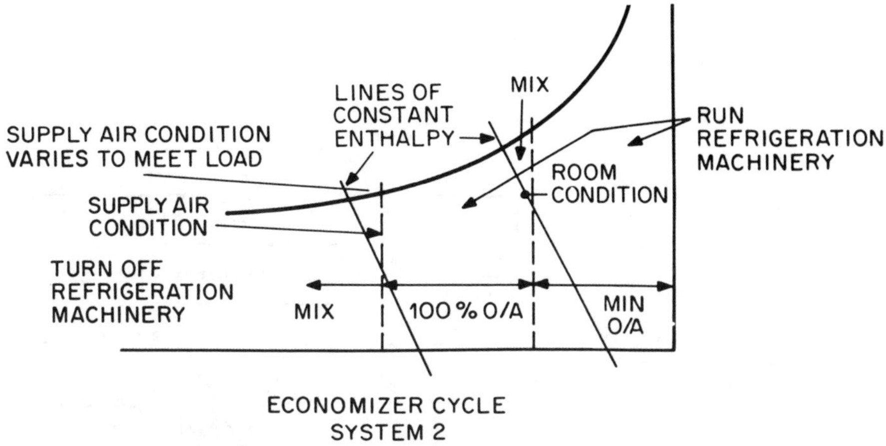

Fig. 9-23 Enthalpy Economizer Cycle

9.3.7 Economizer Cycle (Free Cooling)

Air-handling systems that have access to 100% outside air can provide full cooling without the assistance of mechanical refrigeration whenever the outside temperature is lower than the required supply air temperature. This so-called *air-side economizer cycle* (Figure 9-21) is progressively more effective in northern latitudes, saving up to 70% of mechanical refrigeration energy. In southern areas, such as Florida, the air-side economizer is seldom used. This is because the number of hours during which the outside temperature falls below the controlled space temperature is insufficient to justify the investment in the relief fan, air-mixing chambers, and louvers necessary to dissipate the air pressure caused by supplying 100% outside air.

More energy savings are achieved with a more complicated economizer cycle when:

- The outdoor air temperature is lower than the supply air temperature required to meet the space-cooling load; compressors and chilled water pumps are turned off; and outdoor air, return air, and exhaust air dampers are positioned to attain the required space temperature.

- The outdoor air temperature is higher than the required supply air temperature but is lower than the return air temperature; compressor and chilled water pumps are energized; and the dampers are positioned for 100% outside air.

- The outdoor air temperature exceeds the return air temperature; the dampers are positioned to bring in the minimum outdoor air required for ventilation.

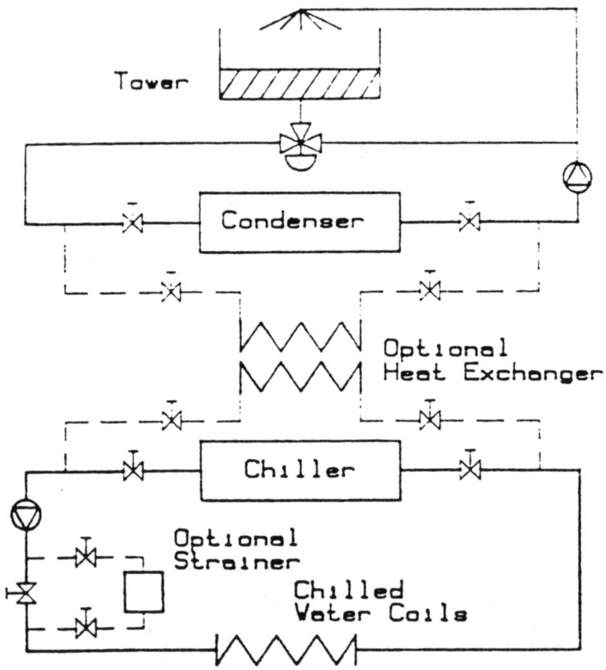

Fig. 9-24 Water-Side Free Cooling Options

To be truly effective, air-side economizer cycles should not be based on dry-bulb temperature but on enthalpy, as illustrated in Figures 9-22 and 9-23 (Dubin and Long 1978). Enthalpy controls are so unreliable, however, that most air-side economizer cycles have been scheduled to revert to a minimum setting for ventilation when the dry-bulb temperature reaches 80°F or an even lower value.

Compartmented air-handling systems that lack the potential for 100% outside air may adopt a winter free-cooling concept for chilled water by interconnecting the chilled water circuit with the cooling tower. This can be done without the benefit of a heat exchanger but at some risk of pipe corrosion and more expensive water treatment. The strainer shown in Figure 9-24 can be as effective as the screening in removing airborne dirt from the tower discharge, but oxygen, which is added to the water by the tower action, is potentially damaging to the chilled water piping circuit. If you do not want to accept this risk, add a heat exchanger to separate the chilled water circuit from the cooling tower circuit. This adds capital cost for the heat exchanger and reduces the effectiveness of water-side cooling because of the additional Δt of the heat exchanger. Water-side free cooling may be as little as one-half as energy conserving as air-side free cooling, depending on climate.

Another form of free cooling involves purging the conditioned areas with cool night air prior to occupancy the following morning. This can avoid the cooling energy necessary to overcome the heat buildup from cleaners' lights, personal computers, and heat flowing back to the conditioned space from concrete floor slabs overheated during the day and from radiant heat from the back of light fixtures. This purging cycle is highly effective in dry climates with low nighttime temperatures, such as in the southwestern United States.

9.3.8 Cold Air Supply Systems

Conventional systems are based on a supply air temperature of 55°F as required to ensure a relative humidity of 50% at a room temperature of 75°F. The advent of ice storage and the ever-increasing cost of air distribution have alerted designers to the potential of supplying less air at a lower temperature. Supply air temperatures can be as close as 5°F to the primary coolant temperature. In the case of water chillers, 40°F water can supply 45°F supply air. In the case of an ice-based storage system, the 36 to 38°F coolant can produce air supply temperatures of 41 to 43°F. Design details for cold air with ice storage are given in Chapter 39, Thermal Storage, in ASHRAE *Applications* (1991). The cold supply air must be tempered with room air to prevent cold drafts. If this is done in fan-powered boxes that operate continuously, some of the energy savings are lost. If the cold supply air causes surfaces to be cooled below the ambient dew point, they must be insulated; this includes air handlers, ducts, and terminal boxes. Since leakage from cold air ducts aggravates any condensation problems, the designer should specify methods of reliably sealing all such ducts. Condensation problems can be reduced during system start-up by lowering the supply temperature gradually.

According to the ASHRAE Comfort Chart (Figure 3-1), the lower relative humidity generated with cold air means that space dry-bulb temperature setpoints can be raised up to 1°F for the same comfort conditions. To ensure full cooling and dehumidification, face velocities at cooling coils supplying cold air should be in the 350- to 450-fpm range, with an absolute maximum of 550 fpm (Dorgan 1989). In cases of high sensible heat ratios, cold air systems may produce undesirably low relative humidity values. A system to remedy this situation uses cold air to satisfy the latent cooling load and supplies additional sensible cooling through 55 to 60°F chilled water coils at the mixing boxes. Operating a chiller to supply chilled water at that temperature is most efficient, and the need for additional chilled water piping can be eliminated by using the sprinkler piping to supply the chilled water coils. This has been permitted by NFPA *Standard* 13, Standard for the Installation of Sprinkler Systems, since 1978 (Meckler 1988).

The use of temperatures lower than 55°F for supply air permits the use of smaller ducts and fans or pipes and pumps. The reduced ceiling space required for the smaller ducts results in significant building height savings in high-rise buildings. Perform an analysis to select the appropriate discharge air temperature, based on a comparative economic analysis of the entire building, not the HVAC system only. That analysis must consider savings in first cost from smaller equipment, a smaller distribution system, and a lower building; increases in first cost for a chiller that operates at a lower temperature; increased operating costs from powered mixing boxes; and reduced operating cost from lower air and water mass flow rates, and, possibly, a slightly lower space setpoint temperature (Dorgan 1989).

9.3.9 Water-Side Conservation

Savings are available from variable flow pumping of chilled water and condensing water and variable speed tower fans. Water chillers can benefit from variable speed compressor motors and variable supply water temperatures when air-handling systems operate at partial loads. Chillers can also be designed to accept a lower condensing water temperature than the old standard of 85°F, when available, to reduce the thermodynamic head on the compressor. Check with the chiller manufacturer to determine the lowest acceptable condenser water temperature and specify control accordingly. The combination of compressor motor, condensing water pump motor, and tower fan motor has been assumed in the past to require one kilowatt input

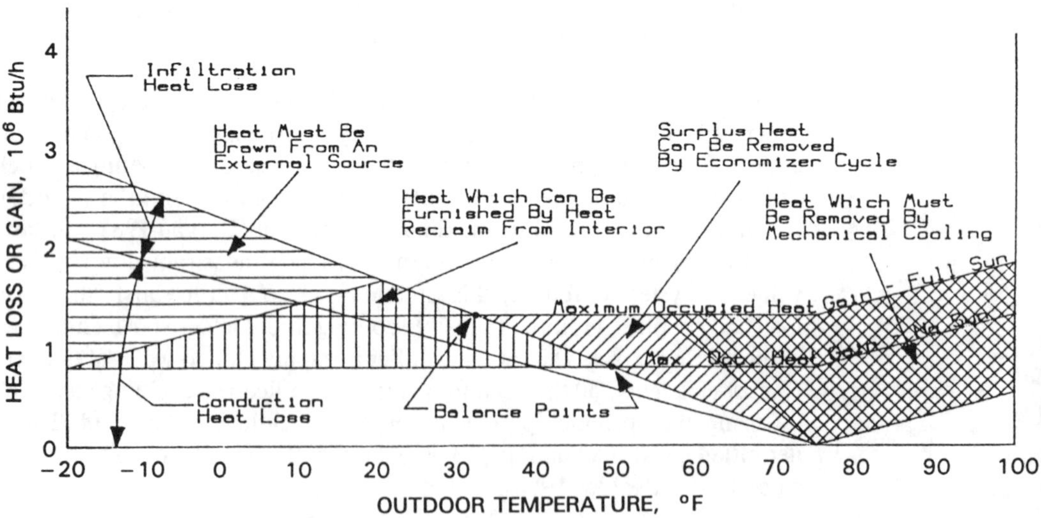

Fig. 9-25 Heat Balance in Perimeter Space

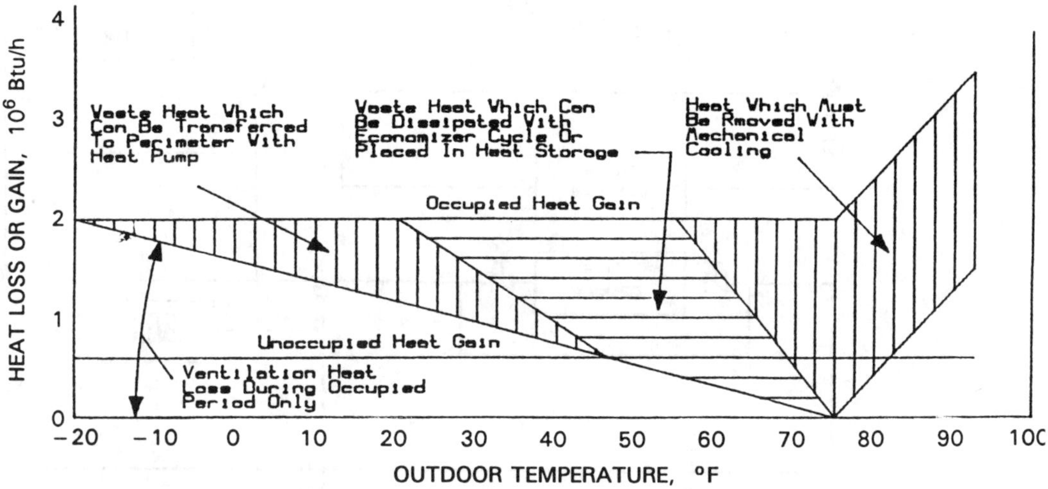

Fig. 9-26 Heat Gain and Heat Loss for Interior Space

for each ton output. Variable temperature and variable speed controls have now enabled this value to be virtually halved for average cooling performance.

9.3.10 Bootstrap Heating

Another form of low-cost heating can take place in winter by operating the chiller to remove the surplus heat in interior spaces and by using the warm condensing water thus generated to offset perimeter heat losses. It might seem counterproductive to use any form of electrical heating when fossil fuels are known to be less expensive. However, if properly operated, a water chiller can be operated as a heat pump with a COP of up to 6. At this rate, heat reclaim may be less than one-half as expensive as the next cheapest fuel. Additional piping and controls are necessary to interface heating and cooling circuitry and to add additional condenser tubing for "clean" heat exchange. Consider these extra costs in the context of savings available from bootstrapping waste interior heat. For this reason, the potential savings must be estimated in advance.

Take care to see that the heat pump chiller is operated only for the heat required by the system. If the chiller must be false-loaded at low capacities, much of the economy will be lost. The control of the heat pump chiller has to be compatible with an economizer cycle. In other words, the mixed air temperature, which determines the percentage of outside air, should be raised only enough to permit the cooling coil to extract the heat required by the perimeter circuits. The balance of the free cooling is

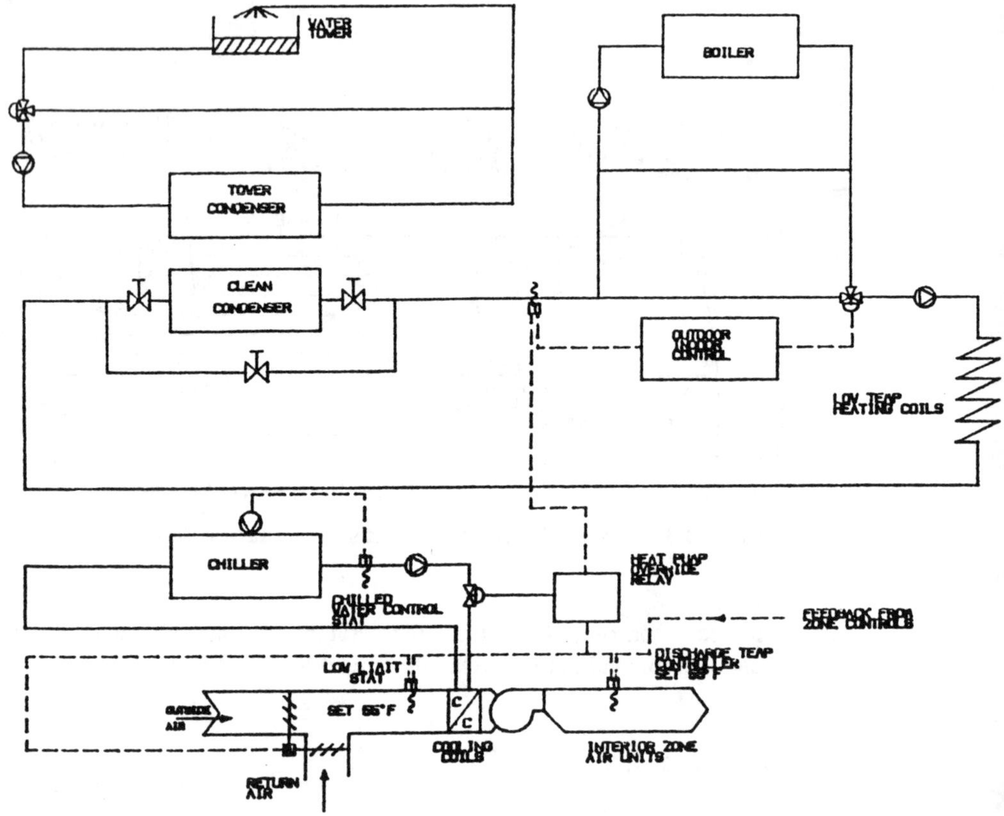

Fig. 9-27 **Control Schematic for Water Chiller/Heat Pump to Use Heat from Interior Zones to Satisfy Heating Demands in Perimeter Zones**

then performed with outside air, while the heat pump boosts the heat taken from the cooling coil to a useful heating temperature.

Buildings have occupied and unoccupied periods. For each, there is a balance point at some outside temperature where heat gains, including the electrical input to the heat pump, balance heat losses. Figures 9-25 and 9-26 indicate how such balance temperatures can be determined. They also illustrate the temperature regions in which heat can be salvaged and usefully transferred and where supplementary heat may be required. The bulk heat that can be bootstrapped at low cost can be analyzed and aggregated with hour-by-hour or bin-type energy analysis programs. The result can then be compared with the premium investment in hardware required to execute the concept.

Figure 9-27 illustrates a form of control that allows the water chiller/heat pump to draw heat from interior zone cooling coils to the extent required to satisfy heating loads. Condensing water temperatures should be arranged in the 90 to 105°F range for heating. Sufficient heating surface needs to be provided so that such low-temperature water can satisfy the heating loads. This seldom creates an investment penalty when the coils are serving alternately for cooling, as with office induction units and hotel room fan-coil units.

The chiller size required for reclaim heating is often less than one-third the size required for cooling. This may validate a two-chiller selection with a one-third size chiller operating for heating and both units operating for maximum cooling. Centrifugal chillers experience some difficulty operating at low capacity with higher condensing temperatures. Thus, chillers for heat pump duty should be selected with care and kept as fully loaded as possible.

9.4 REFERENCES

ASHRAE. 1991. *Handbook—Applications.*

ASHRAE. 1992. *Handbook—Systems and Equipment.*

ASHRAE. 1993. *Handbook—Fundamentals.*

Bhansali, A. and D.C. Hittle. 1990. Estimated energy consumption and operating cost for ice storage systems with cold air distribution. ASHRAE *Transactions* 96(1).

Dorgan, C.E. 1989. Cold air distribution makes cool storage the best choice. ASHRAE *Journal*, May.

Dubin, F.S. and C.G. Long, Jr. 1978. *Energy conservation standards.* New York: McGraw-Hill.

Meckler, G. 1988. Designing and applying super-cold air HVAC systems. *Consulting/Specifying Engineer*, October.

Chapter 10 CONTROLS

10.1 INTRODUCTION

This chapter presents a discussion on fundamentals and illustrates principles for providing reliable and easily maintainable control systems for air-conditioning systems. Such control is required because the systems generally do not operate at full capacity all the time. The simplest control is an on/off control typical of residential applications where, as the room temperature rises, the air conditioner is switched on and as the temperature falls it is switched off. However, most large HVAC systems use modulating controls, *i.e.*, controls that vary the output of the equipment being controlled, to provide the control of temperature and humidity in various spaces in the building from a central air-handling unit.

10.2 TERMINOLOGY

The legend for all figures in this chapter is given in Figure 10-1. Figure 10-2 shows a simple cooling coil control scheme. The basic elements of the system include a *sensor*, that senses the temperature of the control variable—in this case, the air temperature in the duct. The controller compares the sensed value with the *setpoint*. The difference between the measured variable and the setpoint is called the *error*. This error signal is processed to vary the output from the controller. In this case, the output goes to a valve that modulates the flow of water in the coil, thereby influencing the temperature of the air leaving the coil. The process by which the controller output is influenced by the sensed variable that it controls is called *feedback*.

Two types of feedback control strategies commonly used in HVAC applications are *proportional control* and *proportional plus integral (PI) control*. With proportional control, the controller output varies in proportion to the error. In mathematical terms, the controller output is:

$$O = A + K_p \, e$$

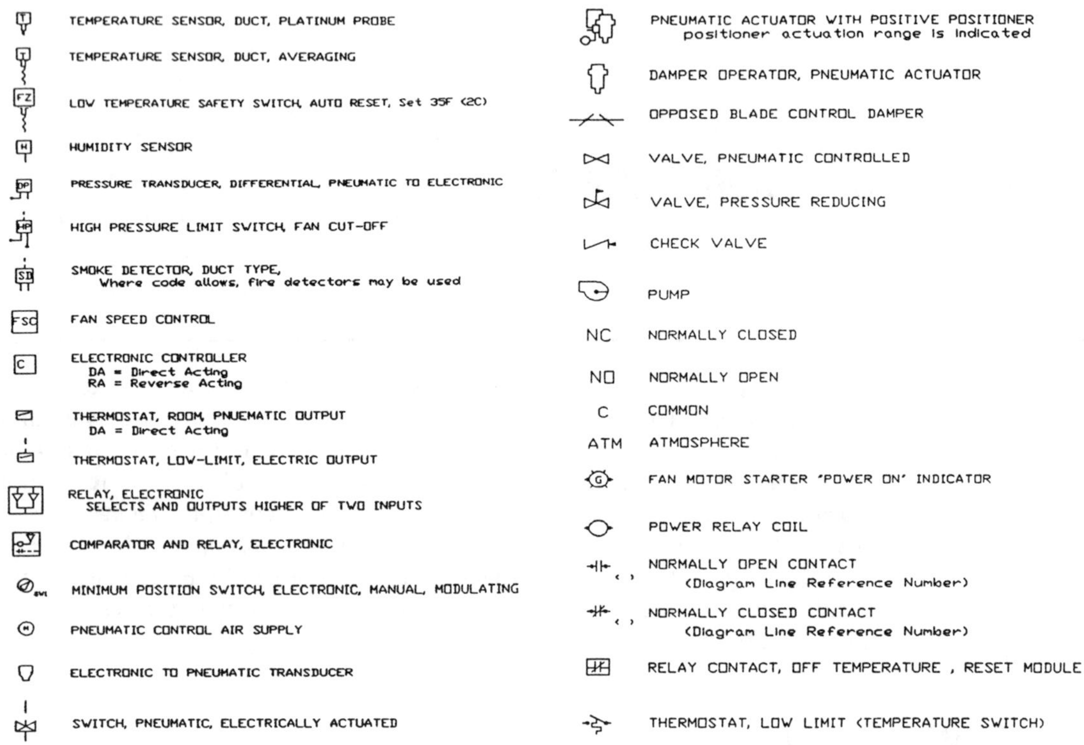

Fig. 10-1 Legend for Control Systems

where

A = constant defining what output should be when error is zero
K_p = proportional gain
e = error
O = output

For example, if a proportional controller were installed to control the temperature leaving a chilled water cooling coil and if the output ranged from 0 to 100%, then A might be 50 and K_p might be 10. Under these conditions, the output from the controller would be 100% when the leaving coil temperature was 5° above the setpoint, 50% when the temperature and setpoint were the same, and 0% when the temperature was 5° below the setpoint. It would take a 10° change in the leaving coil temperature to cause a 100% change in the controller output. This is called the

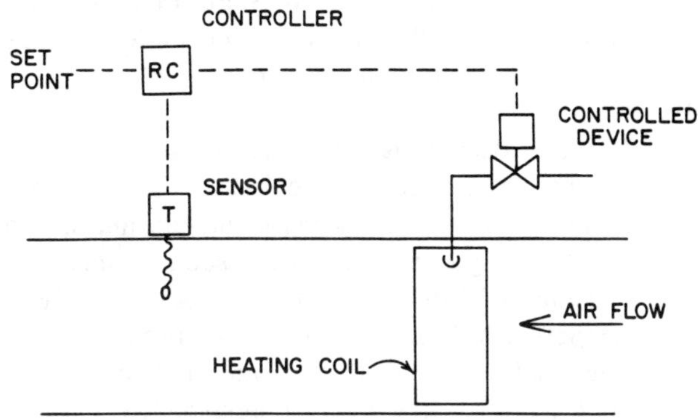

Fig. 10-2 Elements of Control System

throttling range of the controller. As this example illustrates, the controlled variable is only rarely exactly at the setpoint when proportional control is used.

As the name implies, proportional plus integral (PI) control adds integral control action. In mathematical terms

$$O = A + K_p e + K_i \int edt$$

where K_i is the integral gain. By integrating the error signal over time, the output from the controller is increasing or decreasing any time the error is not zero. This forces the error to zero as steady state is reached.

10.3 CONTROL TYPES

Three types of control hardware are used in heating and air-conditioning systems—pneumatic, analog electronic, and direct digital control. *Pneumatic* controls have been used in most building HVAC applications. They are reasonably well understood by designers and maintenance personnel, itself a compelling reason for their use. A second advantage is that pneumatic valve and damper actuators, besides being inherently modulating, are inexpensive and reliable when compared to the electric motors and gears required to produce the same control force with electric actuation. However, most pneumatic control equipment requires a very clean source of supply air. The air must be dry and free of oil. While it may not be difficult to install a system with clean air, one failure or mistake, such as overfilling a compressor with oil, can permanently foul the entire pneumatic system. Another potential disadvantage of pneumatic controls is that they are not particularly precise. Usually, field calibration is required in order to obtain acceptable accuracy, and periodic recalibration of the pneumatic instruments is required.

Modern *analog electronic* electronic control equipment has several advantages. Power is provided electrically, and there are no compressors, driers, or filters to maintain. Signals flow very quickly, and controllers essentially have an instant response (signal processing is instant; temperature sensors and motors still have time constants). Electronic devices can be extremely accurate and free of drift. Resistance temperature detectors made of platinum, nickel, or other metals manufactured to tolerances of 0.5°F are relatively inexpensive, do not require field calibration, and are drift free. They are also inexpensive, and it is easy to display electronic signals using modern digital voltmeters, providing for ready diagnostics. Another advantage is that it is relatively easy to implement proportional plus integral (PI) control electronically. It will be shown later that PI control has important energy-saving advantages.

On the negative side, electronic control systems have often been more expensive than pneumatics, partly because of the high cost of electric valve and damper motors. Another problem with electronic control hardware is that many different types of systems are in use. For example, some companies use inexpensive thermistor temperature detectors while others use platinum or nickel. One company uses pulsed voltages to control the actuator, while others use 0-9 or 2-10 volts. Some HVAC control companies have equipment that follows industrial control practice (platinum resistance temperature detectors and 2-10 volt and 4-20 milliamp control signals). This makes components interchangeable and gives designers access to a variety of high-quality industrial sensors and transducers.

Direct digital control (DDC) may become the standard of the future. Large integrated systems are now common. Small-scale digital control units for single-fan systems have been introduced for new systems and retrofit applications. These units accept analog electronic inputs from appropriate sensors, convert them to digital, use these digitized signals to produce digital outputs via software control algorithms resident in a local microcomputer, and, finally, convert the digital output signals to analog electronic signals to drive control actuators that may be mechanical or pneumatic. Figure 10-3 presents a schematic of a direct digital control concept.

Perhaps the greatest single advantage of DDC is the flexibility it provides because control is implemented through software rather than hardware. The implementation of time sequencing, reset control, PI loops, economizer cycles, and a variety of yet-to-be-invented schemes is accomplished by programming a microcomputer, not by modifying or adding hardware. Of course, with this newfound power comes the designer's responsibility to avoid unnecessary complexity. Another benefit of direct

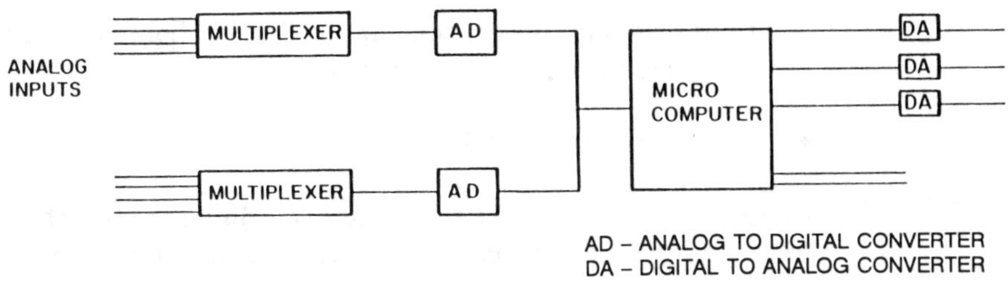

Fig. 10-3 Direct Digital Control Concept

digital control is a further increase in precision and the complete absence of controller drift. For example, once a temperature setpoint is entered into a digital controller, it remains in the microcomputer's memory unambiguously until it is changed by the operator or until the controller fails. The concept of recalibration is simply not part of digital hardware, and maintenance should be required only when the hardware fails.

However, all problems do not vanish with direct digital control. Analog sensors may still require periodic recalibration. Furthermore, direct digital control is a technology that is still not well understood. A serious drawback of its equipment is the proliferation of programming languages (each manufacturer has its own), making DDC system commissioning and maintenance difficult, particularly since the user interface is often not user friendly.

10.4 CONTROL STRATEGIES

Formerly, many details of the control system were left to the discretion of the construction contractor. As the costs of building operation have escalated (largely as a result of increases in energy costs) and control systems have become more complicated, this is no longer true. Today, HVAC designers must be intimately familiar with the thermodynamic, mechanical, pneumatic, and electric/electronic details of control systems and must be acutely aware of how their design features affect operation and maintenance. Building owners and operators want reliable and easily maintainable control systems. To meet these requirements, designers must provide complete plans and specifications for the control system. This includes

- The control sequence of operation
- A control schematic
- Ladder diagrams showing the interaction between the various parts of the system

- Commissioning instructions that allow the building and its air-conditioning system to be put in operation and checked
- A maintenance checklist

Control systems for a number of systems are discussed in the following sections, and examples of control sequences are given in Appendix G. The schematics usually assume that electronic temperature sensing and control is to be used but that it is interfaced with pneumatic actuators. Designers who choose different control hardware must make the appropriate substitutions.

10.4.1 VAV Systems

Figure 10-4 shows a complete control system for a VAV system consisting of the following subsystems: individual room air temperature control, supply air temperature control, mixed air temperature control, and duct static pressure control. Although many designers no longer use return fans (see Section 6.7.2.4), their use is indicated in some applications when return fan system control is also required. Room air tepperature control is achieved by using VAV boxes controlled by room thermostats. Electronic or pneumatic thermostats and actuators are used to modulate the air volume flow rate in response to room temperature. Some VAV boxes also incorporate flow sensors to allow the box to maintain the controlled rate of flow under varying upstream duct pressures.

System-powered or self-powered (deriving motive power from the duct pressure only) VAV boxes are also available. However, some of these devices require excessive duct pressure to operate properly. If they are being considered, it may be advisable to require independent tests before approving them for installation. Heating is provided where required by reheat coils in the duct or by baseboard convectors. They are controlled either by the room thermostat or by outdoor temperature, as described in Appendix C. Baseboard convectors or small fan-coil units (at floor level or above the ceiling) are the preferred approach for heating exterior zones, since their use permits lower minimum airflow settings on the VAV boxes and allows building warmup and night setback heating without turning on the air-handling system.

Supply air temperature control should be accomplished by using a temperature sensor, a proportional plus integral (PI) controller, and appropriate transducers and a pneumatic two-way throttling or a three-way mixing valve on the cooling coil, as shown in Figure 10-4. Commissioning instructions should include detailed descriptions of how to test the control system for stable operation under low and high airflow conditions and cooling loads. Temperature sensors should be located far enough

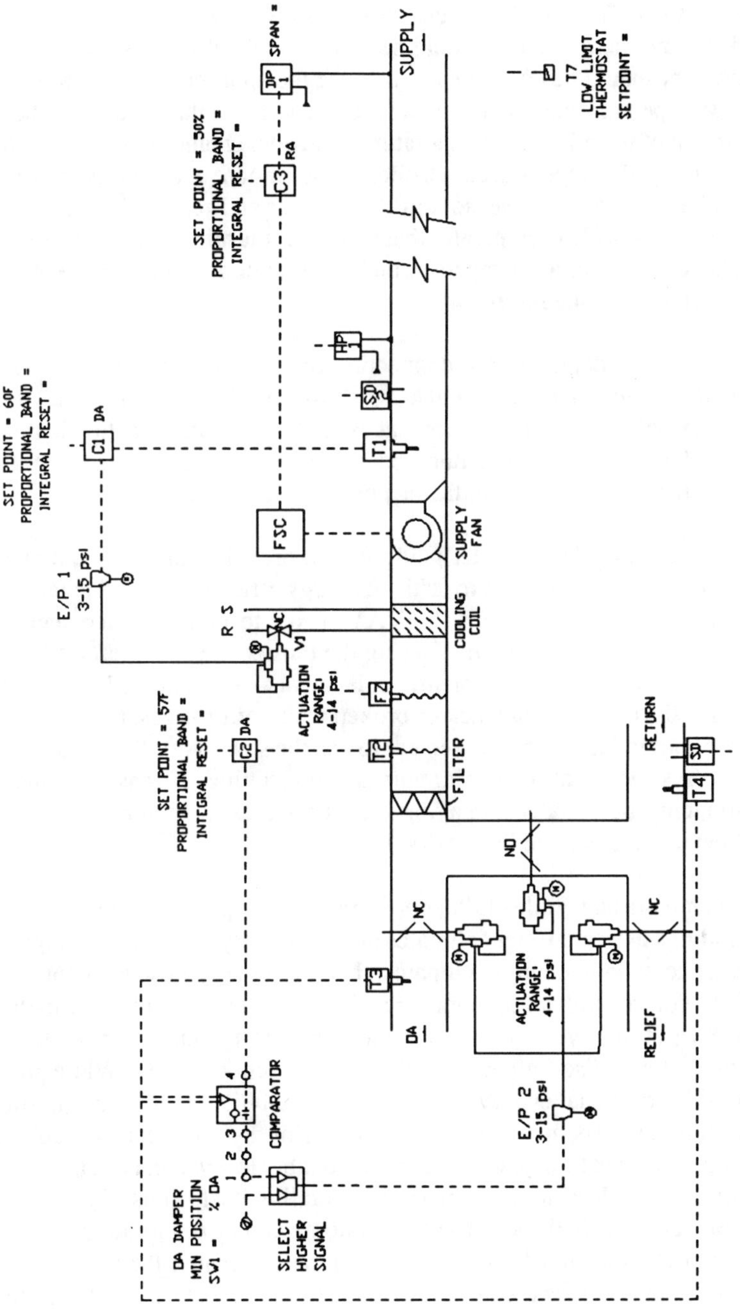

Fig. 10-4 Control Scheme for VAV System

downstream from the fan discharge to ensure adequate mixing (but not further than 25 ft), and their location should be shown on the drawings. Outdoor and return air dampers should be controlled to achieve the appropriate temperature. This is done by a temperature sensor connected to a PI controller that modulates the dampers. The setpoint of the mixed air temperature controller should be lower than the setpoint of the supply air temperature controller by the amount of the temperature rise across the fan. Two temperature sensors in the outdoor and return airstreams can be used in conjunction with a comparator that compares the temperature signals and returns the outside and relief air dampers to their minimum positions whenever the outside air is warmer than the return air.

Construction details of the connection of outdoor and return air ducts to the air-handling equipment should also be shown on the drawings. The designer should specify baffles or other appurtenances to minimize stratification in this duct system. Complete mixing of the outdoor and return air is necessary if the mixed air temperature sensor is to indicate a meaningful value.

Modulation of the fan inlet guide vanes, propeller pitch, or fan speed is usually required in VAV systems to achieve energy efficiency, avoid potentially damaging duct pressures, and permit the VAV boxes to perform adequately. Consider PI control for this application. Varying the fan speed is the preferred control method. Not only is it the most efficient, but it can also modulate to lower airflow rates and avoids the problems of stuck or broken vanes or linkages. Commissioning instructions should include a requirement to test the stability of the control system with all the VAV boxes at their minimum airflow setting. If fans have the capability of producing potentially damaging duct pressures, a high-pressure limit switch or blowout panel should be provided.

Mutammara and Hittle (1990) compared the energy consumption in high-rise office buildings in five different climates under a variety of control strategies. They found that, except in hot, humid climates, the most energy-efficient control strategies are (1) PI control with an economizer and (2) cold deck reset for perimeter zones. Cold deck reset involves the use of room thermostat signals to reset the discharge air temperature (discriminating control, see Section 9.3.3). While this system can provide additional energy conservation, reliability is a serious question. Failure of an individual room thermostat or tampering by an occupant could cause the air temperatures to be adjusted to an unreasonably low value. Fan energy also increases, which might lead to an increase in operating cost. Thus, the simpler fixed deck temperature with PI control and an economizer may be more attractive. Economizers are not recommended for hot, humid climates because the number of hours during which they could be used is small. Instead, only a fixed amount of minimum outside air is provided for ventilation.

Return fans should be avoided in VAV systems because good control of return fans is difficult and involves sophisticated hardware. Return fans also require more energy than relief fans, as shown in Section 6.7.2.4. However, if a return fan is absolutely essential for a particular application, it must be controlled. The proper return fan control minimizes pressure fluctuations and increases energy efficiency. The simplest method is to use the output from the supply duct static pressure controller to also modulate the return fan. Figure 10-5 shows this scheme. It is most effective if the fans are properly set up at full flow to provide the right difference between supply and return airflow and if both fans have similar part-load performance characteristics. A more effective but more complicated scheme is to measure the supply and return airflow rates with a flow-measuring device and use these flows as inputs to the return fan and controller, as shown in Figure 10-6. This control requires careful setup. Both supply fan and return fan flow measurements must be normalized to their respective full-flow conditions. Even in the best circumstances, flow measurements may then be only roughly correct.

Conventional flow-measuring stations are often used to control return fans. Pitot tube racks provide a pressure signal that is transformed to an electronic or pneumatic control signal by a pressure transducer. Unfortunately, at low flow rates, the velocity

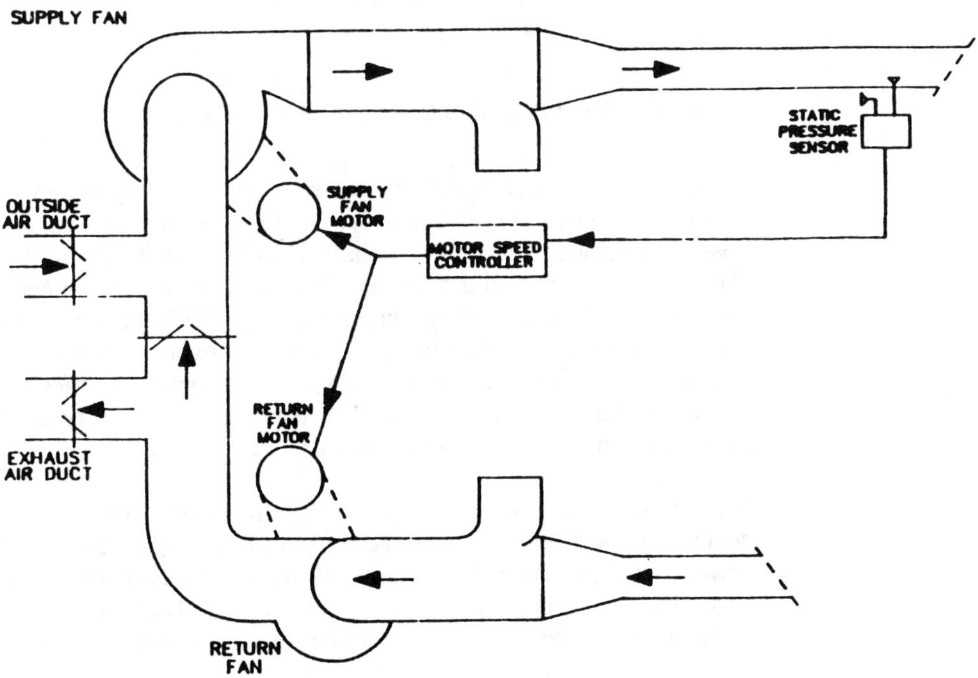

Fig. 10-5 Simple Return Fan Control

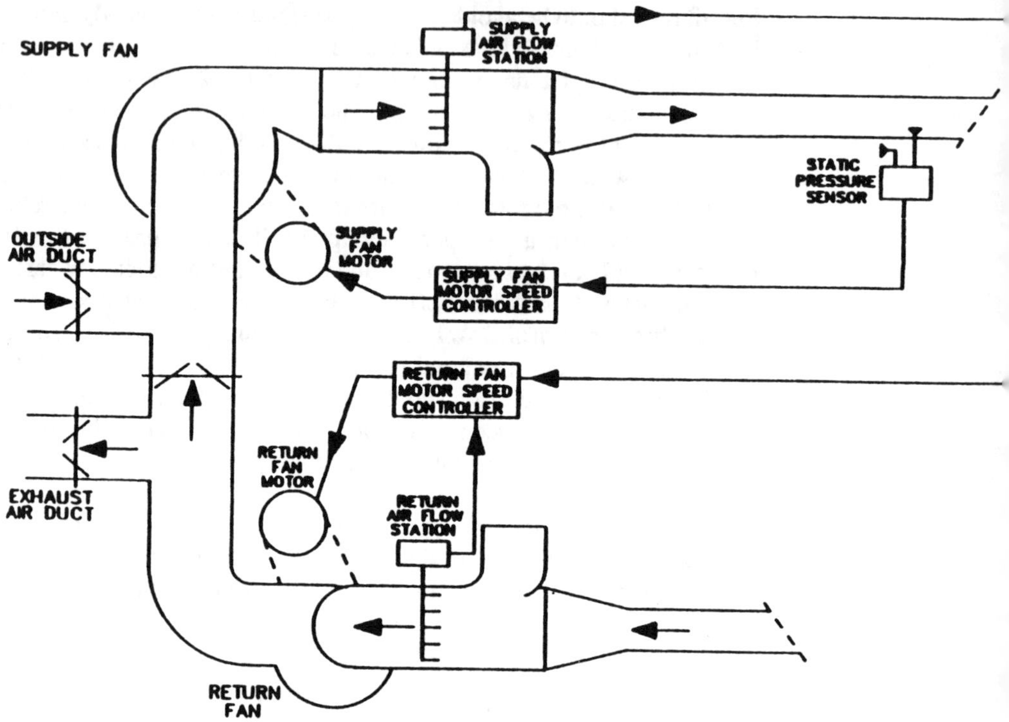

Fig. 10-6 Flow-Matching Control of Return Fans

pressure is so low that it is almost impossible to sense this pressure, especially with low-cost transducers. One solution is to build a special duct section with a reduced cross-sectional area. The flow stations can be placed in these reduced cross sections where the velocities would be higher. One manufacturer recommends a minimum velocity of 700 fpm. Reduced cross sections should be sized to maintain this minimum velocity. Alternatively, if space permits, a venturi can be built into a straight section of both the supply and return ducts with the difference in upstream and throat static pressure converted to flow velocity. A very accurate pressure transducer will be required with either scheme.

Another flow-measuring technique is the use of hot-wire anemometers. If this technique is used, appropriate air guides must be provided so that the anemometers measure only translational flow along the duct and not rotational or turbulent flow caused by local disturbances. An assembly of properly connected hot-wire anemometers to provide average duct velocities must be specified in detail.

For most comfort applications, individual rooms can be controlled with proportional thermostats with local (in the room) setpoint adjustment. The temperature control profile for the VAV dampers and heating coil or baseboard heaters should be as shown in Figures 10-7 and 10-8.

Room thermostats should not be located on outside walls, in dead corners, or in other unsuitable locations where they are likely to be affected by equipment, people, or sunlight. The best locations are those that provide some air motion and allow thermostats to sense the average room temperature. The location of all thermostats should be shown on the plans. If add-on, tamperproof covers are used, they should be the type constructed of widely spaced metal rods to allow free air movement over the thermostats. Metal or plastic box-type security enclosures with air slits should not

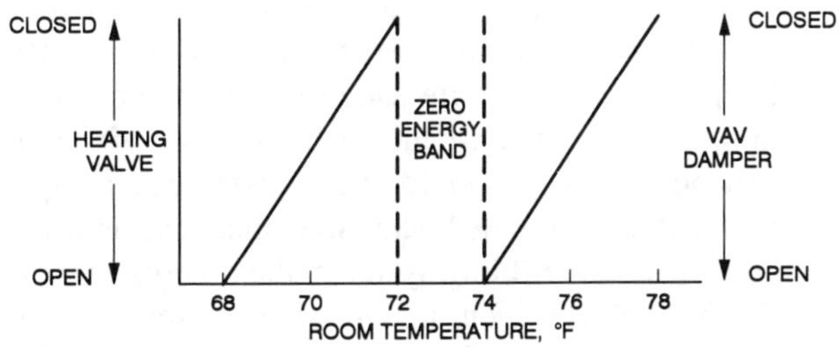

Fig. 10-7 **Room Temperature Control Profile for "Zero Energy Band" Thermostat (Temperatures shown are merely suggested values.)**

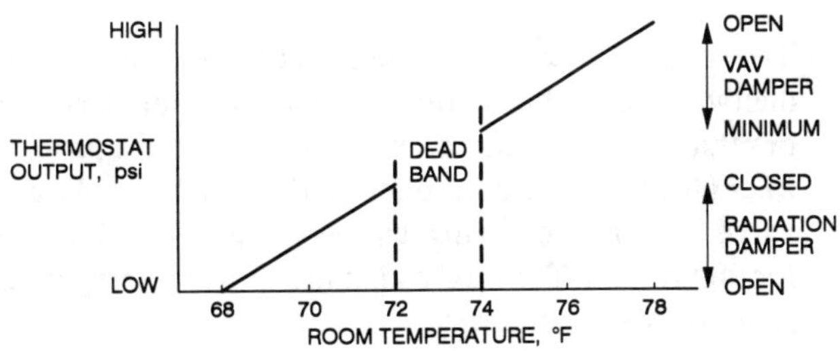

Fig. 10-8 **Room Temperature Control Profile for "Dead Band" Thermostat (Temperatures shown are merely suggested values.)**

be used, since they do not provide enough airflow over the thermostat for it to respond rapidly to changing room conditions.

10.4.2 Single-Zone, Constant-Air-Volume Systems

Very small single-zone air-handling units are thermodynamically equivalent to room fan-coil units or residential air-conditioning systems and can be controlled by simple two-position electric or electronic thermostat/valve systems common in this type of unit.

For larger single-zone systems, one of the two schemes illustrated in Figures 10-9 and 10-11 can be used. Figure 10-9 shows the minimum configuration, which may provide inadequate control and should be used only on closely coupled systems, *i.e.*, systems with relatively short ductwork and high air volume flow rates compared to the room volume. In this scheme, a room or return air temperature sensor is connected to a single proportional controller, which modulates the heating coil valve, mixing dampers, and cooling coil valve in sequence. Their temperature control profiles should be as shown in Figure 10-10. The outside and return air temperature sensors are used to return the outside and return air dampers to their minimum position whenever the outside air is warmer than the return air. This economizer cycle should have a hysteresis of up to 2°F, *i.e.*, the cycle should be initiated or terminated when the outside air is 2°F cooler or warmer than the return air. For small building systems, the controllers may modulate electronic valve and damper actuators, but pneumatic actuators are often used on larger systems. The proper sequencing of the valves and dampers is achieved by carefully adjusting the pilot positioners on the pneumatic equipment. Instructions for making the needed adjustments should be in the commissioning instructions. The room temperature sensors should not be located in the return air duct if the return air is drawn through lighting fixtures or in any other case where the return air would be at a different temperature than the room air. Under these circumstances, a wall-mounted room thermostat should be used.

The scheme shown in Figure 10-11 is to be used in applications that require longer ductwork and tighter control, or that serve very large rooms. This scheme uses a room or return air temperature sensor connected to a proportional controller to reset three PI controllers. One controller modulates the cooling coil valve, one the heating coil valve, and one the mixing dampers. The reset scheme of Figure 10-12 should be used for this system. The mixed air and supply air temperature sensors should be mounted as in the VAV system.

If very tight temperature and humidity controls are required, one of the systems shown in Figures 10-13 and 10-14 should be used. The system shown in

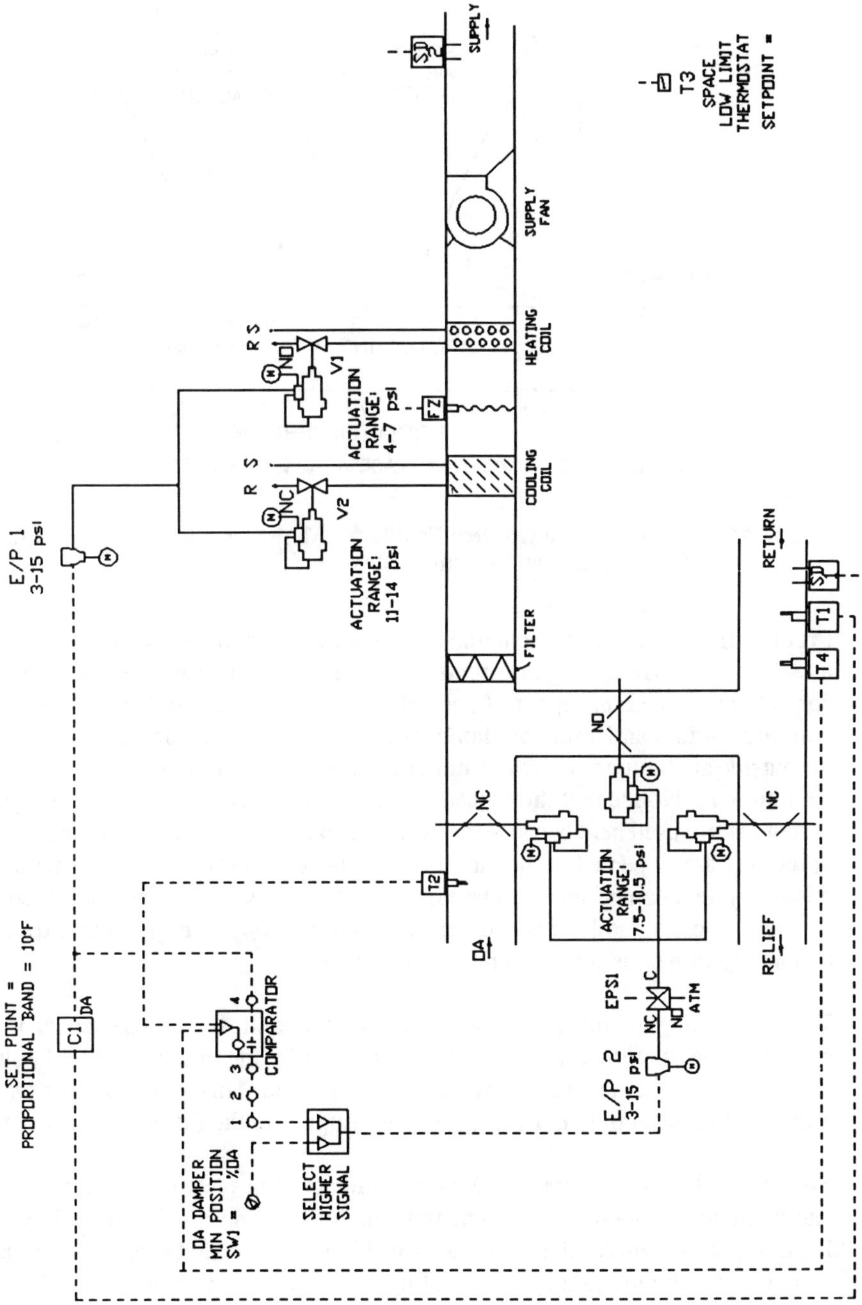

Fig. 10-9 Minimum Single-Zone Control System

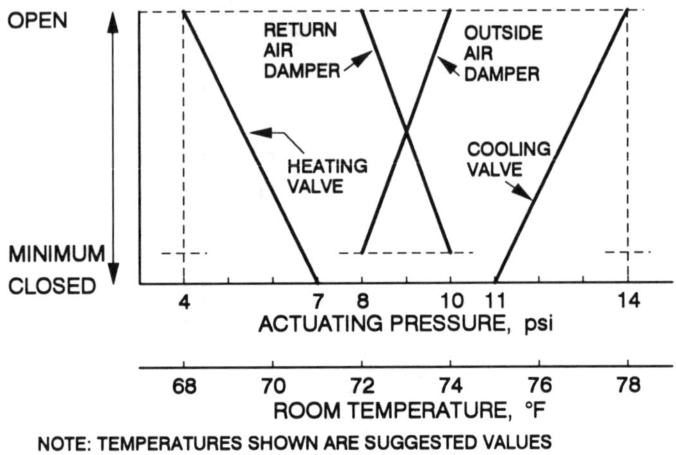

Fig. 10-10 Temperature Control Profile for Minimum Single-Zone Control System Shown in Figure 10-9

Figure 10-13 is more efficient and can be used where loads are fairly constant and mostly sensible (reheating is not required for humidity control). Room temperature control is achieved through a PI controller that modulates the heating coil, cooling coil, and outdoor and return air dampers in sequence. Air temperature sensors return the outside air dampers to their minimum position when the outdoor air is hotter than the return air. Notice that the cascade control system recommended for single-zone systems is not used here because the system is expected to run continuously and load variations are expected to be small and to occur slowly. Hence the temperature control system can be set up to be fairly sluggish. In systems controlling computer rooms, no heating coil is needed because there is always a requirement for cooling and plenty of heat is provided by the computer.

The humidifier is controlled with a proportional controller whose sensor is downstream of the humidifier. This controller is reset by the output of the PI controller whose sensor is in the return airstream or in the room. This avoids adding excessive moisture to the supply airstream, causing dripping in the discharge duct.

Figure 10-14 shows a more extensive humidity control system. While it is also more energy intensive, it ensures sufficient dehumidification under more widely varying loads or in cases where latent heat gains are higher. A constant supply air temperature is maintained by modulating the cooling coil as well as the outdoor and return air dampers in exactly the same way and using the same components as in the VAV system described in Section 10.4.1. A PI controller sensing return or room air

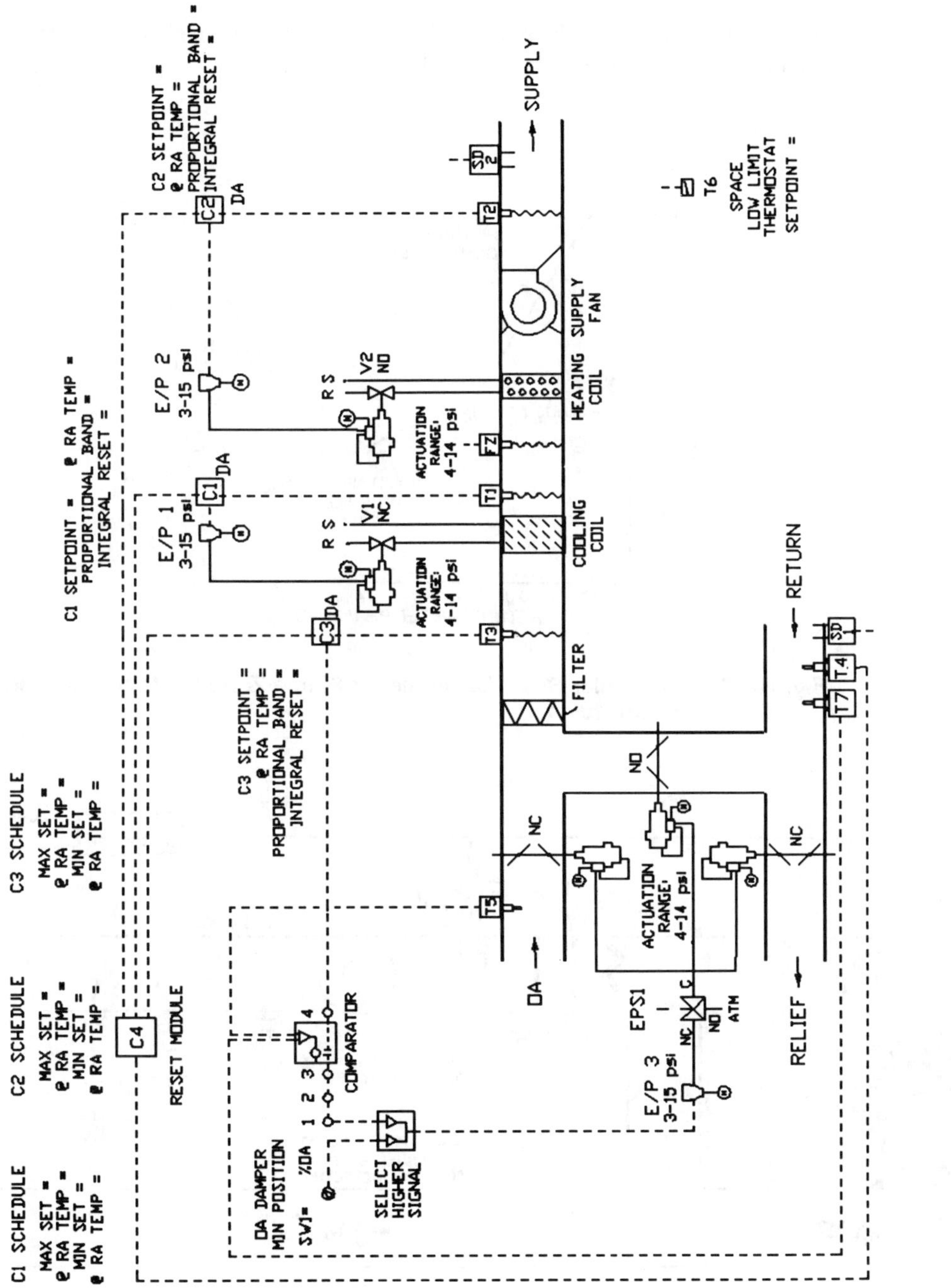

Fig. 10-11 Cascade Control for Single-Zone System

temperature modulates the reheat coil valved to control the room temperature. The humidifier is controlled as shown in Figure 10-13.

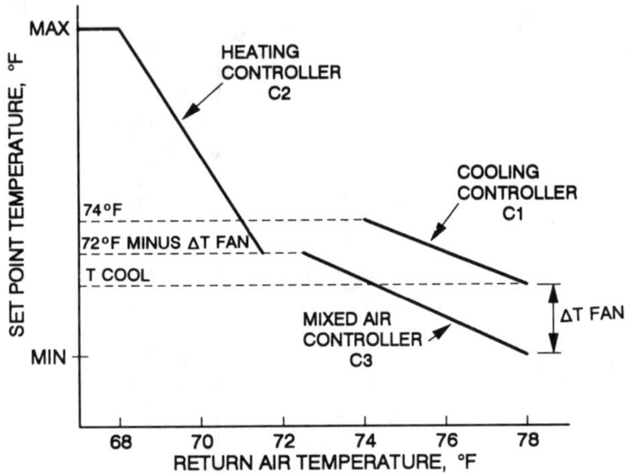

Fig. 10-12 Controller Reset Schedule for Single-Zone System Shown in Figure 10-11

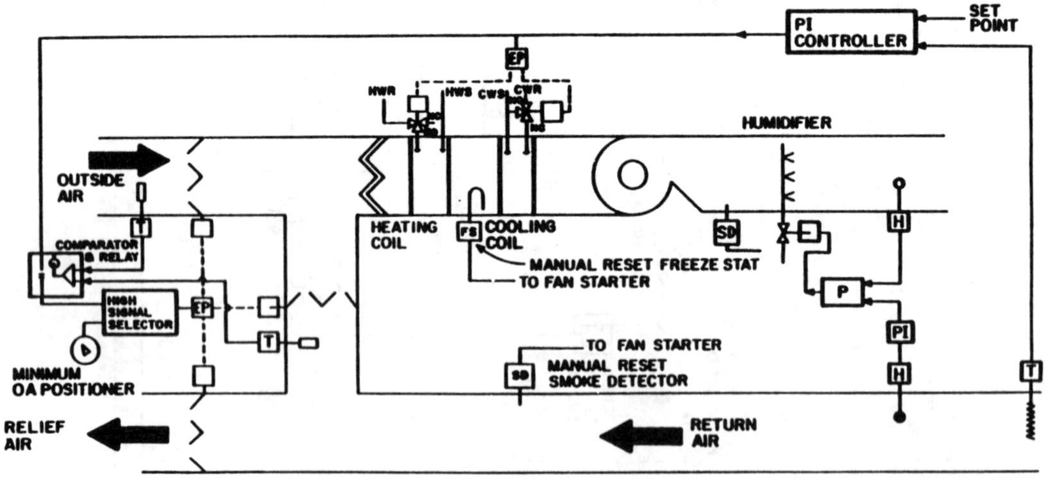

Fig. 10-13 Single-Zone System with Precise Temperature Control

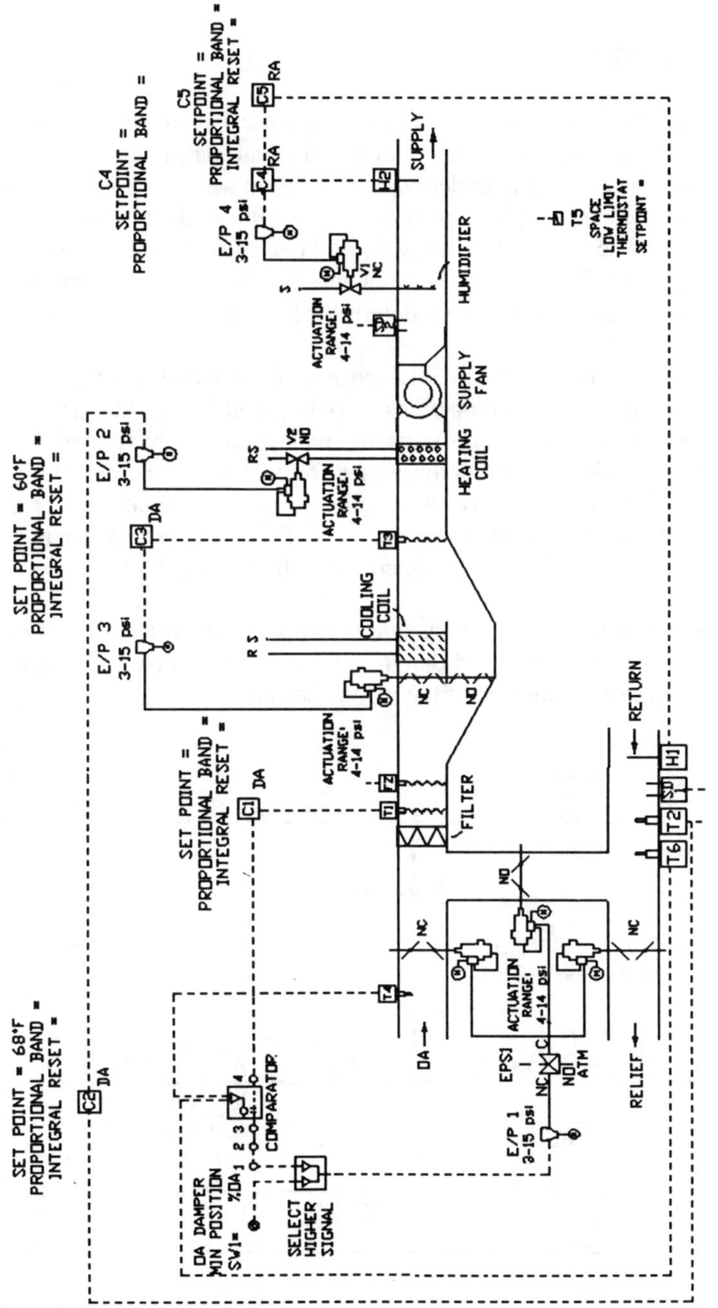

Fig. 10-14 Single-Zone System with Precise Temperature and Humidity Control

10.4.3 Dual-Duct Systems

An automatic control arrangement is shown in Figure 10-15. It is adequate for most installations and can be easily modified to suit particular applications and requirements. The minimum outdoor damper is open when the fan is running, except when all areas are unoccupied during warm-up. The hot duct temperature is controlled by the submaster hot duct thermostat, which operates the heating coil valve (or valves) in sequence. The control point of this thermostat is reset by the outdoor master thermostat according to a predetermined schedule.

The air in the cold duct is maintained at the desired dew-point temperature by the cold duct thermostat, which operates a cooling coil valve. At a predetermined outdoor temperature, the system is switched to natural cooling by the two-position switchover thermostat. When outdoor air is used for cooling, the cold duct thermostat controls the operation of the maximum outdoor damper, exhaust, and return air dampers through the use of a three-way air valve. This thermostat also controls the operation of the cooling coil when mechanical cooling is required.

The control of the central system is simple, since it is required only to serve as a source of warm and cold dehumidified air. Of course, there are refinements for operating economies, ventilation, and so forth.

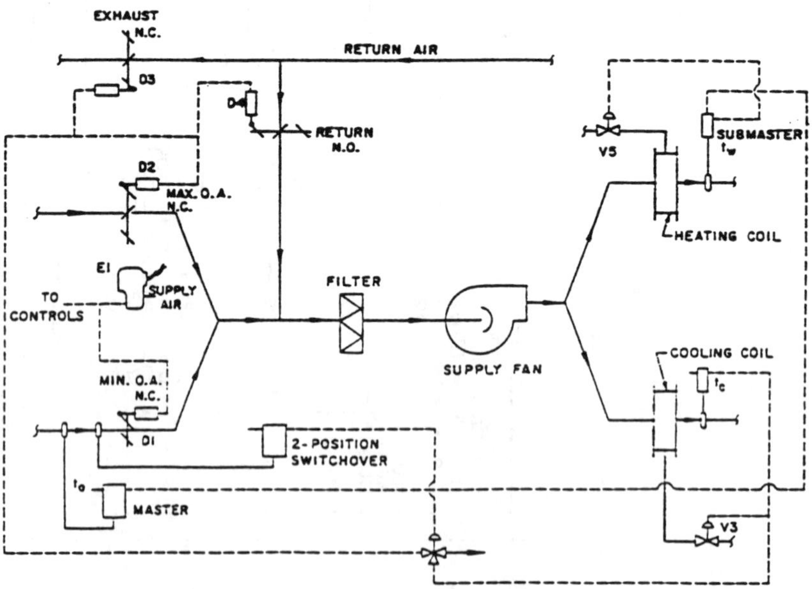

Fig. 10-15 Simple Automatic Control for Dual-Duct System

Dual-duct systems with mechanical or automatic flow regulators do not need fan or system static pressure regulation, since each terminal unit compensates itself for any change in backpressure. However, to take full advantage of fan energy savings at reduced flow conditions in dual-duct VAV systems, fan regulation is recommended. This also helps to avoid high-pressure drop noise effects in VAV terminal units and is a more important consideration in dual-fan systems.

10.5 COMPONENTS

Control valves must be sized to provide the proper control stability and authority. *Authority* is the ratio of the pressure drop through the wide-open valve to the combined pressure drop through the valve and coil or valve and heat exchanger. Line-size valves are almost always too large. Valve sizes and valve coefficients should be shown on a schedule on the plans. Equal percentage valves (see Section 5.5.2) with a turndown ratio of 20:1 or better are usually needed for most HVAC applications. The authority of control valves should usually be 0.3 or higher. Additional information on valves is presented in Section 5.5.

Dampers are often provided with air-conditioning systems and are frequently too large. For control purposes, opposed-blade dampers or a mixture of opposed- and parallel-blade dampers usually provide the best performance. In most cases, dampers should be smaller than the return or outdoor air duct. The wide-open velocity through a damper that is to control airflow should preferably be 1500 fpm or higher. Dampers should be shown on a schedule on the plans. Experience has shown that copper, nylon, or thermoplastic bushings on damper shafts may fail prematurely. Since dampers do not receive routine maintenance in most HVAC systems, the use of sealed ball bearings at the ends of shafts is recommended.

Positive positioners should be specified for most pneumatic valves and actuators, except for the small actuators on VAV boxes, reheat coils, or baseboard heaters. A pilot positioner is required to reduce hysteresis, to improve actuator response, and to permit adjusting the gain of the controlled device.

10.6 REFERENCES

Mutammara, A.W. and D.C. Hittle. 1990. Energy effects of various control strategies for variable-air-volume systems. ASHRAE *Transactions* 96(1).

Appendix A PROPOSED MECHANICAL AND ELECTRICAL SYSTEMS FOR OFFICE BUILDING WITH RETAIL STORES

A-1 SUMMARY OF RECOMMENDATIONS

1. A single fan room shall be installed on each office floor containing a complete factory-packaged combination air-handling unit and self-contained DX cooling coil and refrigeration compressor section, as follows:

   a. The air-handling system shall consist of two 11,000-cfm supply air fans with inlet vortex dampers, pleated (flat) filters, cleanable precooling coils, DX-type cooling coils, and controls.

   b. Each DX unit shall be rated at 35 tons and shall consist of two separate compressors arranged for sequential start and simultaneous use when the load requires.

   c. The DX units on each floor shall be connected to a 2,500-ton-capacity cooling tower located on the roof and pumps located in the basement. An auxiliary pair of condenser water risers shall be installed for future tenants' use.

   d. The condenser water pumps and the auxiliary risers shall be piped such that it shall be possible to isolate one of the tower cells with the auxiliary risers for winter operation.

2. The air distribution system on each floor shall consist of the VAV fan, medium-pressure ductwork, conventional VAV boxes serving the interior zone from a common interior/perimeter air distribution main, and fan-powered VAV boxes serving the perimeter zones. Heating shall be provided by means of electric

resistance coils placed in the low-pressure supply air ductwork downstream of the individual terminal boxes.

3. Ventilation air for the individual fan rooms shall be provided via a vertical, fan-pressurized air shaft. The fan shall contain a heating section for preheating the outside air to 40°F to prevent freeze-ups at the individual floor units. The shaft shall be sized to provide up to 15% (3,300 cfm) fresh air to each floor's air-handling system.

4. Two vertical exhaust shafts shall be connected to exhaust fans on the roof, capable of exhausting the equivalent of 8 to 10 air changes per hour from the largest floor in the building.

5. Stair pressurization fans shall be capable of pressurizing the individual stair shafts with outdoor air during alarm conditions.

6. A complete fire standpipe system serving sprinkler loops on each floor shall be connected to two fire pumps piped in a cascading arrangement.

7. Local electric domestic hot water heaters shall serve the tenant floor toilets three or more floors at a time. Toilets shall have floor drains as required by code.

8. Electric service at 480/277 volts shall be obtained from the public utility transformer vault, constructed under the sidewalk or roadway adjoining the building, and distributed throughout the building via bus ducts run vertically through electric closets on each floor.

9. An emergency distribution system and an emergency generator shall be installed to provide power to life-safety and other critical equipment in the event of a failure of the normal power supply.

A-2 DESIGN CRITERIA

1. 1987 BOCA Model Energy Conservation Code

2. Design Conditions

Summer	Outdoor	94°F dry-bulb (ASHRAE 1%) 77°F wet-bulb (ASHRAE 1%)
	Indoor	78°F dry-bulb (State Energy Code)
Winter	Outdoor	14°F dry-bulb (ASHRAE 97.5%)
	Indoor	70°F dry-bulb
Humidity control		None except in designated computer rooms
Occupancy load		1 person per 100 ft^2
Lighting load		2.0 W/ft^2
Receptacle load		2.0 W/ft^2
Cooling tower		Tower and auxiliary condenser water system risers to be sized for an additional PC and receptacle load of 3.0 W/ft^2 or 500 tons of equivalent cooling load.

3. Energy Code Requirements

Overall exterior wall	$U_o = 0.35$
If facade contains 40 – 50% glass, recommended wall value	$U_w = 0.10$
Assumed value for glass	$U_g = 0.30$
Assumed glass shading coefficient	SC = 0.30 (coated)
Roof/ceiling assembly	$U_r = 0.08$
Overall thermal transfer value	OTTV = 33.5
Assumed glass shading coefficient	SC = 0.30 (coated)

4. Ventilation Criteria

Office spaces. In response to the growing concern about the lack of proper ventilation in office buildings (sick building syndrome), a minimum of 15 cfm per person outside air will be supplied. Provision shall be made at each air-handling unit control panel to remotely vary the quantity of outside air intake into each unit via control of the outside air damper position. Toilets and slop sinks shall be exhausted at the rate of 75 cfm per fixture.

Electrical spaces and elevator machine rooms. In general, a minimum of 8 air changes per hour shall be mechanically exhausted. Positive supply shall be utilized for all areas containing transformers or other heat-dissipating equipment that require larger amounts of exhaust.

Main telephone room and elevator machine rooms. 85 to 90°F ambient air temperature shall be provided with fully recirculating local air-conditioning units plus a minimum amount of ventilation exhaust.

5. Acoustical Treatment

The following design goals shall be used for the occupied spaces:

Offices	NC-35 (except where adjacent to local floor fan room)
Corridors	NC-40
Conference rooms	NC-35
Lobby	NC-45

6. Air Filtration

Outside air and recirculated air shall be filtered at the air-conditioning units by means of pleated filters.

7. Domestic Hot Water

System shall be designed in accordance with ASHRAE *Systems and Equipment* (1992), Chapter 12. Electric storage-type heaters providing 105°F to building plumbing fixtures shall be provided every three floors.

8. Provisions for Retail Areas

- Metered electricity for each tenant's use
- Domestic cold water service
- Sanitary drainage and vent connections
- Sprinkler connections
- Valved condenser water outlets
- Outside air for tenants' ventilation needs, if introduced at store front, must be a minimum of 10 ft above street level

A-3 ESTIMATED LOADS

The following estimates are based on 625,000 ft^2 of gross office space plus lobby and two below-grade floors:

	Demand	Installed
Electric power	7600 kVA	From preliminary load study (not shown)
Refrigeration	1830 tons	2100 tons
Space heating	3350 kW	5200 kW
Domestic hot water heating	150 kW	8 × 9 kW + 78 kW for future capacity

The maximum refrigeration demand and installed capacity was determined as follows:

Base building, 600,000-ft^2 offices (from computer printout, not shown): 16,484,000 Btu/h	=	1374* tons
Adjustment for additional 25,000-ft^2 office	=	60*
Additional 1.0 W/ft^2 on office floors	=	170
Allowance for additional corner glass	=	190
Lobby plus two lower levels	=	200
5% of above for miscellaneous interior loads	=	90
Maximum demand with 40% — 0.5/0.5 glass	=	2000* tons
Maximum demand with 40% — 0.3/0.3 glass	=	1880 tons

* Check for reasonableness:
600,000/1,374 = 437 ft^2/ton
25,000/60 = 417 ft^2/ton

The typical office floor shall demand 65 tons of cooling if the maximum building demand is shared equally among the 25 typical floors. Due to the expected load variation between floors, it is recommended that each floor be provided with a minimum of 70 tons of cooling capacity under a floor-by-floor air-conditioning system concept. Therefore, the minimum projected installed capacity would be 25 × 70 + 300 tons for lobby, etc. = 2050 tons, and the proposed cooling tower capacity is 2500 tons.

Note: The calculations for the heating load and installed capacity are not shown here.

The supply air quantity for each typical office floor has been calculated as follows:

Base building, with 3 W/ft^2 and 0.5/0.5 glass	=	18,400 cfm
Additional 1 W/ft^2	=	3,900
Allowance for additional corner glass	=	2,000
Subtotal	=	24,300 cfm
Deduction for 0.3/0.3 glass	=	2,600
Maximum operating air for typical floor	=	21,700 cfm

A-4 ENERGY SOURCE AND UTILITY ANALYSIS

The available energy sources at this site and their current estimated average price per unit of consumption are as follows:

Electric power, normal use	9.0¢/kWh
Electric power, heating	7.6¢/kWh
Natural gas, firm	47¢/therm
Natural gas, interruptible	32¢/therm
No. 2 fuel oil	60¢/gal

Electric power prices are for delivered energy at the job site, with perhaps 3 to 5% inefficiency or line losses prior to consumption. The fossil fuel prices, on the other hand, are for raw fuel, which, by means of a boiler plant, has to be converted into heat energy.

When depreciated over a 20-year period (at a 10% interest rate) and adjusted for owning and additional labor cost, the heating energy generated in a fossil fuel plant is more costly than purchased electric energy, especially where there is a reduced demand charge for electrically heated buildings.

The following is a comparison of overall energy production costs, per million Btu delivered to system, between the electric and the fossil fuel-heated building.

Fuel	$ per 1000 Btu	Plant Labor	Plant Cost	Total
Electric power	23.40	0.00	0.00	23.40
Firm gas	6.20	6.70	13.00	25.90
Interruptible gas	4.20	6.70	13.00	23.90
Oil	5.70	6.70	13.00	25.40

Notes: Plant labor is based on an additional crew of 1.5, requiring $60,000 per year more than the all-electric system. Owning or plant amortization cost is based on a one-million dollar extra construction cost attributable to the boiler plant plus the penalty to be paid for the "wet" system with hot water heating boilers. Therefore, conclusions on the various available energy sources are as follows:

1. Electric energy is more expensive in terms of its unit price and annual heating energy cost, but it will prove to be more economical when used in connection with a "dry" resistance heating system. This is because a fossil-fueled plant, using No. 2 heating oil or natural gas and distributing steam or hot water through a "wet" piping system, will raise the overall construction cost of the building by $1.0 to $1.2 million.

2. In the case of a refrigeration plant for this project, the recommended source of energy is electric power. Other types of drives cannot presently compete with electric motor-driven chillers. Over the past decade, input to electric-drive chillers has been reduced from 0.85 to 0.90 kW/ton to 0.65 to 0.70 kW/ton, a 23% reduction, while input requirements to steam-driven chillers have remained essentially the same.

3. Although the utility or fuel cost of the gas-fired heating system is lower than that of the electrically heated building, the first cost of the gas-fired boiler cannot be economically justified during the lifetime of the building.

A-5 CENTRAL VERSUS DECENTRALIZED HVAC SYSTEMS

For location of the air-handling systems, two possible options have been evaluated:

Option 1. A central fan room located at the lowest floor (above the main lobby) and at the building penthouse serving the entire building.

Option 2. One fan room per floor.

The choice between central and local floor fan room schemes depends on the following:

1. Structural reasons, such as column transfer levels, etc., encouraging the use of a central fan room.

2. Major tenants' staffing and preferred mode of servicing and maintaining the mechanical and electrical systems.

3. Building height restrictions, if any.

4. Anticipated overtime or off-hour use of the building systems on different floors.

5. Method of tenant utility metering and billing.

6. First cost, operating cost, and rental space considerations.

Space Requirements

1. **Central fan rooms (top and bottom)**

Outdoor cooling tower	3,000 ft^2
Penthouse mechanical and electrical room	9,000
Lower mechanical and electrical room	7,000
Chiller plant	4,000
Duct area through typical floors	5,000
Total	28,000 ft^2

2. **Individual fan room per floor**

Outdoor cooling tower	3,000 ft^2
Typical floor fan room, 20 ft by 30 ft (25 floors)	15,000
Outdoor air intake shaft	1,500
Smoke shaft	375
Lobby AC room	1,200
Basement AC room	1,200
Total	22,275 ft^2

In this project, the proposed structural system does not require column transfer space; therefore, there is no unused secondary space suitable for mechanical equipment installation. We know of no tenant preference for either option, and there are no height restrictions. Frequent overtime and off-hour use of the building HVAC systems is anticipated. There appears to be no problem in proportionately billing the tenants for overtime use of the system without installing individual electrical meters on each floor.

The overall installed costs of the central fan room option and the individual fan room per floor option are approximately the same. From an operating cost standpoint, there are two issues to be considered:

1. The use of an air-side economizer particular to the central fan room option favors this concept by approximately $23,000 per year in terms of energy cost, as shown in the summary. This is because the water-side economizer associated with the floor-by-floor fan room scheme is less efficient at or near outdoor temperatures when the changeover from the cooling to the heating mode takes place. The air-side economizer can provide "free cooling" at temperatures below 55°F outdoors, whereas with the water-side economizer, free cooling will not be available until the outdoor air temperature reaches 45°F.

2. The requirement for off-hour operation of the system favors the individual fan rooms per floor system since complete shutdown of systems is possible with this scheme whenever a floor is not in use. With a central fan room scheme, although system turndown ratios of 6 to 1 are possible, the resulting energy savings will not be proportional to the turndown ratio.

A-6 COMPARISON OF CENTRAL VERSUS INDIVIDUAL REFRIGERATION SYSTEMS

1. Capital costs (not included in this appendix)

2. Operating costs

Scheme	Energy Use, Btu/ft^2/yr	Utility Costs, $		
		Electricity	Gas	Total
Air-side economizer	48,900	731,400	29,200	760,600
Water-side economizer (with plate heat exchanger)	50,400	754,600	28,900	783,500

The operating costs of the central chilled water plant will be lower during normal working hours since the larger machines are more efficient and can also take advantage of the considerable diversity of loads in a large building. It is during the extended periods of use beyond normal working hours that the individual air-conditioning units on each floor will be much more economical to operate than a lightly loaded central chiller. If an additional licensed refrigeration engineer would be required for the central chiller plant, the operating cost of the central plant might be even higher during normal working hours.

3. Maintenance Costs

The maintenance costs of the central chilled water plant will obviously be less than the costs of maintaining a multiplicity of individual smaller chiller packages. It is also apparent that replacement costs for the individual units will be higher than for the central plant. The return on the initial cost savings should more than make up for the increased maintenance costs, particularly in the first five years, which should be relatively maintenance-free. The likelihood of maintenance problems is much higher for the individual air-conditioning units, since regular maintenance can be carried out during the off-season on the main chillers, even during the water-side economizer operation, while preventive maintenance is generally not practiced on installations with a number of air-conditioning units.

4. Leasing Considerations

From a tenant's standpoint, the concept of total control of one's office environment, day and night, is favorably looked upon by the people responsible for leasing the spaces. Nor does there seem to be a drawback in apportioning the additional cost of overtime use to the tenants using the equipment, whether they are full-floor or part-floor tenants.

5. Space Requirements

Our study confirms that the individual fan room per floor need not be enlarged to accommodate the addition of the DX chiller section, whereas the central chiller plant will require 4,000 ft^2 of space in the basement.

6. Flexibility, Reliability, and Redundancy

The central chiller plant concept, almost by definition, is more flexible and more reliable and possesses a greater degree of redundancy than the concept involving individual cooling packages in each fan room. On the other hand, if one of the main chillers broke down during a peak-load day, the entire building would be affected, whereas with the individual units, only the respective floor would be involved.

7. Energy Considerations

The project, as designed, will be in compliance with applicable energy codes. Energy-efficiency enhancement techniques, such as heat recovery, economizer cycle, and thermal storage, are feasible with the central plan concept. These alternatives have been studied and have not proved to be economically viable. The greatest degree of energy savings will be attained as a result of the decision to use high-performance glass with a low (0.30) U-value and a low (0.30) shading coefficient.

8. Recommendation

Based on the foregoing, we recommend that individual packaged DX units be incorporated within each of the individual floor fan systems in lieu of a central chilled water plant. This recommendation is made with the understanding that the potential noise and vibration problems inherent in these units will be addressed by an acoustical consultant and that the unit specified and installed will have been successfully used in prior installations.

A-7 PERIMETER HVAC SYSTEM OPTIONS

Details of the heating energy calculations are not reproduced here. Three perimeter heating schemes were investigated:

- VAV + Perimeter baseboard heat
- VAV + Perimeter fan-powered VAV (FPVAV)
- VAV + Perimeter four-pipe fan-coil units

The following two tables summarize the results of our computer analysis on the percentage and type of glass in the facade and the overall effectiveness of the three types of perimeter systems studied.

Glass Properties				
U-Value	**Shading Coefficient**	**Percent Glass**	**Energy Use, Btu/ft$^2\cdot$yr**	**Annual Utility Cost, $**
0.3	0.3	40	46,300	737,400
0.4	0.4	40	48,900	760,600
0.5	0.5	40	51,800	782,500
0.4	0.4	50	51,100	782,500

Scheme	Estimated First Cost, $	Energy Use, Btu/ft^2•yr	Utility Costs, $			Net Present Worth, $
			Electric	Gas	Total	
VAV + Perimeter Baseboard	13,593,750	64,400	739,200	65,400	804,600	15,489,800
VAV + Perimeter Fan-Powered VAV	13,750,000	47,000	737,300	24,000	761,300	15,425,300
VAV + Perimeter Four-Pipe Fan-Coils	14,375,000	53,700	688,500	38,400	726,900	15,739,500

The following conclusions were drawn:

1. The utility cost of the baseboard heating scheme is the highest because of its uncontrolled heat delivery mode in winter. If, in order to obtain control of the heat output, the individual baseboard lengths were to be controlled in sequence with the VAV box volume control, then the first cost of the baseboard system would become prohibitive.

2. The utility cost of the four-pipe fan-coil-unit system is the lowest because more than 50% of the floor's cooling is being accomplished through the chilled water system piped to the fan-coil units. Compared to the all-air VAV systems, therefore, the fan-coil unit system uses much less fan energy to achieve the same amount of perimeter cooling.

3. Despite the operating cost advantages of the fan-coil unit system, the perimeter fan-powered variable-air-volume (FPVAV) system displays the lowest overall cost. This is due to its lower first cost and its reasonably good energy performance compared to the fan-coil unit system.

The major reasons for recommending the FPVAV system are:

1. Its first cost is more accurately predictable because of its wide application in this area. It is definitely less expensive to install than a comparable four-pipe fan-coil unit system.

2. Its annual energy is 14% less than that of the fan-coil unit system and 37% less than that of the baseboard heating system. The FPVAV system's ability to use the "heat of light" available in the hung ceiling cavity for winter heating is the main reason for its superior performance.

3. The FPVAV system would permit the floor-to-ceiling glass (with 8- to 12-in. sill height) contemplated for this building, whereas standard fan-coil units require higher sills. The lower profile units are not produced competitively.

4. The heating elements provided with the FPVAV boxes can provide warm-up (in winter) without running the air-handling system fans. Similarly, they can provide temporary heat prior to occupancy if installed without ductwork in their permanent locations.

A-8 PROPOSED AIR DISTRIBUTION SYSTEM FOR TYPICAL FLOOR

Under an FPVAV system, the 15-ft perimeter zone will receive conditioned air from overhead ductwork. One fan-powered variable-air-volume control box in the hung ceiling space will be assigned to each 450 ft^2 of floor area and will be controlled by a room thermostat. Boxes will have electric resistance heating coils downstream in the low-pressure ductwork. The perimeter zones will be fed through a continuous slot between columns via 4-ft boots.

The interior zones of the floor will have conventional boxes, one per 1000 ft^2 of floor area, connected to the common supply air duct main serving the entire floor. VAV boxes will be controlled by return air or by room thermostats. Supply air may be fed through conventional 2 by 2 air diffusers suitable for installation in a 2 by 4 lay-in ceiling through light/air troffers or through 12 by 12 perforated face diffusers. Space air will be returned through air slots incorporated within the light fixtures. A certain number of interior VAV boxes will have duct heaters as required by the floor slab or roof exposure of the space served or as required by special zoning or warm-up needs of the space. Controls will be pneumatic/electronic with local DDC panels capable of handling desired local functions as well as forming part of a central building.

A-9 BUILDING AUTOMATION SYSTEM

A central and expandable building automation system (BAS) is recommended. The system will be composed of a programmable central monitoring control and supervisory system that incorporates and integrates the following functions from the various DDC field panels:

1. Automatic temperature control system readouts and status of equipment.

2. Energy management functions, such as logging of equipment run-time and scheduled start-stop.

3. Lighting control systems.
4. Intercom between operating personnel.
5. System operation and maintenance scheduling.
6. Functions interfacing with the fire alarm system.

Appendix B DESIGN CALCULATIONS FOR SINGLE-ZONE BUILDINGS

B-1 INTRODUCTION

The sample calculations in this appendix and in Appendix C illustrate the application of the procedures presented in Chapter 6, All-Air Systems. Any conclusions you draw from comparisons of systems apply only to the specific application presented and do not necessarily apply universally. A step-by-step procedure is given for each air system's load components accompanied by a corresponding psychrometric chart plot. Several off-season operating conditions are also analyzed to illustrate the ability of each system to maintain design conditions under a wide range of loads.

The building analyzed here and in Appendix C is a multiexposure, multistory building with a roof at one level. Any specific treatment of the roof level as to loads or control zones is ignored, and all vertical separations are assumed to be ceiling-to-floor assemblies without any roof. All decks are considered floors separating conditioned spaces; therefore, deck losses are a heat loss from, not a heat gain to, the ceiling plenum.

All supply and return ducts are assumed to be outside the conditioned spaces, so that they do not affect ceiling temperature equations. This assumption is made for simplicity but, if the supply ducts in the return air ceiling are insulated and there are only small runs of return air stub ducting in the ceiling, the resulting error is small. It is also assumed that the peak airflow for all rooms is governed by room sensible heat.

All fan heat gains are assumed as a temperature rise at the fan discharge even though the velocity pressure component of the fan energy occurs along the ducts. This introduces a small psychrometric error when draw-through and blow-through systems are compared. The velocity at the fan discharge is assumed to be 2,500 fpm or 0.39 in. of water velocity pressure (VP). In a system with 6 in. of water total pressure (TP), this VP represents only 6.5% of the total pressure (or energy) of the fan. The error is accepted for simplicity in this presentation.

B-2 BUILDING LOADS

B-2.1 Physical Data

Total conditioned area		160,000 ft^2 (10 stories, 126.5 by 126.5 ft)
Perimeter area		70,000 ft^2 based on 16-ft depth
Interior area		89,300 ft^2
U-factors for ceiling assembly		
Ceiling	=	0.5-in. acoustical lay-in tiles; heat flow down
U	=	$1/(0.23 + 1.19 + 0.92) = 0.43$ Btu/h·ft^2·°F
Floor	=	2-in. concrete slab, tile covered; heat flow up
U	=	$1/(0.23 + 0.40 + 0.05 + 0.61) = 0.77$ Btu/h·ft^2·°F

Return-air hung ceiling plenum with return air grilles in ceiling at 150-ft^2 module locations, resulting in the plenum temperature being midway between room temperature and the temperature of the air entering the duct.

B-2.2 Design Conditions for Full Load

Room design summer condition	76°F db, 45% rh
Room design winter condition	76°F db, 20% rh
Outdoor design summer condition	95°F db/75°F wb
Outdoor design winter condition	0°F db, 20% rh
Minimum outdoor air	0.2 cfm/ft^2 = 32,000 cfm

B-2.3 Summer Design Transmission Loads (Solar and Conduction)

Assume no roof (for simplicity, see Section B-1); therefore, loads are from walls and glass only, distributed over the 16-ft perimeter area that is partitioned from the interior area. The following figures are assumed to be per square foot of perimeter floor area.

Exposure	N	E	S	W
Btu/h·ft^2 block load	5.88	6.90	7.92	39.9
Btu/h·ft^2 max. in each exposure	5.88	32.2	18.2	39.9

Each exposure of 10 stories has 17,675 ft^2 of floor area. Total loads for perimeter area are, therefore:

Block load	104 + 122 + 140 + 705 = 1,071,000 Btu/h
Σmax load	104 + 569 + 322 + 705 = 1,700,000 Btu/h

Conduction load per square foot of perimeter floor area is 5.45 Btu/h·ft². The remainder is solar load:

Loads (1000 Btu/h)	Total	Conduction	Solar
Block	1071	385	685
Σmax	1700	385	1315

B-2.4 Winter Design Block Transmission Load (Conduction Only)

Load is 1,540,000 Btu/h or 21.8 Btu/h per ft², of perimeter floor area. Loads do not include transmission from ceiling and floor cavities to room area. These loads are given for completeness only. Winter conditions are not analyzed in this appendix.

B-2.5 Lighting and Miscellaneous Electric Loads

The following tabulation includes ballast, office equipment, and table lamps shown as design values with and without a 10% diversity allowance. Only systems with true VAV characteristics in multizone applications are allowed diversity reductions, since constant-volume systems must provide capacities for Σmax, the sum of the individual peaks. When the ceiling is a return air plenum, only part of the lighting emissions to the ceiling are retransmitted to the room as a heat gain. However, with a stagnant ceiling in a ducted return air system, all of it is retransmitted.

	Σ max Full Load		Load with 10% Diversity	
	W/ft²	1000 Btu/h	W/ft²	1000 Btu/h
65% heat of lights emitted directly to ceiling plenum	3.04	1660	2.74	1496
35% heat emitted directly to room	1.63	890	1.47	803
Miscellaneous room electric loads	0.56	306	0.50	273
Total electric load	5.23	2856	4.71	2572
Ceiling lighting alone	4.67	2550	4.21	2299

B-2.6 Occupancy Loads (Assumed as High-Occupancy Building)

Full load of 2,378 occupants; 67 ft^2 per occupant. If design permits diversity, use 2,140 occupants, 75 ft^2 per occupant. Use 240 Btu/h sensible and 210 Btu/h latent heat per occupant.

B-2.7 Design Data Full Load

Supply fan total pressure	6 in.	
Return fan total pressure	1 in.	
Cooling coil leaving air dry-bulb temperature (summer)		50°F
Minimum room supply air temperature (winter)		55°F

B-2.8 Intermediate Season Loads

Outdoor temperature	=	65°F db/65°F wb
Indoor minus outdoor temperatures	=	11°F
Cooling coil temperature	=	48°F minimum; control droop 2°F lower
Transmission load	=	–223,000 Btu/h (heat loss) at –11°F Δt without solar load
Heat gain from lights		
at 75% full load	=	2,143,000 Btu/h
with 10% diversity	=	1,927,000 Btu/h
People load		
at full load	=	570,700 Btu/h
with 10% diversity	=	513,600 Btu/h

B-2.9 Special Room Loads

Table B-1 lists standard load components for typical rooms under various operating conditions. The first set of calculations in this appendix will help you in understand the numbers in the table.

B-2.10 System Calculations

For simplicity, all overall system calculations are based upon either the block loads or Σmax loads, assuming that summer sensible heat governs all room air volumes. In unusual cases, certain room air volumes and room peak calculations may indicate that ventilation, humidity, or winter sensible heat may govern, but these conditions rarely occur and are ignored in the present analysis. In all cases, particular stress is given to system calculations as they affect room conditions, particularly humidity.

B-3 LOAD CALCULATION FOR CONSTANT-VOLUME SINGLE-ZONE SYSTEM

Although the nature and the magnitude of these design criteria were chosen to analyze and compare the results for fairly sizable buildings with a requirement for zoning flexibility, they are used here as the loads for a simple, single-zone system to illustrate the basic procedures. Even though the type of system analyzed here is not intended as a recommended design for this application, the results serve to illustrate that the use of more sophisticated air systems might result in appreciable increases in air, refrigeration, and heating capacities to achieve the desired effect. Figure B-1 represents the system cycle with the state of air at all locations shown in Figure 6-1a. The corresponding points on the psychrometric chart and the component heat quantities absorbed or liberated between points are illustrated in Figures 6-1b and 6-1c. Assume that the simple, single-zone load is identical to the simultaneous peak load of the entire building, as if there were just one large area with diversity.

B-3.1 Cooling Cycle Calculations at System Peak

Step 1. Find trial room sensible heat gain; it cannot be finalized until heat gains from ceiling and floor cavities are determined after finding the average plenum temperature. Assume that 40% of ceiling light heat will be transmitted to the room, down through the ceiling, and up through the floor below (see Figure B-1).

Heat gain from ceiling and floor	=	0.4 × 1,496,000 = 598,400 Btu/h
Lighting and miscellaneous heat gain	=	803,000 + 273,000 = 1,076,000
Simultaneous maximum transmission heat gain	=	1,071,000
People load	=	2,140 occupants @ 240 = 513,600
Sensible heat load (trial)	=	3,259,000 Btu/h

Table B-1 Special Room Loads (Basic Form)

	1 Lecture and Projection	2 Interior Clerical	3 Perimeter Executive Office	4 Perimeter Conference Room	5 Perimeter Office
Area, ft^2	5,000	320	400	1,800	320
(1) Full load electric W/ft^2	4.0	6.0	5.75	5.0	4.5
Full load occupancy	250	3	6	36	4
Location	Interior	Interior	South	North	West
Room Peak Cooling, 95/75 day, 100% Lighting and Occupancy					
(2) Conduction load	0	0	2,180	9,800	1,740
(3) Solar load	0	0	5,100	774	11,024
(4) Plenum load					
Occupancy sensible load	60,000	720	1,440	8,650	960
Room sensible load					
Occupancy latent load	52,500	630	1,260	7,550	840
Room internal load					
SHF = Room sensible heat factor					
(5) Cooling at Part Load, 65/65 day	**@ 0% Lighting**	**@ 65% Lighting**	**@ 80% Lighting**	**@ 25% Lighting**	**@ 80% Lighting**
(6) Conduction load (solar = 0)	0	0	(5) –1,260	(5) –5,690	(5) –1,010
Plenum load					
Occupancy sensible load	60,000	720	1,440	8,650	960
Room sensible load					
Occupancy latent load	52,500	630	1,260	7,550	840
Room internal load					
SHF = Room sensible heat factor					

Notes for Table B-1

All loads are in Btu/h.

Only the basic criteria that apply to any system design are given in this table. Blanks are left where data will vary from one system to another. No diversity in individual rooms is allowed at peak.

(1) Electrical loads in specific areas vary from the average of 5.23 W/ft^2, depending on wall and floor outlet usage or nonstandard ceiling lighting. Loads are intended to be actual demands, not connected loads.

(2) Loads are pure wall and glass conduction with no allowance or proration for roof.

(3) These are solar loads over and above conduction loads. Items (2) and (3) for each area correspond to the Btu/ft^2 maximum loads from Section B-2.3, Summer Design Transmission Loads. Only room 3, south, is given at its noon peak; all other rooms peak individually but simultaneously in late afternoon.

(4) Direct emission from light heat to any room is considered to be the same for all systems, but indirect transmissions through floor and ceilings from the return air plenum are a function of the average ceiling temperature, which varies with the return air volume from the room and the temperature above the deck or below the floor adjacent to unconditioned spaces.

(5) The percent of full load lighting indicated is for each room only. Systems lighting for all 65/65 outdoor conditions is 75%. For interior spaces, 65% of full load lighting is assumed a realistic minimum in a preplanned, modular lighting system for an area without business machines and table lamps, since local occupancy switches are rarely provided. For a perimeter office with a light switch, 80% lighting load is assumed when occupied and 0% when unoccupied; for conference rooms, 25% when occupied and 0% when unoccupied.

(6) The conduction values shown constitute calculated values at the Δt for the indicated outdoor temperature. They are applied as internal loads only in specific systems tabulations, for air volume calculations, to the extent that they are not balanced by auxiliary heating systems, *e.g.*, perimeter radiation, terminal heating coils.

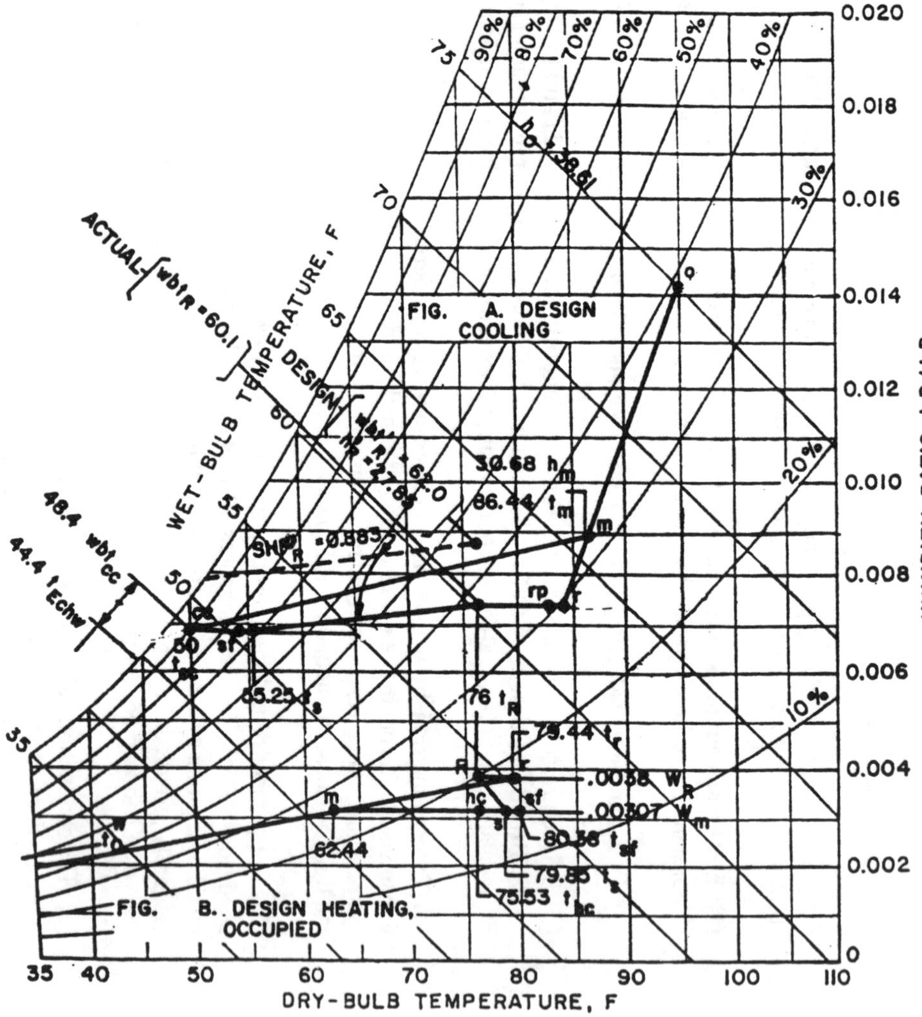

Fig. B-1 Psychrometric Analysis for Single-Duct, Single-Zone System

Step 2. Find the supply air temperature and the trial supply and return cfm. With a draw-through coil and the fan and drive completely enclosed in the fan plenum, all the heat of the supply fan, motor, and drive, as well as the supply duct transmission loss, becomes a temperature rise in the supply air after the cooling coil.

$$\text{Fan temperature rise} = \frac{0.363 \times \text{Total pressure in inches}}{\text{Fan efficiency} \times \text{Motor efficiency} \times \text{Drive efficiency}} \quad \text{(B-1)}$$

Fan temperature rise: 0.363 × 6/(0.6 × 0.9 × 0.95) = 4.25°F, independent of cfm.

Assuming a 1°F temperature gain from that portion of the supply duct that passes through 90°F unconditioned surroundings, the supply temperature becomes 50 + 4.25 + 1 = 55.25°F.

Using Equation (1) in Chapter 6, the trial supply air for sensible cooling is obtained as 3,259,000/[1.1(76 – 55.25)] = 142,800 cfm.

Recirculated air rate equals supply air rate minus outdoor air rate.
Recirculated air rate: 142,800 – 32,000 = 110,800 cfm.

Now check the assumed heat gain from the ceiling and floor. Using the appropriate equation from the chapter Duct Design of ASHRAE *Fundamentals* (1993), the difference between room and return air plenum temperatures is found from

(Lighting heat in plenum)/(0.5 × Floor and ceiling conductance + Return air cfm)

$$1{,}496{,}000/[0.5 \times 160{,}000(0.77 + 0.43) + 1.1(110{,}800)] = 6.78°\text{F}$$

The difference between room and average return air plenum temperatures is one-half of that value, and the heat gain from the ceiling and floor plenums then becomes

$$160{,}000(0.77 + 0.43)6.78/2 = 650{,}880 \text{ Btu/h}$$

Step 3. A revised try for cfm, repeating steps 1 and 2 with new values gives the following:

Heat gain from ceiling and floor		=	650,880 Btu/h
Lighting and miscellaneous heat gain	803,000 + 273,000	=	1,076,000
Simultaneous maximum transmission gain		=	1,071,000
People load	2,140 occupants × 240	=	513,600
Sensible heat load (revised)		=	3,311,500 Btu/h
Revised supply air rate		=	145,000 cfm
Revised recirculated air rate		=	113,500 cfm
Temperature difference	4,496,000/(96,000 + 1.1 × 113,500)	=	6.77°F
Return air plenum temperature		=	82.8°F
Heat gain from ceiling and floor	160,000 (0.77 + 0.43)(6.77/2)	=	650,880 Btu/h
Return plenum heat gain	1.1(113,500)(6.77°F)	=	845,230 Btu/h
Total direct lighting emission to plenum	650,000 + 845,230	=	1,496,110 Btu/h

This is within 0.007% of direct emission allowed in the breakdown criteria. Note that this is 44% instead of the 40% lighting load fraction assumed. The total heat gain contribution of direct and indirect heat of light is 1,076,000 + 650,880 = 1,727,000 Btu/h or 67% of the total electric load.

Step 4. Find the room sensible heat factor and check the obtainability of the design room temperature. Use Figure B-1a. Plot room design air state 76°F db, 45% rh.

Room sensible heat load	3,321,100 Btu/h
Room latent heat load	2,140 occupants × 210 = 449,400 Btu/h
Room total internal load	3,770,500 Btu/h
Sensible heat factor	3,321,100/(3,770,500) = 0.88

Draw a line with this slope through the design state and note that the required supply air temperature of 55.25°F, after a 5.25°F rise from the 50°F coil temperature, could not possibly fall anywhere within this process line (see dotted line in Figure B-1a). A reasonable coil temperature must be assumed before the actual room humidity ratio and total system load can be determined, since the latter is a function of actual, not design, humidity.

A final adjustment to the calculated loads must always be made based upon a given room design condition, when the state of air leaving the actual selected coil is low enough in dew point to depress the room humidity below the selected design condition. If a reconciliation is not made between the calculated load and the air load from the psychrometric chart conditions, the calculated load may be too low. Also, since the final leaving coil air condition cannot be determined until the total system load is known, a rule of thumb will suffice to zero in on a reasonable coil condition. It is based upon the average coil surface temperature, which may be approximated closely as the intersection of the cooling line with the saturation curve (assumed straight), in this case, line m-cc extended in Figure B-1a. The rule may be expressed as:

$$(t_{ccs} - t_{Echw})/(wbt_{cc} - t_{Echw}) = K_{cc} \qquad \text{(B-2)}$$

where

t_{ccs} = coil surface temperature as defined previously
t_{Echw} = entering chilled-water temperature
wbt_{cc} = leaving wet-bulb temperature of air off coil
K_{cc} = chilled water coil constant

K_{cc} is a function of the ratio of extended surface to prime surface of a coil. It varies from about 0.5 for circular fin coils with 8 fins-per-in. spacing to 0.7 for those with 14 fins per in., and from about 0.6 to 0.75 for continuous plate fin coils. The overall effect for a given t_{Echw} is to raise the leaving coil dew-point temperature as the extended surface ratio increases. The designer should always check the assumed cooling coil temperature against that obtainable with final coil selection.

Step 5. Find the dry-bulb temperatures and humidities of return air and mixed air. These humidities can only be found by trial and error using Equations (B-1) and (B-2). From Equation (B-1), the return fan temperature rise equals 0.363 × 1 in./(0.6 × 0.9 × 0.95) = 0.71°F. Assume a return duct transmission gain (in the unconditioned space) of 0.29°F. Then the return air temperature becomes 76 + 6.77 + 0.29 + 0.71 = 83.8 (point *r* in Figure B-1a). Also plot point *o*, the state of outdoor air. From Equation (5a) in Chapter 6, and the developed 22% of outside air ratio, the mixed-air temperature is found to be 83.8 + 0.22(95 – 83.8) = 86.26°F.

Assume a coil with $K_{cc} = 0.5$. Graphically, start with a lower room humidity than the design condition and plot in sequence the temperatures of Equation (B-2) such that $K_{cc} = 0.5$. This occurs when $wbt_{cc} = 48.4°F$, $t_{ccs} = 46.4°F$, and $t_{Echw} = 44.4°F$. Figure B-1a shows this at a room condition of 76°F db/60°F wb (rh = 38.7%), which is considerably below the 45% design relative humidity. At this point, coil selection should be checked with actual manufacturer's data to verify that conditions are attainable with an optimum selection of coil area, rows, gpm, t_{Echw}, and water temperature rise.

Step 6. Find the final total system cooling load from load component summation and psychrometric chart values.

Room sensible heat	3,321,100 Btu/h
Room latent heat	449,400
Room total heat	3,770,500 Btu/h
Light heat in return air	845,230
Return duct and fan 1.1 × 113,500(1°F)	124,850
Supply duct and fan, sensible 1.1 × 145,500(5.25)	840,260
Outside air, sensible 1.1 × 32,000(95 – 76)	669,000
Outside air, latent 4840 × 32,000(0.01410 – 0.00752)	1,019,000
Total calculated load	7,268,800 Btu/h

Step 7. Find the total coil load from the enthalpy difference between mixed air and coil leaving air.

Coil load	4.5 × 145,500(30.56 – 19.43)	7,287,400 Btu/h

Step 8. Find the total load from temperature and humidity ratio differences across the coil.

Sensible coil load	1.1 × 145,500(86.26 – 50)	5,803,410 Btu/h
Latent coil load	4840 × 145,500(0.00885 – 0.00683)	1,422,520
Total coil load		7,225,900 Btu/h

B-3 COMMENTS

a. Steps 6, 7, and 8 illustrate a 0.83% reconciliation for three calculation methods, including the psychrometric chart analysis to synthesize the total load.

b. Note that the entire ceiling plenum load becomes a system load only when all the air from the room is returned to the supply fan. In the extreme case, if all the return air were dumped after the return air fan, the heat of the ceiling plenum in the return air never becomes a system cooling load. (Conversely, for winter heating, the plenum heat that is cast off by the return air fan before blending with outdoor air cannot be credited to the heating coil or system.)

c. The full load with a 5.23 W/ft^2 electric load requires only 145,500 cfm/160,000 ft^2 = 0.91 cfm/ft^2 because of the heat of lights in the return air, even with a moderate supply air temperature difference of 20.75°F. If the ceiling were not used as a return air plenum, this load of 845,230 Btu/h would become a room sensible load requiring approximately 37,000 more cfm, or a total supply of 1.14 cfm/ft^2.

d. Figure B-1b shows the psychrometrics for winter conditions. For an explanation of winter conditions, see Appendix C.

Appendix C DESIGN CALCULATIONS FOR SINGLE-DUCT VAV BUILDING WITH PERIMETER RADIATION

C-1 INTRODUCTION

This is the same building as the one described in Appendix B. Data not shown here are taken from that appendix. The following description defines the building in more detail.

1. A continuous perimeter hot-water radiation system is assumed with scheduling of the supply water temperature from the outdoor temperature. It is sized throughout for adequate auxiliary heat to maintain a 76°F room by balancing the heat loss from conduction plus any infiltration. Scheduling options are illustrated.

2. Full-closing VAV units are assumed, but means for maintaining a minimum acceptable air motion and ventilation level without minimum positioning are illustrated.

3. Multiple-zone flexibility is illustrated by examining interior and perimeter exposed areas of different uses.

4. Full diversity benefit is taken.

5. Table C-1 completes the basic form of Table B-1 for this particular system.

C-2 CALCULATIONS AT SUMMER PEAK

The calculations and psychrometric chart for the simple, single-zone cooling application with a draw-through fan may be considered identical to this VAV application for the solution of average system conditions at summer design. Figure

Table C-1 Special Room Loads

	1 Lecture and Projection	2 Interior General Offices	3 Perimeter Executive	4 Perimeter Conference	5 Perimeter Office
Area, ft^2	5,000	320	400	1,800	320
(1) Full load electric, W/ft^2	4.0	6.0	5.75	5.0	4.5
Full load occupancy	250	3	6	36	4
Location	Interior	Interior	South	North	West
Room Cooling Peak on 95°F db day, 100% Full Load Lighting and Occupancy					
(2) Conduction load	0	0	2,180	0.800	1,740
(3) Solar load	0	0	5,100	774	11,024
(4) Plenum load	55,300	4,657	5,260	22,056	3,375
Occupancy sensible load	60,000	720	1,440	8,650	960
Room sensible load	115,300	5,377	13,980	41,280	17,099
Occupancy latent load	52,500	630	1,260	7,550	840
Room internal load	167,840	6,007	15,240	48,830	17,939
SHF = Room sensible heat factor	0.69	0.9	0.92	0.85	0.95
cfm/ft^2 at $t_s = 55.25°F$	1.01	0.74	1.85	1.0	2.34
Room relative humidity, %	46.6	38.6	40.0	40.0	37.6
(5) **Cooling at Part Load, 65/65 Day**	@ 0% F.L. Lighting	@ 65% F.L. Lighting	@ 80% F.L. Lighting	@ 25% F.L. Lighting	@ 80% F.L. Lighting
(6) Conduction load offset by radiation (solar = 0)	0	0	0	0	0
(4) Plenum load	19,200	3,409	4,441	10,595	2,887
Occupancy sensible load	60,000	720	1,440	8,650	960
Excess radiation	0	0	0	0	400
Room sensible load	79,200	4,129	5,881	19,245	4,247
Occupancy latent load	52,500	630	1,260	7,550	840
Room internal load	131,700	4,759	7,141	26,794	5,087
SHF = Room sensible heat factor	0.60	0.87	0.82	0.72	0.83
cfm/ft^2 at $t_s = 58.1°F$	0.80	0.66	0.75	(Exh) 0.65	0.65
Room relative humidity, %	55.0	47.7	45.5	48.5	44.5

Table C-1 Special Room Loads *(Concluded)*

Cooling at Heating Peak on 0°F Day

(6) Conduction load (reference only)	0	0	–8,730	–39,300	–6,990
Peak solar load	0	0	8,800	300	5,250
Maximum plenum load	55,300	4,657	5,260	22,056	3,375
Maximum occupancy sensible load	60,000	720	1,440	8,650	960
Excess radiation	0	0	4,150	0	4,100
Maximum room cooling load	115,300	5,377	19,650	31,006	13,685
cfm/ft² at t_s = 55.25°F	1.17	0.74	2.15	0.76	2.17
Minimum room cooling load at t_s = 61.5°F	15,000	2,620	4,150	10,595	4,100
cfm/ft² at t_s = 61.5°F	1.01	0.05	0.65	(Exh) 0.65	0.65
	0	0	0.65	0.65	0.65

Notes for Table C-1

All loads are in Btu/h.

(1) Electrical loads in specific areas vary from the averge of 5.23 W/ft², depending on wall and floor outlet usage or non-standard ceiling lighting. Loads are intended to be actual demands, not connected loads.

(2) Loads are pure wall and glass conduction with no allowance or proration for roof.

(3) These are solar loads over and above conduction loads. Items (2) and (3) for each area correspond to the Btu/ft² maximum loads from Section B-2.3, Summer Design Transmission Loads. Only room 3, south, is given at its noon peak; all other rooms peak individually but simultaneously in late afternoon.

(4) Direct emission from light heat to any room is considered to be the same for all systems, but indirect transmissions through floor and ceilings from the return air plenum are a function of the average ceiling temperature, which varies with the return air volume from the room and the temperature above the deck or below the floor adjacent to unconditioned spaces.

(5) The percent of full load lighting indicated is for each room only. Systems lighting for all 65/65 outdoor conditions is 75%. For interior spaces, 65% of full load lighting is assumed a realistic minimum in a preplanned, modular lighting system for an area without business machines and table lamps, since local occupancy switches are rarely provided. For a perimeter office with a light switch, 80% lighting load is assumed when occupied and 0% when unoccupied; for conference rooms, 25% when occupied and 0% when unoccupied.

(6) The conduction values shown constitute calculated values at the Δt for the indicated outdoor temperature. They are applied as internal loads only in specific systems tabulations, for air volume calculations, to the extent that they are not balanced by auxiliary heating systems, *e.g.*, perimeter radiation, terminal heating coils.

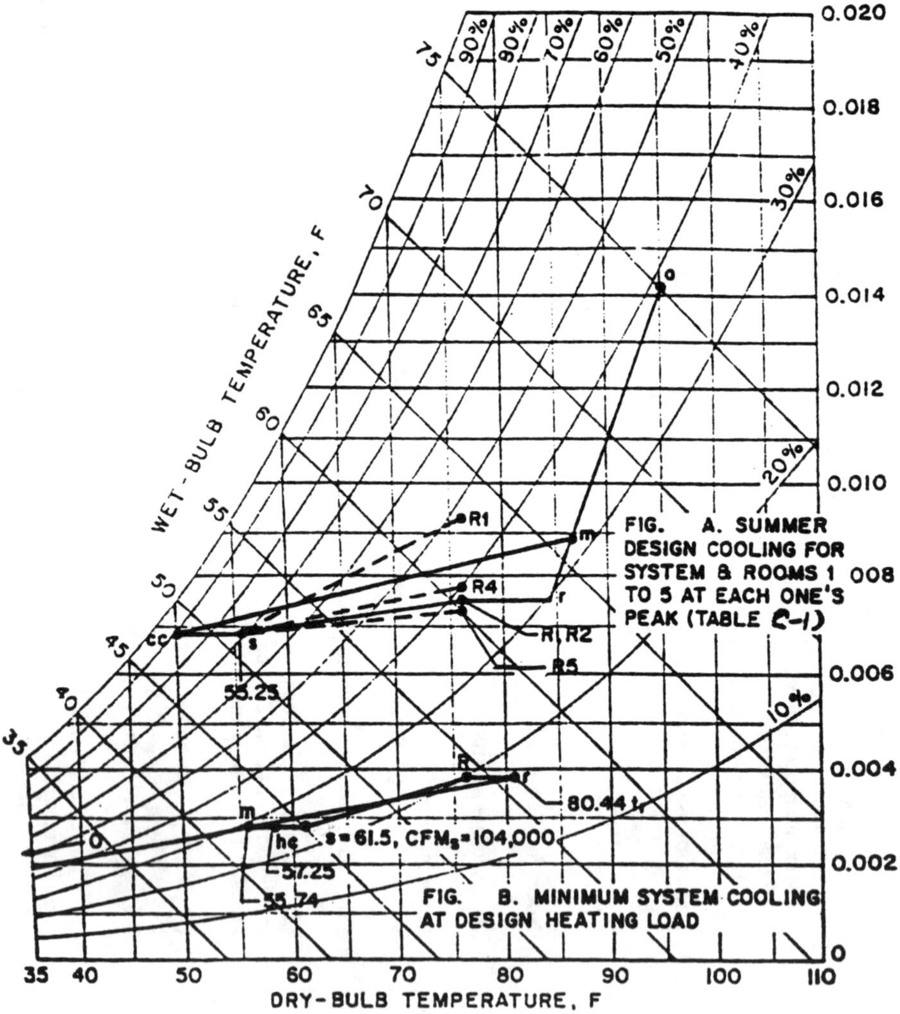

Fig. C-1 Psychrometric Analysis for Single-Duct VAV System with Separate Perimeter Radiation at Design Loads

C-1a therefore shows, as solid lines for system averages, the identical summer plot as Figure B-1, while the dotted processes are for special rooms defined in Table C-1. The room state subscript numerals correspond to the column numbers in the table.

Footnote 4 of Table C-1 shows the direct light heat emission to each room from ceiling lights and miscellaneous room electric loads plus the indirect heat as a function of the building's average ceiling temperature rather than as a percentage of the individual maximum direct light heat emission to the ceiling. The plenum temperature is therefore a function of the percent of the system's entire full-load ceiling lighting, independent of the specific ceiling light intensity for any specific room.

Room 1: Lecture and Projection, Peak Load (see column 1, Table C-1)

Take ceiling lighting alone as 3 W/ft^2, miscellaneous electrical loads as 1 W/ft^2, and the temperature difference between plenum air and room air as 3.39°F (from Appendix B). Again, as in Appendix B, assume 35% of the lighting load enters the room. Room loads are then as follows:

Electrical load	(1 + 0.35 × 3)(5,000)(3.41)	=	35,000 Btu/h
Load from plenums	5,000(1.2)(3.39)	=	20,340
Occupant sensible	250 × 240	=	60,000
Room sensible			115,340 Btu/h
Occupant latent	250 × 210	=	52,500
Room total internal			167,840 Btu/h

Sensible heat factor	SHF = 115,300/167,840 = 0.69
Room supply air rate	115,300/1.1(76 - 55.25) = 5,051 cfm = 1.01 cfm/ft^2

Draw line s-R1 through point s in Figure C-1a at a slope of SHF = 0.69 for graphical solution of room state at 76°F db, 46.6% RH_1. This is not as low as the design condition but is considered acceptable; therefore, no design reheat or equivalent is required for this mode.

Columns 2 through 5 in Table C-1 are calculated in a similar manner. Only numbers that are not readily apparent are explained next.

Room 2: Interior Clerical, Peak Load (see column 2, Table C-1)

Electrical	(1.5 W/ft^2 + 0.35 × 4.5 W/ft^2 ceiling)(320)(3.41)	=	3,355 Btu/h
Plenums	(320 ft^2)(1.2 × 3.39) Btu/(h·ft^2)	=	1,302
Sum			4,657 Btu/h

Supply air rate 5,377/1.1(20.75) = 236 cfm at full load = 0.74 cfm/ft^2
Draw line s-R2 at SHF = 0.9 for room RH_2 of 38.6%

Room 3: Perimeter Executive Office, South-Facing, Noon Peak (column 3, Table C-1)

From Assumed Loads, Appendix B, Paragraph 3:

Conduction	400 × 5.45 Btu/h·ft^2	=	2,180 Btu/h
Solar	400(18.2 – 5.45)	=	5,100
Lighting	(1 + 0.35 × 4.75)(400)(3.41)	=	3,632
Plenums	400(1.2 × 3.39) Btu/h·ft^2	=	1,628
Sum			5,260 Btu/h

Room supply air rate 13,980/1.1(20.75) = 741 cfm = 1.85 cfm/ft^2
Draw line s-R3 at SHF = 0.92 for room RH_3 of 40% (this line not shown in Figure C-1a)

Room 4: Conference Room, North-Facing, Peak Load (column 4, Table C-1)

Conduction	1,800 × 5.45	=	9,800 Btu/h
Solar	1,800(5.88 – 5.45)	=	774
Electrical	(1 + 0.35 × 4)(1,800)(3.41)	=	14,730
Plenums	1,800(1.2)(3.39)	=	7,326
Sum			22,056 Btu/h

Room supply air rate 41,280/1.1(20.75) = 1,808 cfm = 1.0 cfm/ft^2
Draw line s-R4 at SHF = 0.85 for room RH_4 of 40%

Room 5: Perimeter Office, West, Late Afternoon Peak (column 5, Table C-1)

Solar	320(39.9 – 5.45)	=	11,024 Btu/h
Electrical	(0.5 + 0.35 × 4)(320)(3.41)	=	2,073
Plenums	320(1.2 × 3.39)	=	1,302
Sum			3,375 Btu/h

Room supply air rate 17,099/1.1(20.75) = 749 cfm = 2.34 cfm/ft^2
Draw line s-R5 at SHF = 0.95 for room RH_5 of 37.6%

C-3 CALCULATIONS FOR PART-LOAD COOLING

C-3.1 System Analysis for Outdoor Conditions 65/65°F, No Sun, 75% Electric and 100% Occupancy Loads

With true VAV, only the following changes on the psychrometric chart have an effect on room conditions (see Figure C-2):

1. The cooling coil temperature will suffer a natural control droop from the discharge thermostat that controls it as well as from the controller of the refrigerated medium. For part-load humidity control, it is desirable to permit this to occur, and some designs even have a desired droop programmed into their control cycle. It is to be noted, however, that this tends to increase the throttling ratio for VAV-controlled zones.

2. The temperature rise from supply fan heat and duct gains may change as a result of the drop in system air volume and different ceiling plenum balance and temperatures. Although Equation (B-1) shows the fan heat temperature rise to be independent of cfm, it is an inverse function of fan efficiency at any given operating condition. This may vary in any given fan, as the constant static pressure control operates the various devices and may rise or fall, depending upon the position of the full-load operating point in the efficiency curve, the type of control, and the part-load value. For simplicity in comparisons, it is assumed that variable-inlet vanes are used and, within the range of 100 to 50% of system volume, the bhp is taken to be proportional to the percent of full-load volume. Therefore, the temperature rise is 48 + 5.25 = 53.25°F. This tends to keep the fan Δt rise constant within the stated range regardless of volume. Note, however, that the fan efficiency for variable-speed drives remains constant in the range typical for VAV systems.

3. For any given room latent load, the room SHF decreases as the room supply volume is decreased to balance reduced room sensible loads.

4. At constant room air cfm, the temperature rise from the ceiling heat of lights is proportional to the percent of full lighting load. However, this percentage varies inversely with room air supply volume. Repeat the trial-and-error solution for this quantity, illustrated in steps 1 through 3 of the cooling load calculations in Appendix B, for 75% lighting and miscellaneous electric loads, full occupancy, and a greater ratio of lighting loads into the room than the previous 44% found there. With a substantially lower room air volume, assume a ratio of 57% because of the inverse relationship.

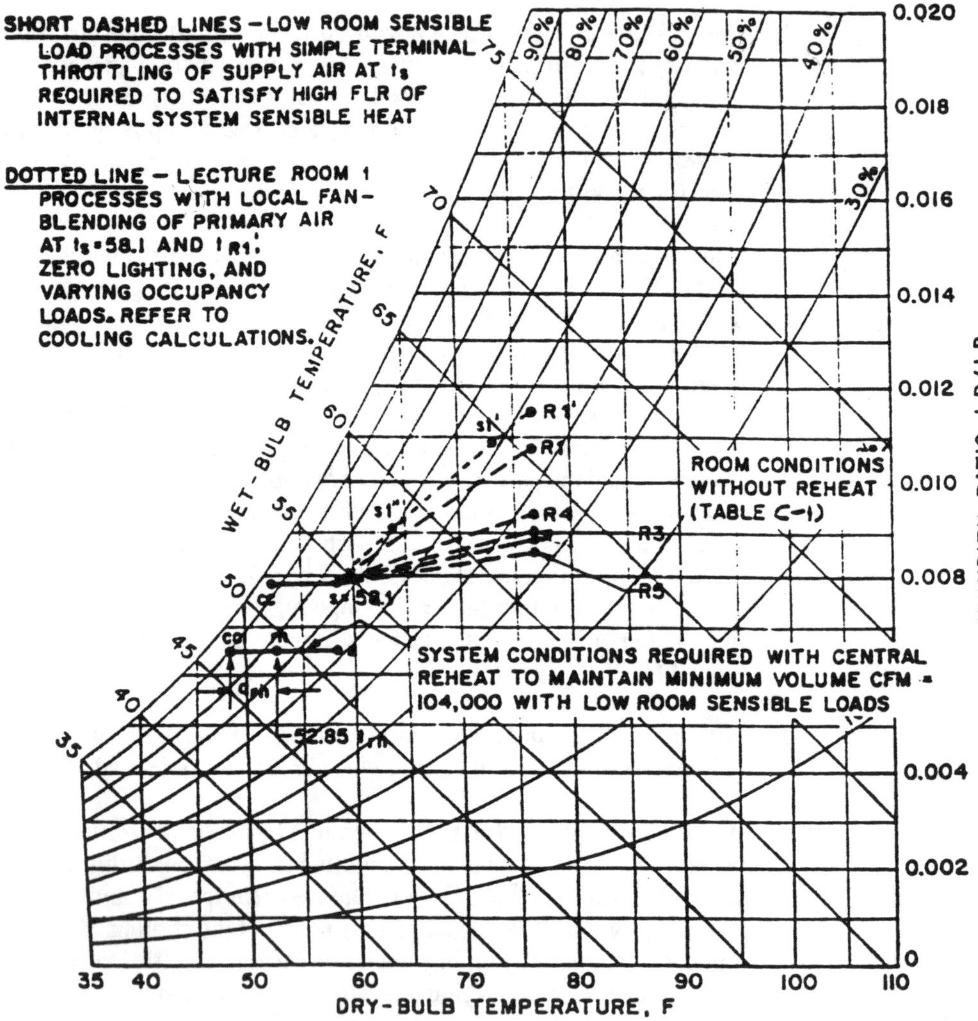

Fig. C-2 Psychrometric Analysis for Single-Duct VAV System with Separate Perimeter Radiation at Outdoor Conditions of 65/65°F

Direct lighting to room			
and miscellaneous electrical loads	0.75 × 1,076	=	807,000 Btu/h
Plenum loads	0.57(0.75 × 1,496.1)	=	639,500
Sensible occupant load			
at full load with diversity			513,600
Transmission load (balanced by radiation)			0
Total room sensible load			1,960,100 Btu/h

Supply air of 78,300 cfm is at a temperature of 53.25°F, allowing for a rise to room air of 22.75°F. This represents 54% of the full-load air supply and an average of 0.49 cfm/ft^2. Assume, however, that 0.65 cfm/ft^2 is the minimum acceptable value (see Section C-3.2.5) for a total of 104,000 cfm. In actual operation, when a controller senses a system reduction to this desired minimum point, the supply air temperature can be scheduled to rise to some tolerable point that still permits adequate cooling at this minimum volume. In this case, a Δt of 17.9°F, *i.e.*, a supply temperature of 58.1°F, would be necessary with a cooling coil temperature of 52.85°F as shown in the upper cycles of Figure C-2. If this coil temperature were too high or if the load were lower, the basic outdoor air radiation schedule could be elevated as a second step, after the maximum acceptable supply air temperature has been reached—all from the same flow volume-sensing device. You may assume that 58.1°F is satisfactory until its use is checked for desired performance in the special rooms, and the 104,000 cfm may be used for the first try in the modified ceiling temperature rise equation, as before:

$$\frac{0.75\,(1{,}496{,}110)}{0.5\,(160{,}000)\,(1.2)+1.1\,(104{,}000-32{,}000)} = 6.4$$

There is no need to check the heat load from the plenums, since it was predicated on an arbitrary return air rate of 104,000 – 32,000 = 72,000 cfm. However, the trial-and-error cycle must be closed by checking the amount of lighting heat directly emitted to the plenum against the assumed value of 0.75 × 1,496,000 Btu/h = 1,122,000 Btu/h, which checks very closely as shown in the following calculation.

Load from plenums to room	160,000(1.2)(3.2)	=	614,400 Btu/h
Load from plenums to return air	1.1(72,000)(6.4)	=	506,900
Total		=	1,121,300 Btu/h

5. Were it not for the need to raise the supply air temperature above the design point to control air volume, the system's percent of full load operation would have no effect on the room relative humidity, since the latter is strictly a function of that temperature, not the system load. For this particular system load, and a supply

air temperature of 58.1°F required, simple chilled-water temperature or flow control may be considered first if most or all room conditions can be satisfied with the higher dew-point temperature at the cooling coil. Terminal reheat, ceiling induction, or local fan recirculation with ceiling plenum mix may be a more economical solution (in both dollars and energy use) than main air system reheat to take care of especially difficult rooms (see lower left of Figure C-2). You can now check all room conditions from Figure C-2 without the need to plot average system conditions. Only cooling coil and supply air temperatures are relevant.

C-3.2 Room Analyses for Outdoor Conditions 65/65°F, No Sun, Electric and Occupancy Loads as Noted in Table C-1

The calculations follow the previous pattern. Special conditions are highlighted below.

1. *Room 1 at zero lighting and full occupancy.*

Room electrical load during projection from isolated booth		=	0
Plenum loads	5,000 $ft^2 \times 1.2 \times 3.2$	=	19,200 Btu/h

Completing the summation in Table C-1 as before with SHF = 0.60 requires a location for the coil temperature to find the supply air temperature. With a rise of 5.25°F (assuming that Δt will stay substantially the same with decreased air volume and a smaller temperature difference between plenum and supply air), then the coil temperature is 58.1 – 5.25 = 52.85°F. With a lower air velocity across the coil, the leaving air will be more saturated than at full load; so, in Figure C-2 the upper cycle is plotted at 52.85/51.6, and the supply air will be at 58.1/53.0. For SHF = 0.60, RH_2 = 55%.

This result warrants an interesting observation that is seldom considered by the designer when examining part-load and no-load conditions. Unless a particular room is under a roof, or over an unconditioned space, or its ceiling cavity is completely isolated from the remaining ceiling cavities on the floor, or unless all building lighting is suspended within the room, the room is never under a true no-load condition, even with no lights or occupancy. Thus, there is always a substantial plenum load when a building is occupied and, if the perimeter conduction loss is adequately treated, the floor and ceiling heat gains constitute a year-round heat gain and also act as radiant heating panels. Without this effect, this crowded lecture room without lights would have an SHF of 0.535 and a relative humidity of 59.5% (see

Figure C-2, line s-R1′). This effect is even more pronounced in lowering room humidity when recessed lights are used in a *dead* ceiling with a higher plenum temperature.

2. *Room 2 at 65% electric and 100% occupancy loads.*

The table is self-explanatory. In order to avoid overloading Figure C-2, the psychrometric conditions for this room are not shown in that figure.

3. *Room 3 at 80% electric and 100% occupancy loads.*

The conduction load is neutralized by radiation and is therefore zero.

4. *Room 4 at 25% lighting and 100% occupancy loads.*

Similarly, the conduction load is zero. The room air rate is 977 cfm, which equals 0.54 cfm/ft^2. Since this is less than the 0.65 cfm/ft^2 criterion (see Section C-3.2.5, although it meets cooling requirements) and it has an infrequent type of room use, an expedient that takes care of such a room without penalizing the entire system is worth considering. One of the simplest ways to provide a minimum of air motion in a conference room, if temperature and humidity control is not a problem, is to employ a smoke-exhaust fan at the minimum desired cfm/ft^2 value with make-up from neighboring spaces through relief grilles. This permits the job to be satisfied without any additional cost, since an exhauster is usually advisable in this type of room in any event. Psychrometrically, the 977 cfm mixes with the extra 198 cfm of relief air at room conditions when the room is occupied. This yields the same room relative humidity with a satisfactory volume of total air movement.

5. *Room 5 at 80% lighting and 100% occupancy loads.*

Similarly, this room needs 188 cfm or 0.59 cfm/ft^2. Obviously, the intense sun effect at design condition places too much of a throttling demand in a west perimeter zone during sunless periods, with a similar but less intense situation for the east exposure. During weather below 76°F, this can be taken care of with separate zoning of east and west radiation and selective reset of the radiation schedule for excess radiation, when appropriate, to maintain the minimum airflow in the shaded rooms. Thus, room 5 would need only 400 Btu/h excess radiation to maintain 0.65 cfm/ft^2. If moving shadows result in simultaneous sunlit and shaded rooms on the same exposure, with the shaded room set for radiation to maintain 0.65 cfm/ft^2, the sunlit room will have excess cooling available to neutralize a sun effect in each module (16 ft deep, 1 ft wide) in a west exposure of 1.1(2.34 – 0.65) cfm/ft^2 (76 – 58.1) = 33.3 Btu/h·ft^2, which is obviously more than required to neutralize the excess radiation and the sun effect in a sunlit room. During weather above 76°F, the problem is nonexistent because the additional conduction load brings up the sensible load and the room air rate within the criteria limits.

Since room types 2, 3, and 5 constitute practically the entire building and can be satisfied to 45.5% relative humidity or less with a supply temperature of 58.1°F, the designer should explore raising this temperature and even lowering the room temperature somewhat below 76°F in order to raise the room air rate. In the illustration given, there is no doubt that the west- and east-side exposures could be brought up to 0.65 cfm/ft^2 with supply air scheduling alone, leaving radiation sequence reset only for more stringent requirements; the tabulation is shown with excess radiation.

With effective distribution, 0.65 cfm/ft^2 can produce in excess of 10 fpm air motion in the occupied zone. Simple VAV systems under certain part-load conditions are operating satisfactorily with considerably less than that value, while others have encountered occupant complaints. The designer must use judgment in appraising these results, as well as the effect of air quality control and code restrictions during part-load operation, upon the design. Diffusers with a high induction ratio may be helpful in avoiding potential low circulation problems.

C-4 COOLING CALCULATIONS FOR WINTER PEAK LOAD, OUTDOOR CONDITIONS OF 0°F (FIGURE C-1B)

C-4.1 System Load Calculations at Peak Cooling, 100% Lighting, Occupancy, and Solar Loads

When the entire perimeter conduction and infiltration heat loss is balanced by auxiliary radiation, there is no transmission heat gain from the wall, only a solar gain.

The entire air system is on a year-round room cooling cycle except during an interior warm-up period or for interior spaces with exposed roof or floor. Therefore, no main system coil heat is required unless the minimum outdoor air mixture is lower than needed to maintain the desired supply air temperature after fan and supply duct heat gains have been added to the mixed-air temperature.

Since a high-limit humidity control problem is nonexistent below 58°F outdoor air temperature, the supply air temperature may be determined on the basis of providing enough cooling for all spaces but not so low as to create air movement or cold draft problems.

For maximum cooling conditions, take full lighting and occupancy loads and an assumed block solar gain of 530 MBh. This load has not been calculated but can be shown to be approximately 77% of the 685-MBh summer block peak.

Solar load	530,000 Btu/h
Plenum load (assumed same as on 95/75 day)	650,900
Direct lighting and electrical loads	1,076,000
Occupants sensible load	513,600
Total room sensible load	2,770,000 Btu/h

This requires 140,680 cfm at 58.1°F, the same supply temperature as that used for the 65/65°F condition. Hence, for practical purposes (since the full-load volume was 145,500 cfm), take all plenum, lighting, electrical, and occupancy values the same as for full load, and the above loads for the 95°F day can be used for the 0°F day in the Table C-1 summary. Room conditions may be examined for the extreme conditions of maximum and minimum cooling in each room without regard to room humidity.

C-4.2 Room Calculations at 0°F for Maximum and Minimum Room Cooling under Maximum System Cooling Conditions

1. *Rooms 1 and 2, Interior*

At maximum cooling, Table C-1 indicates that any interior space would require the same air quantitites and temperatures as for peak summer conditions. As a practical matter, wide experience with interior constant-volume cooling systems has shown that the supply air temperature must be several degrees higher in the winter than in the summer to avoid complaints. Therefore, assume that ordinary interior rooms, such as room 2, can be satisfied with 58.1°F supply air and the same air volumes as for the peak summer condition. Consequently, the cfm/ft^2 values for such interior offices are left at peak summer values, even though the higher supply temperature calculates out to 1.17 cfm/ft^2. The lecture room is left at the calculated higher 1.17 cfm/ft^2 peak volume because of the high occupancy concentration and its effect on indoor air quality.

For minimum cooling, the worst type of situation that could occur in general office areas would be for local air distribution to be designed for greater modular flexibility with a peak capability of handling 6 W/ft^2 lighting for full load in the Table C-1 criteria, including a 1 W/ft^2 allowance for miscellaneous electrical loads and a 60 ft^2/occupant concentration. If a particular tenant had no miscellaneous electrical load and only 4 W/ft^2 ceiling lighting with no occupancy, the only cooling load is the direct lighting load plus a reduced plenum load if the situation were to exist over the entire floor without air volume reduction. The designer is left with many choices, such as (a) accepting lower minimum air volumes, (b) raising system volume with higher coil and supply temperatures to preserve cooling capability in perimeter areas, (c) terminal reheat or induction reheat for low part-load areas, or (d) local recirculation. To carry through the simple VAV approach, it is assumed here that a combination of higher supply air temperature (controlled from system volume) and lower room temperature will be programmed to maintain a minimum 0.65 cfm/ft^2 of supply air at 61.5°F. These results are shown in Table C-1 as the two extremes with 58.1°F supply air for maximum cooling and 61.5°F for minimum cooling.

The lecture room with an assumed minimum load of no lights, 25% occupancy, and a substantial load from a common ceiling plenum would have a load of only 15,000 + 20,340 = 35,340 Btu/h. In the worst case, however, such a room would be isolated for fire protection and remain with no heat load except the 15,000 Btu/h from occupancy, shown for minimum cooling in room 1. A local fan unit, which mixes VAV system supply air and recirculated air to produce a constant 1.01 cfm/ft^2, can handle all situations (in lieu of overairing). The processes for such a unit are shown as a dotted line in the upper cycle of Figure C-2. With occupancy as the only load, process line s-R1′ is at an SHF of 240/450 = 0.53. The supply air state s1″ at 64.3°F is the mixture required for full occupancy, and s1′ at 73.1°F for 25% occupancy, both yielding room air of 76°F db, 59% rh. If this type of room suffered any heat losses (*i.e.*, exposure), then reheat should be considered to keep relative humidity below 60%.

2. *Room 3*

At minimum cooling with 61.5°F supply air, a minimum volume of 0.65 cfm/ft^2 needs a sensible load of 4,150 Btu/h to maintain a room temperature of 76°F, or 3,300 Btu/h to maintain 73°F. The former can be accomplished with 80% lighting and one occupant, while the latter can be satisfied with about 70% lighting and one occupant, both without any excess radiation or reheat. When a building facade has moving shadows, it is usually impractical to provide individual room control of radiation output or facade zoning to avoid excess radiation in the sunlit areas. It is easier to design on the basis of all areas receiving enough radiation to neutralize the minimum VAV at some supply temperature, such as this 61.5°F, when any area is sunlit without any internal load. Thus, the shaded areas will receive enough excess radiation to keep all south-facing VAV controls at the minimum air volume, while rooms in the sun require some margin of air to neutralize this radiation and maintain control on a year-round cooling cycle. Therefore, 4,150 Btu/h is used in Table C-1 and is assumed to have come from some combination of lighting and excess radiation. For example, a shaded south-facing room could conceivably be unoccupied, unlit, and without solar gain, while the remainder of the building is generally occupied. Assuming individual room VAV control, such a room would have to be heated by radiation only to handle the conduction loss plus the minimum air volume of 0.65 cfm/ft^2 tempered from 61.5 to 76°F. To design for this worst condition requires 4,150 Btu/h of excess radiation, which becomes a cooling burden in a sunlit, fully loaded identical room on the same radiation riser without individualized room radiation control. This is the reason for the 4,150 Btu/h of radiation load (excess) for room 3 and 4,100 Btu/h for room 5.

At maximum noon cooling, the winter design condition in the south governs the peak room cfm but, with true VAV, only the south-facing branches and diffusers need be sized to handle the peak, not the system fan, since air not required in other areas is probably available. With supply air at 58.1°F and the 4,150 Btu/h excess radiation for the worst-case scenario, this room needs 2.15 cfm/ft^2 to maintain 76°F instead of the summer peak of 1.85 cfm/ft^2, an increase of 0.3 cfm/ft^2. If all south-facing offices were the same, an additional volume of (17,675 ft^2 of south-facing perimeter) × 0.30 cfm/ft^2 = 5,302 cfm would be required at the period of south peak. Other calculations for full-load cooling on a 0°F day (not shown) indicate a diversity reserve of 145,000 – 121,400 = 24,100 cfm several hours later, which translates into a much greater reserve during the south peak, especially when occupancy and lighting diversity is allowed for the entire southern facade, because most of the offices have a lighting load of 4.5 W/ft^2 rather than 5.75 W/ft^2.

3. *Room 4, North Conference Room*

The north, winter, maximum cooling load without conduction gain, being considerably lower than the summer load, still has enough load to permit 0.76 cfm/ft^2. No excess radiation is required, since all north-facing rooms would receive only enough radiation to balance the conduction losses.

The minimum cooling load with 25% lighting and no solar and no occupancy needs only 0.37 cfm/ft^2 from the VAV system at 61.5°F supply air, but air movement can be kept to the minimum 0.65 cfm/ft^2 with the exhaust fan and relief from an adjacent area.

4. *Room 5*

The cooling load can vary from zero to 13,685 Btu/h with full solar, occupancy, and lights. A maximum 0°F day cooling load needs 2.87 cfm/ft^2 of 55.25°F air, while a no-load room needs 4,100 Btu/h excess radiation to provide the minimum 0.65 cfm/ft^2 with 61.5°F air. This excess radiation may be treated as described for room 3.

5. *Condition at Zero Lighting*

Although there might be an occasional occupied perimeter space without lights or plenum gains, a practical solution is excess radiation for normal rooms, such as 3 and 5, and exhausters for special rooms. It is assumed that general office interior areas are not occupied unless lights are on, and, if the entire interior were unlit and unoccupied, a full-throttling VAV terminal could go to complete shut-off. Occasional interior spaces without lights would have a plenum gain of 3.4 Btu/h·ft^2 if surrounded by spaces with an average of only 65% full-load lighting. This would result in a reduced room temperature of 66.3°F with 0.65 cfm/ft^2 of 61.5°F supply air. However, this is more tolerable than the lower temperatures encountered with a constant-air-volume system and less energy-consuming than adding reheat to the system.

Appendix D ALL-WATER SYSTEM

D.1 A PROJECT STORY

The project was a new high-rise apartment house located in a large city on the East Coast of the United States. The developer already owned several other apartment buildings in the area. He was targeting this project primarily toward childless business couples in the middle income bracket. Most apartments were to have one or two bedrooms. The engineer presented several system alternatives in the concept development phase of design. Although not the designer's first choice, the developer opted for a basic two-pipe fan-coil system. Since the developer already owned and operated such systems in other buildings, he was aware of and comfortable with the system's foibles. Tenants in those buildings chiefly complained about stuffiness and lack of air conditioning in spring and fall but not enough to affect the occupancy rates. The leases stipulated that air conditioning (cooling) would be provided from May 15 to September 15. Secure in his knowledge of system performance and costs, the client did not want to spend more money for a four-pipe system or one with water-source heat pumps. The engineering project manager reluctantly accepted the decision, but he convinced the owner to bid an alternative design that would provide positive ventilation. If the cost of this alternative was close to the base bid, the owner would accept it.

The designer was concerned that there would be inadequate cooling during spring and fall in the living areas along the southern exposure. Assuming 60 cfm of infiltration through the windows for the heat balance equation (heat gain equals heat loss), he checked the changeover point for a south-facing living room:

$$\text{Hourly heat gain, solar + Lights + People} \\ = 1.1\ (\text{Infiltration cfm})\ (t_r - t_{co}) + UA\ (t_r - t_{co})$$

Solving for the changeover temperature t_{co}:

$$t_{co} = t_r - [\text{Hourly heat gain, solar + Lights + People}]/[1.1\ \text{Infiltration cfm} + UA]$$

For September 15 at 1 P.M., he estimated the following values:

$$t_{co} = 76°F - \frac{5000\,(\text{solar}) + 300\,(\text{lights}) + 250\,(\text{people})}{1.1 \times 60 \text{ cfm} + 0.6 \times 100 \text{ ft}^2 + 0.08 \times 100 \text{ ft}^2}$$

$$t_{co} = 35°F$$

The record low for September 15 at that location was 43°F, which occurred early in the morning. Thus, if cooling were discontinued any time near this date, discomfort from excessive temperature in the living quarters was a near certainty. At this point, the engineer explained his concerns to the owner, suggesting they provide separate zoning for southern exposures to allow cooling at outdoor temperatures below those normal for September, plus the addition of supplementary heat to permit heating when the solar load would not be present but with the zone still indexed to the cooling mode. The owner rejected the recommendation because of the complexity and additional expense. The engineer confirmed in writing his concerns, his recommendation, and the owner's decision.

Fortunately, this story did not end in tragedy because the engineer proposed an alternative bid. The engineer designed a ventilation system for the apartments combined with the corridor air-conditioning system, which was part of the base design. The design included heat recovery provisions from the central toilet exhaust system to temper the ventilation supply in a rooftop air-conditioning unit. Since alternative unit selections were not burdened with the ventilation cooling load, some money was saved by selecting smaller fan-coil units and reducing pipe sizes and pump capacity. A credit was thereby obtained against the increased ventilation system cost. The additional cost for the alternative design submitted by the low bidder was $23,000, which was accepted by the owner. The engineer also received the fee for the additional design effort.

Epilogue: Three years later, the developer undertook another, larger apartment project with a new architect but requested the engineer who designed his previous project. This time, apartment ventilation was included in the base design. Supplemental electric heat was considered as an alternative.

D.2 DESIGN EXAMPLE

D.2.1 Design Parameters

The following design example concentrates on two aspects of the all-water system design—unit selection and method of providing ventilation air.

Apartment house living room, intermediate floor

Conditioned area	20 ft × 15 ft = 300 ft^2
Gross wall area	20 ft × 10 ft high = 200 ft^2
Window area	5.5 ft × 18.2 ft = 100 ft^2
Glazing and shading	Insulating glass, each light 1/4-in. thick
Heat-absorbing outer light, clear inner light, light-colored venetian blinds	
SC = 0.36, U = 0.60 Btu/h·ft^2·°F (estimated for both summer and winter)	
Wall	Insulated masonry U = 0.08 Btu/h·ft^2·°F
Peak coincident internal heat gains	
Lights, fan, and equipment	300 W
Two people, very light work	
Sensible heat gain	230 Btu/h each
Latent heat gain	190 Btu/h each

Design Criteria

Ventilation	15 cfm/person or 60 cfm to make up for kitchen exhaust
Infiltration	Assumed to be 60 cfm without ventilation supply, 0 if separate ventilation supply is included
Ventilation supply (alternative)	60 cfm at 65°F db, 62°F wb at summer peak 65°F db at other times
Design Conditions	
Outdoors	Cooling 95°F db, 75°F wb; heating 10°F db
Indoors	Cooling 76°F db, 63°F wb; heating 70°F db

D.2.2 Unit Selection

Units selected are for a two-pipe, dual-temperature system using floor-mounted, under-window fan-coil units with three-speed fans. Units must be selected for similar pressure losses so that the water distribution system will be self-balancing. A safety factor of 10% is desired in the capacity of the units.

Comments on Unit Selection (refer to Tables D-1 and D-2)

1. If the sensible heat factor (SHF) of the room is lower than the sensible heat factor of the coil, the space humidity may be slightly higher than the design value. The sensible heat factor of both room and coil will adjust until a balance is struck. The room latent load may decrease slightly as humidity rises, and the coil will condense more moisture as the dew point of the entering air increases. Under most circumstances, this is not a design concern.

Table D-1 Summary of Apartment House Living Room Loads Fan-Coil Unit Selections[a]

		N	E	S	W
Cooling Peak Loads					
Time of room peak		July 4 P.M.	July 11 A.M.	Sept. 2 2 P.M.	July 6 6 P.M.
Q_{cond} =	Conduction load	1,300	600	800	1,300
Q_{sol} =	Solar load	900	4,700	4,800	5,200
$Q_{s,\ occ}$ =	Occupancy[b] sensible load	1,500	1,500	1,500	1,500
$Q_{s,\ inf}$ =	Infiltration sensible load	1,300	600	800	1,200
Q_s =	Room sensible load	5,000	7,400	7,900	9,200
$Q_{l,\ occ}$ =	Occupancy latent load	400	400	400	400
$Q_{l,\ inf}$ =	Infiltration latent load	1,200	1,200	1,200	1,200
Q_l =	Room latent load	1,600	1,600	1,600	1,600
Q_t =	Room total load	6,600	9,000	9,500	10,800
RSHF =	Room Sensible Heat Factor	0.76	0.82	0.83	0.85
Fan-Coil Unit Selection		4	6	6	6
Airflow, cooling, high fan speed, cfm		400	620	620	600
Rows of coil based on 10°F WTR[c]		3	3	3	4
Sensible cooling capacity		6,700	9,100	9,100	11,800
Total cooling capacity		8,200	9,500	9,500	13,100
Catalog flow rate, gpm		1.6	1.9	1.9	2.7
Pressure drop, ft		6	2	2	4
Adjusting flow rate for more equal pressure loss					
Flow rate, gpm		1.6	2.7	2.7	2.7
Pressure drop, ft		6	4	4	4
Adjusted sensible cooling capacity		6,700	9,600	9,600	11,800
Adjusted total cooling capacity		8,200	10,800	10,800	13,100
Coil sensible heat factor		0.82	0.89	0.89	0.90
Heating loads without credit for internal and solar heat gain					
$Q_{cond,\ wall}$		500			
$Q_{cond,\ window}$		3,700			
$Q_{infiltration}$		4,100			
Q_t =	Room heat loss	8,300	8,300	8,300	8,300
Coil capacity at cooling, gpm		11,400	16,200	16,200	18,700

[a] All loads and capacities in Btu/h. [b] Lighting, equipment, and people. [c] Water temperature rise.

Table D-2 Summary of Apartment House Living Room Loads[a] Fan-Coil Unit Selections (If ventilation air is supplied by separate system)

	N	E	S	W
Cooling Peak Loads				
Time of room peak	July 4 4 P.M.	July 11 A.M.	Sept. 2 2 P.M.	July 6 P.M.
Q_s = Room sensible load	3,000	6,100	6,400	7,300
Q_l = Room latent load	1,000	1,000	1,000	1,000
Q_t = Room total load	4,000	7,100	7,400	8,300
RSHF = Room Sensible Heat Factor	0.75	0.86	0.86	0.87
Unit Selection	2	4	4	4
Airflow, cooling, high fan speed, cfm	175	400	400	390
Rows of coil based on 10°F WTR[b]	4	3	3	4
Sensible cooling capacity	4,000	6,700	6,700	8,000
Total cooling capacity	4,600	8,200	8,200	9,100
Catalog flow rate, gpm	1.0	1.6	1.6	1.9
Catalog pressure drop, ft	2	6	6	11
Adjusted flow rate, gpm	1.8	1.8	1.8	1.8
Adjusted pressure drop, ft	6	8	8	9
Adjusted sensible cooling capacity				7,900
Adjusted total cooling capacity				8,800
Heating Loads				
Q_t = Room heat loss	4,500	4,500	4,500	4,500
Coil capacity at cooling, gpm	7,700	15,000	11,500	13,000

[a] All loads and capacities in Btu/h. [b] Water temperature rise.

2. Few unit manufacturers design their coils for consistency of pressure drop and inherent self-balancing. By adjusting the flow values obtained from catalog ratings, a fairly tight range of unit coil pressure drops can be achieved. While ratings are presented in catalogs according to water temperature rise, there is no need to select units for those rises and the flows indicated. In actuality, the rise through any coil will be determined by the load imposed on it. The return water temperature for the system will be based on the total sum of unit flows and the block system cooling load. If only a few flow values are assigned to all the units on the entire project, the design procedure for the water distribution system is greatly simplified.

3. A ducted ventilation supply offers superior performance over the infiltration or open-window source. It adds to the project cost. Yet, as the alternative selections indicate, there is some offsetting cost benefit resulting from smaller

unit sizes and, due to the lower flow requirement, smaller piping and reduced pumping. If the ventilation air is adequately dehumidified, very little latent cooling is required of the fan-coil units. If the water supply to the units can be designed for a temperature of 50°F or higher, little condensation on the coils should occur. This is desirable to reduce dependence on the drain pans and condensate removal piping, which are vulnerable to clogging. Moreover, a wet coil is going to collect more dirt, which then becomes a moist medium for biological growth, odor collection, and subsequent emissions into the airstream.

4. Note that the factor of safety for the west-facing living room following unit flow adjustment is less than the desired 10%. You have several options: increase the unit size, accept a lower factor of safety, increase water flow for all units, or lower the entering water temperature. The answer must reside with your judgment, weighing the specific project conditions and the cost implications.

5. The chilled-water supply temperature is tentatively selected at 45°F, based on past experience. A higher temperature generally results in somewhat larger units and higher space relative humidity; however, the refrigeration plant could be smaller and have a higher COP. The opposite results occur with lower temperatures; therefore, a balance is sought. The selected temperature is also well suited to selecting cooling coils and air-handling units that serve other areas of the building.

   As the fan-coil units are being selected, a slight adjustment of the supply chilled-water temperature may produce less costly selections.

6. Heating capacities of the unit coils selected for cooling are usually more than adequate because of the higher temperature difference between water and room. You may elect to evaluate the heating capacity at reduced flow and to consider operation with reduced pumping capacity during the heating mode. Fine-tuning the heating water supply temperature usually results in very limited initial cost reduction for the effort. In this example, 110°F has been chosen for the heating water supply temperature and is somewhat arbitrarily based on past practice.

Appendix E LIQUID DESICCANT SYSTEM WITH ENGINE CHILLER AND FAN-COIL TERMINALS

This appendix describes the procedure for designing a typical liquid dessicant HVAC system, which is shown psychrometrically in Figure E-1 and schematically in Figure E-2. This system and a conventional all-air VAV system is also compared.

The system incorporates a single-stage desiccant unit that dehumidifies 0.2 cfm per square foot of outside air. The desiccant (in this example, a lithium chloride solution) is sprayed over the incoming outside airstream. The dehumidified air is distributed at constant volume to fan-coil terminals throughout the building. Evaporatively cooled exhaust air cools the desiccant in order to remove the heat of sorption. A gas-engine-driven chiller provides 55°F chilled water for sensible cooling at terminals as well as recovered engine heat for desiccant regeneration (summer) and hot water for perimeter terminal heating coils (winter). A hot water boiler supplements the recovered engine heat.

In winter, the dehumidifier serves as a humidifier. Water is added to the desiccant solution. The incoming outside air, preheated to 80°F, absorbs moisture from the desiccant solution spray as it flows through the desiccant unit.

Basic Data

Building type	6-story office building
Dimensions	240 ft × 120 ft
Building glass area	30%
Building location	Southern New Jersey, 40°N
Lighting load	2.0 W/ft^2
Floor power load	1.5 W/ft^2
Occupancy	1 person/100 ft^2
Ventilation rate	20 cfm/person or 0.2 cfm/ft^2
Occupied floor area	143,000 ft^2
Perimeter zone floor area	67,000 ft^2
Interior zone floor area	76,000 ft^2
Gross floor area	159,000 ft^2
Winter design temperature	10°F
Summer design conditions	88°F db, 76°F wb (118 grain/lb)

Indoor design temperature		75°F all year
Indoor relative humidity		Max. 50%, min. 30%
Occupied hours		12 hours/day, 5 days/week
Electric utility tariff		
Demand:	June-September	$9.14/kW
	October-May	$7.94/kW
Use:	On-peak (8 A.M.-7 P.M.)	$0.08757/kWh
	Shoulder (weekends)	$0.07662/kWh
	Off-peak (7 P.M.-8 A.M.)	$0.05848/kWh
Gas utility tariff		
	June-September	$5.06/million Btu
	October-May	$5.45/million Btu

Summer Cooling Loads

			Conventional All-Air VAV System, Btu/h	Desiccant System, Btu/h
1.	Transmission and solar peak coincident envelope load (calculations not shown)		2,088	2,088
2.	Electrical load			
	Lighting load	143,000 × 2.0 × 3.4	972	972
	Equipment load	143,000 × 1.5 × 3.4	729	729
3.	Occupant load	(143,000/100 = 1,430 persons)		
	Sensible	1,430 × 250	358	358
	Latent	1,430 × 200	286	286
4.	Ventilation	(143,000 × 0.2 = 28,600 cfm)		
	Sensible	28,600 × 1.08(88 – 75)	402	402
	Latent (50% indoor rh)	28,600 × 0.69(118 – 65)		1,006
	Latent (40% indoor rh)	28,600 × 0.69(118 – 52)		1,263
5.	Fan heat	(3°F temperature rise)	(to be determined later)	

Conventional All-Air VAV System

6. Total air quantity
   1. + 2. + 3. = 4,147,000/(1.08 × 20) — 192,000 cfm
7. Fan heat
   192,000 cfm × (3°F) × 1.08 — 622,000 Btu/h
8. Total system refrigeration load
   1. + 2. + 3. + 4. + 7. — 6,463,000 Btu/h= 539 tons

Integrated Desiccant System

9. Primary air quantity — 28,600 cfm
10. Primary air load (to space at 80°F, 37.5 gr/lb, see note 3 below)
    Sensible load — 28,600 × 1.08 × (80 – 75) — 154,000 Btu/h
    Latent load — 28,600 × .69 × (37.5 – 52) — –286,000 Btu/h
11. Secondary air quantity
    1. + 2. + 3. + 10. = 4,301 × 10^6 Btu/h/(1.08 × 17) — 234,000 cfm
12. Secondary fan heat gain
    1.08 × (1°F) × 234,300 — 253,000 Btu/h
13. Total system refrigeration load
    1. + 2. + 3. + 10. + 12. — 4,554,000 Btu/h = 380 tons

DESIGN PROCEDURE

1. Establish zoning arrangement for the building and its spaces.

2. Calculate peak heat gain/loss for each zone as well as the building peak coincident cooling/heating loads using ASHRAE recommended procedures (steps 1 through 4).

3. Determine the primary air moisture level required to do the space dehumidification. In most cases, the target humidity level will be 40% rh (which equals 52 gr/lb at 75°F). (For the comparative conventional VAV system, the moisture content of 50% rh at 75°F is equal to 65 gr/lb.) This information is used in step 4. From

$$\text{Btu} = \text{cfm} \times 0.69 \times (\Delta \text{ grain/lb})$$

   the required primary moisture level is

$$52 - [286{,}000/(28{,}600 \times 0.69)] = 37.5 \text{ gr/lb}$$

4. Selection of the liquid desiccant system components is fairly straightforward once some constraints are set. The design outdoor conditions and the desired supply air moisture level are known. Since the desiccant solution will be cooled by exhaust air and not by refrigeration, the capacity of the evaporative fluid cooler becomes the limiting factor. Given the cfm and the conditions of the exhaust air, the desiccant manufacturer will size the fluid cooler, the conditioner, and the regenerator and will provide data on desiccant pump sizes, supply air temperature, and the heat required for regeneration of the dilute desiccant. Be sure to obtain these data for a range of outdoor conditions. The manufacturer will also provide information regarding the heat required at the preheat coil for winter preheating and humidification. Working closely with the manufacturer will ensure consistent supply air conditions throughout the range of outdoor air conditions.

   For this example, the following information was supplied by the manufacturer. Supply air condition: 80°F, 37.5 gr/lb (used in step 10). The fluid cooler delivers desiccant to the conditioner at 75°F and it leaves the conditioner at 99°F. The desiccant concentration is 37.5% in the conditioner and 42% in the regenerator. It requires 3,190,000 Btu/h of regeneration heat and 9,500 cfm of regeneration air at design conditions.

5. Select the zone fan-terminal units. They should be blow-through units. Interior units include cooling coils only, while perimeter zone units include four-pipe

cooling and heating coils. When totaling the sensible cooling loads, remember to include contributions from the terminal fan (1°F temperature rise) and from the primary air (supply temperature from manufacturer). If possible, select coils with less than four rows, using a chilled-water inlet temperature of 55°F and an air-side Δt of 15 to 20°F. For heating, use an inlet temperature of 140°F and an air-side Δt of 20°F.

6. Lay out and size the secondary ductwork using any ASHRAE-approved method, being careful not to exceed the rated external static pressure of the terminal units. Select diffusers and place them according to manufacturer's data.

7. Lay out and size the primary ductwork using any ASHRAE-approved method. The network should be designed as a medium-pressure, high-velocity system to minimize duct sizes. Air delivery to the terminal units can be by hard connection to the inlet end of the terminal or by discharging into the plenum near the inlet end of the box.

8. Select the primary fan(s) based on the pressure drop of the distribution system (step 7) and the pressure drop of the desiccant conditioner (from the manufacturer). Remember to include louvers, filters, and so forth.

9. Lay out and size hot and chilled-water distribution systems as in any other system. The control method of the terminal units should be two-way modulating instead of three-way modulating. This allows the option of not insulating the return chilled-water piping, because low space and plenum moisture levels will not promote condensation on pipes at 60°F or higher.

10. Select an engine-driven chiller based on the system peak refrigeration load (see load calculations) using a condenser temperature of 85 to 90°F, a chilled-water supply temperature of 50 to 53°F (based on 55°F at terminals), and a chilled-water Δt of 10 to 12°F. Consult the manufacturer's data to determine whether the amount of recoverable heat from the engine can satisfy the heat requirements of the desiccant regenerator throughout the range of expected outside air temperatures. Remember that the requirement for regeneration heat remains constant or increases as chilled-water and heat output drop off.

    Because the heat available at design (1,773,000 Btu/h) is significantly less than the heat required (3,190,000 Btu/h) at regeneration, select an air-to-air heat recovery system to reduce the net heat input at design (approximately 15% reduction) and to stabilize the ratio of heat input to moisture removed for all outside (regeneration) air temperatures. Keep in mind that regeneration of the liquid desiccant is not continuous but cyclical. Careful consideration of this batch process will determine whether storage of the cogenerated heat is warranted. (For this example, all the available heat from the chillers could be used if the regeneration process were continuous, but in reality only 45 to 50% of the available heat can be recovered.) The system should be arranged so that engine heat can be recovered to energize the perimeter hot water system and regeneration heat can be supplied by the system boiler.

11. Select a boiler based on the building's peak heating load, using appropriate safety and pickup factors. Check the output versus the heat required by the regenerator. Under all operating conditions, the boiler output must be more than adequate.

12. Select all pumps to handle design flow rates considering friction losses

$$1\ \text{Btu/(lb}\cdot{}^\circ\text{F)} \times 60\ \text{min/h} \times 8.34\ \text{lb water/gal} = 500\ \text{Btu/h} \times (\text{gpm} \times \Delta t)^{-1}$$

This equation is used to determine the required water flow when heat flow and Δt are given.

13. Select an open cooling tower by consulting manufacturer's data or representatives. Use a range of 10 to 15°F and a leaving temperature of 85 to 95°F. Be sure to specify freeze protection, since the tower will be used during the winter for free cooling of the chilled-water loop.

14. Figure E-1 on page E-8 is a psychrometric chart that illustrates this example.

Calculation Summary

EQUIPMENT

Conventional VAV System

Fans		
Supply fans	total 192,000 cfm	106 hp
Return fans	total 180,000 cfm	120 hp
Refrigeration		
Compressors	total 539 tons at 1.1 kW/ton	593 kW
Condenser fans		60 hp

Liquid Desiccant System

Fans		
Terminal unit fans	total 234,300 cfm (1.6 cfm/ft^2)	90 hp
Primary air fan	28,600 cfm at 4.5 in. static pressure	30 hp
Regeneration fan (from desiccant manufacturer)	9,500 cfm at 2.5 in. static pressure	10 hp
Return air fans	28,600 cfm at 3.0 in. static pressure	20 hp

Pumps		
Chilled-water pump, P-1	4,554,000 Btu/h/(500 × 12) = 760 gpm	40 hp
Hot water pump, P-2	2,800,000 Btu/h/(500 × 15) = 375 gpm	15 hp
Condenser water pump, P-3	6,050,000 Btu/h/(500 × 15) = 800 gpm	20 hp
Liquid desiccant conditioner pump, P-4	(from desiccant manufacturer)	10 hp
Liquid desiccant regenerator pump, P-5	(from desiccant manufacturer)	7.5 hp
Hot water recovery pump, P-6	3,190,000 Btu/h/(500 × 20) = 320 gpm	10 hp
Engine chiller		
Output, chilled water	4,554,000 Btu/h (380 tons) at 50°F	
Output, hot water	1,773,000 Btu/h at 240°F	
Input, gas	3,190 ft³/h	
Cooling tower	6,050,000 Btu/h, 800 gpm, 15°F Δ*t*	
Fan	30 hp	
Desiccant subsystem		
Heat input (from manufacturer)	(319 gpm) × (20°F Δ*t*) × 500	3,190,000 Btu/h
Supplemental heat required	3,190 – 1,773 (from chiller)	1,417,000 Btu/h

ENERGY USE AND COST ANALYSIS

	Conventional All-Air VAV System	Desiccant System
Electric demand (kW)		
Summer day peak	897	257
Winter day average	451	240
Annual electric use (MWh)		
Fans	642	605
Pumps	—	342
Refrigeration (including condensers)	680	58
Electric resistance heat	1,388	0
Total	2,710	1,005
Annual gas use (million Btu)		
Chiller	—	4,553
Desiccant regeneration	—	2,966
Space heat	—	6,455
Total	0	13,964
Annual electric demand cost	$61,400	$24,600
Annual electric energy cost	$201,900	$81,900
Annual gas cost	0	$73,200
Total	$263,300	$179,700
Annual utility cost per ft²	$1.66	$1.13

COMMENTS

In this example, the liquid desiccant system permits a reduction in the size of the central refrigeration plant—from 539 to 380 tons—when compared to a conventional all-air VAV system. This reduction is primarily due to removing the latent load from the chiller. The annual energy cost of the desiccant system is \$0.53 per ft^2 less than that of a conventional all-air VAV system.

The desiccant system also permits a major reduction in the size of primary ducts and fans and in the fan energy required for air distribution based on distributing 0.2 cfm per ft^2 of primary air instead of 1.4 cfm per ft^2. However, it requires the addition of chilled-water piping from the central plant to all terminals.

The engine chiller was incorporated because the desiccant system can use both engine heat (for regeneration) and chilled water (for sensible cooling). Alternatively, the system could include a 380-ton electrically driven chiller and a larger hot water boiler. The choice depends upon the balance between the differences in energy cost and first cost and is strongly dependent on the utility rates in effect at the building location.

Point	Description	°F	grains/lb
1	Outside air entering primary fan	88	118
2	Air entering desiccant dehumidifier	91	118
3	Air leaving dehumidifier, supplied to terminals	80	38
4	Mixture of dry air and plenum air	77.5	50
5	Air leaving terminal supply fan	78.5	50
6	Air leaving sensible cooling coil, supply to space	60	50
7	Space conditions (40% rh)	75	52
8	Plenum conditions	77	52

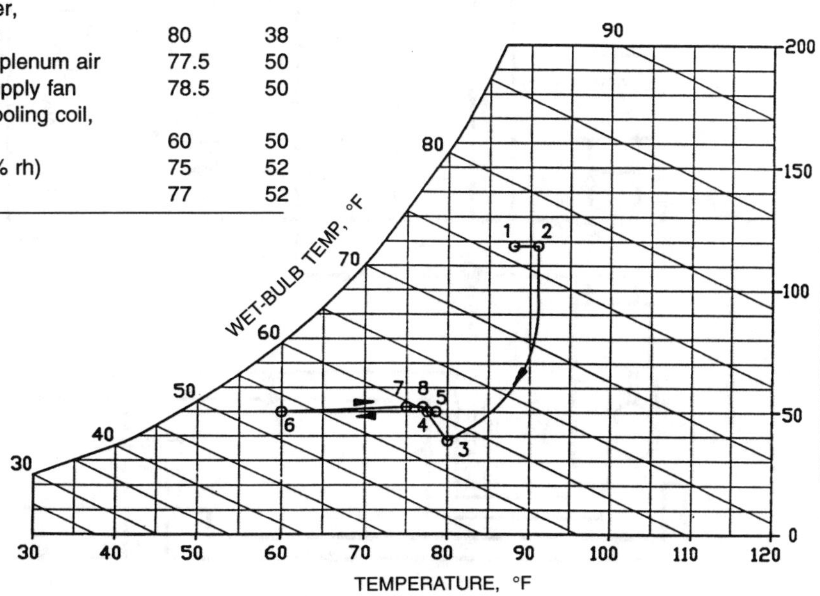

Fig. E-1 Psychrometric Chart for Example

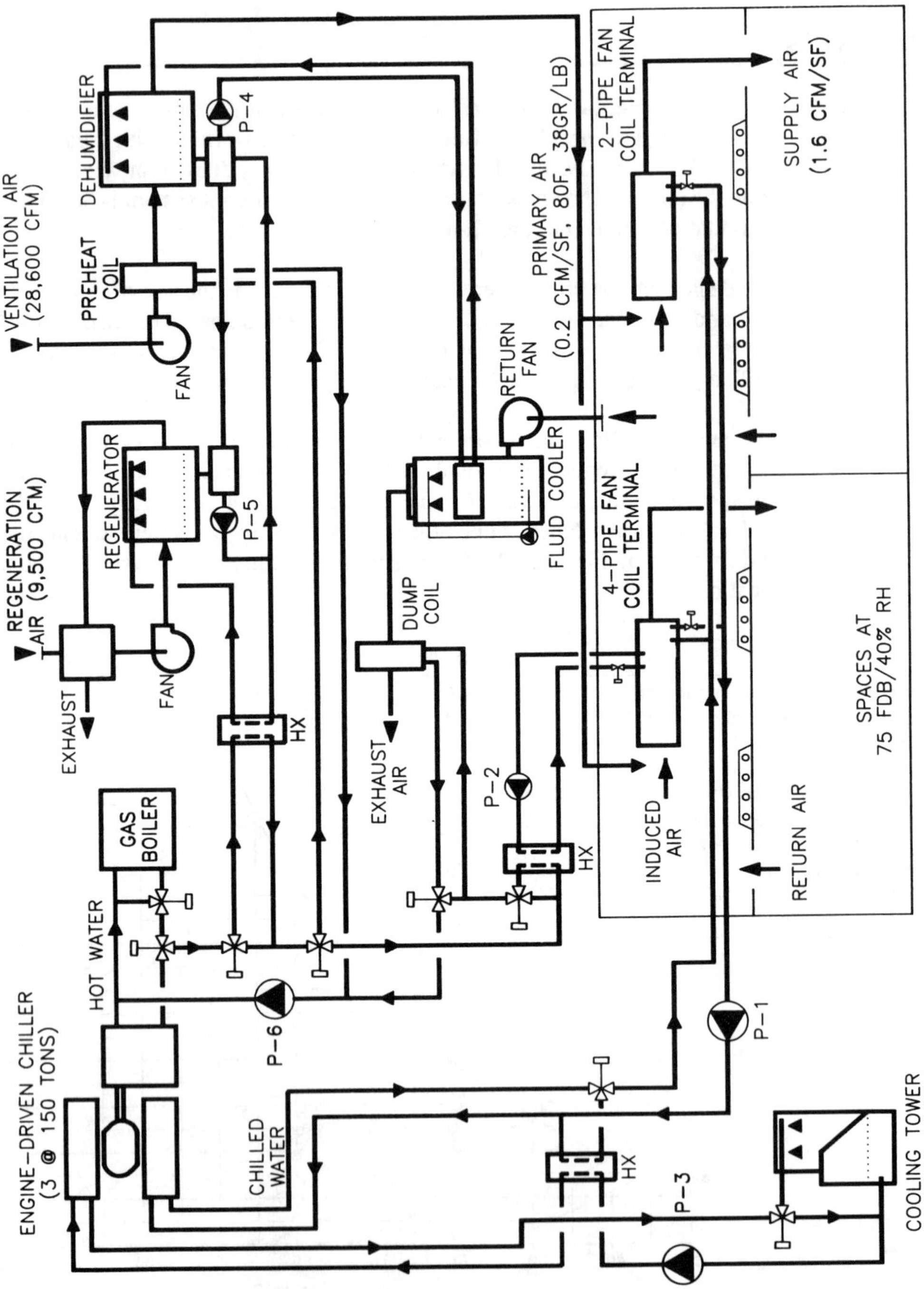

Fig. E-2 Schematic of Liquid Dessicant/Engine-Driven Chiller System

Appendix F PRELIMINARY COMPARISON OF CONVENTIONAL AND ICE STORAGE SYSTEMS FOR FOUR-STORY OFFICE BUILDING

F-1 EXECUTIVE SUMMARY

This is a four-story, 264,200-ft^2 office building designed to operate as a low-energy-use building with an energy budget of 22.7 kWh/ft^2 (77,500 Btu) of floor area per year. Energy costs are estimated to be $237,000 per year or $0.90/ft^2/year (based on Ontario Hydro utility rate schedules, expressed in Canadian dollars).

The energy analysis compared a *conventional variable-air-volume (VAV) system and water chiller* with a *low-temperature-air VAV system and ice storage.* The ice storage system will save $2,200 per year based on current utility rates. The ice storage system's capital cost is estimated to be equal to or less than that of the conventional system, according to the mechanical designers for the building.

A number of design strategies can reduce the energy costs below the budget of $237,800 per year. Among the options are:

- Circuiting of perimeter zone lights for daylighting and peak cooling load control.
- Computerized switching of lighting circuits or low-voltage switching.
- Energy-efficient motors.
- Variable-speed pumping.

- Discriminator control for supply air temperature.
- Building automation system with direct digital control (DDC).

F-2 BUILDING CHARACTERISTICS

The office building structure incorporates east and west office wings linked by a circular rotunda area. The total conditioned floor area is 264,200 ft^2. The conditioned floor area by floor is summarized as follows:

Concourse level	67,100 ft^2
Ground level	53,500 ft^2
Second level	50,400 ft^2
Third level	48,200 ft^2
Fourth level	45,000 ft^2
Total floor area	264,200 ft^2

The building is divided into discrete thermal zones for the computer simulation of energy use. Five zones are used in the energy simulation:

	Floor Area	Roof Area	Wall Area	Glass Area
Rotunda	52,350	11,250	11,064	5,188[a]
North zone	33,900	6,780	10,290	9,016
South East	20,125	4,025	12,000	9,475
South West	20,125	4,025	12,400	8,962
Interior	137,700	27,540	—	—
Totals	264,200	53,620	45,754	32,641

[a] 5,188 ft^2 includes skylight glazing.

The interior zone requires only cooling and ventilation since this zone has no significant envelope losses or gains; the other four zones all require heating and cooling.

Thermal characteristics of the walls, roof, and windows are summarized as follows:

Wall	R20
Roof	R20
Glass	R3 with 0.40 shading coefficient

The occupancy loads for the building are:

Lights	1.5 W/ft^2
Tenant equipment	1.0 W/ft^2
Miscellaneous electric	0.5 W/ft^2
People	1 person/150 ft^2

F-3 PRELIMINARY ENERGY BUDGET

The preliminary energy budget for the building is based on the characteristics and operation schedules predicted for the building. The energy budget is summarized for the ice storage HVAC system and the conventional chiller HVAC system.

Energy End Use	Conventional, kWh/(ft^2·yr)	Percentage, %	Ice Storage, kWh/(ft^2·yr)	Percentage, %
Heating	5.2	23	5.2	23
Cooling	1.3	6	2.1	9
Fans and pumps	2.0	9	1.7	7.5
Electric equipment	6.3	28	6.3	28
Lighting	6.6	30	6.6	29
Domestic hot water	0.8	4	0.8	3.5
Energy budget (elec.)	22.2 kWh/(ft^2·yr)		22.7 kWh/(ft^2·yr)	
Energy cost	$240,000/yr		$237,800/yr	

The utility rates used for the energy simulation are:

Electricity	
Demand	$3.68/kW
Energy	3.69¢/kWh on-peak 2.50¢/kWh off-peak

Peak period is defined as 7 A.M. to 9 P.M. Monday through Friday.

Natural Gas	
Demand	Not applicable
Energy	$4.50 per million cubic feet

F-4 END-USE ANALYSIS

Figure F-1 provides a breakdown of the total annual energy costs into the major energy end-uses: space cooling, heating, water heating, operation of fans and pumps, lighting, and other equipment (*e.g.*, appliances, computers, and process equipment). The cooling costs are calculated for a conventional cooling system. This information is useful for pinpointing opportunities for savings. Detailed summaries of the energy simulation program inputs and printouts are shown in Tables F-1 and F-2. Note that they include *all* energy uses, not just HVAC.

Annual End-Use Breakdown—Conventional

End Use	Cost, $	% Total	Energy Use	Demand
Cooling	17,338	7	346,751 kWh	492 kW
Heating	22,641	9	47,028 therm	42 therm/h
Hot water	3,516	1	7,301 therm	2 therm/h
Fans and pumps	25,703	11	514,038 kWh	310 kW
Electric equipment	83,577	35	1,671,607 kWh	357 kW
Lighting	87,239	36	1,744,862 kWh	396 kW
Total	$240,014	100		

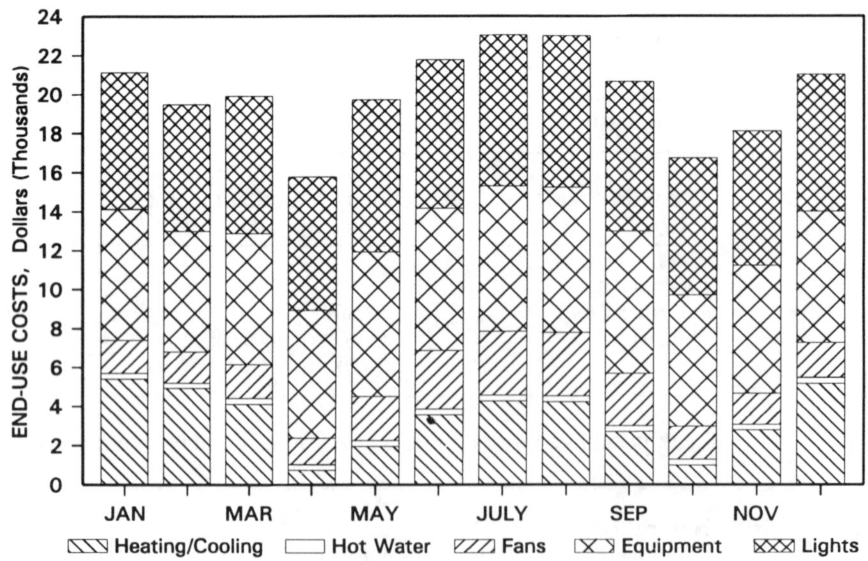

Fig. F-1 End-Use Costs—Conventional

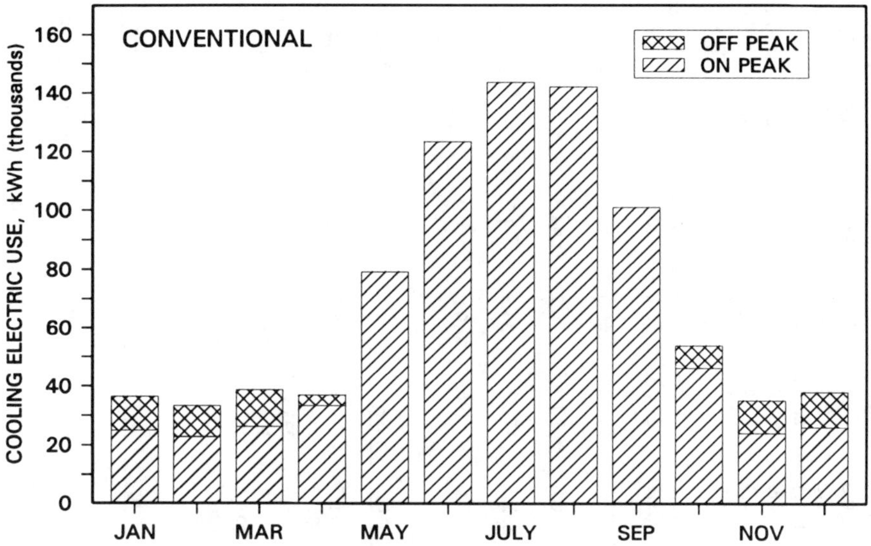

Fig. F-2 Cooling Electric Use—Conventional

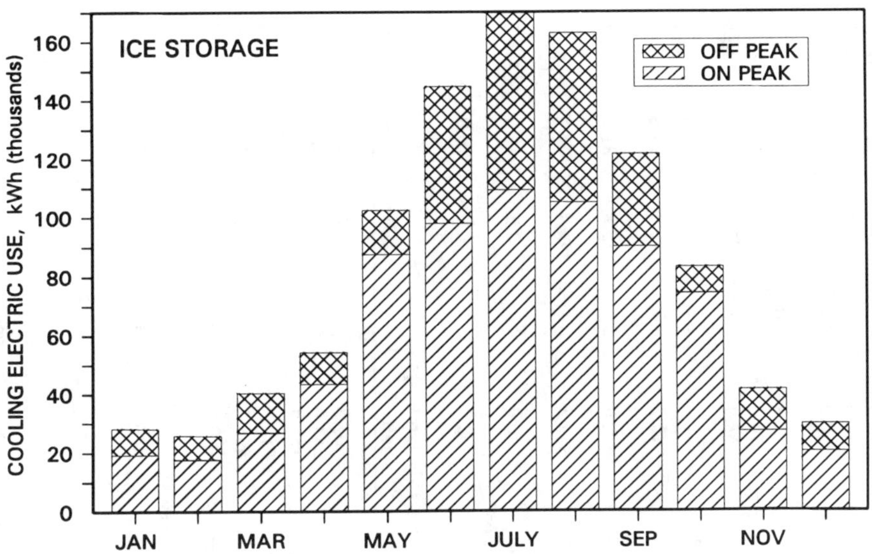

Fig. F-3 Cooling Electric Use—Ice Storage

Table F-1 Conventional System

HVAC Systems Performance (Million Btu or Million cfm · hr)

System #1	Jan	Feb	Mar	Apr	May	Jun	Jul	Aug	Sep	Oct	Nov	Dec	Year
Cooling	0.0	0.0	7.1	29.0	123.2	228.8	277.9	274.4	174.2	63.1	6.0	0.0	1,183.8
Heating	1,114.6	1,017.5	630.1	70.2	15.9	0.0	0.0	0.0	4.5	23.3	557.5	1,069.4	4,702.8
Heat pump	0.0	0.0	0.0	0.0	0.0	0.0	0.0	0.0	0.0	0.0	0.0	0.0	0.0
Fans and pumps	38.6	35.7	36.0	21.9	61.0	97.6	106.4	106.1	86.0	44.7	25.6	39.1	698.3
Hot water	61.8	56.0	62.2	59.7	62.1	59.8	61.9	62.0	60.2	62.0	60.3	62.1	730.1
Supply fan energy	70.3	63.8	72.5	59.6	68.6	74.8	83.9	82.8	67.6	61.0	71.7	74.0	850.6
Supply airflow	58.9	53.3	57.1	50.1	46.9	44.8	48.1	47.9	44.1	45.6	53.6	60.2	610.2
Outstanding airflow	31.6	28.3	34.7	37.7	33.2	22.4	18.6	18.2	27.8	38.8	36.3	32.8	360.8
Return fan energy	15.2	13.7	16.3	14.8	17.4	19.5	22.1	21.8	17.2	15.0	16.7	16.2	206.1
Sensible cooling load	0.0	0.0	15.8	81.0	407.1	819.7	1,032.0	1,021.6	590.2	166.5	12.2	0.0	4,146.4
Sensible heating load	588.4	530.4	428.3	191.5	56.4	0.2	0.0	0.0	20.9	74.2	294.2	546.0	2,730.2
Latent cooling load	0.0	0.0	2.3	18.5	118.6	255.7	337.7	332.5	172.7	34.3	1.3	0.0	1,273.7
Humidification load	2.3	2.7	0.0	0.0	0.0	0.0	0.0	0.0	0.0	0.0	0.0	0.3	5.5

Building Loads (Million Btu)

Item	Jan	Feb	Mar	Apr	May	Jun	Jul	Aug	Sep	Oct	Nov	Dec	Year
Transmission	−502.2	−449.9	−394.0	−255.3	−199.1	−136.8	−108.9	−111.5	−179.3	−248.0	−300.6	−448.6	−3,334.4
Infiltration	−56.0	−50.0	−58.4	−64.7	−64.0	−41.2	−26.3	−32.7	−51.3	−75.6	−60.5	−50.2	−631.0
Solar loads	170.7	154.9	188.5	202.3	228.6	235.7	247.4	246.7	231.0	222.7	192.3	178.2	2,499.1
Equipment gain	314.3	283.8	315.5	303.3	314.3	303.6	314.2	314.5	304.5	314.7	305.2	315.1	3,702.5
Lighting gains	495.2	447.5	497.5	477.9	495.9	478.7	495.4	495.8	480.4	496.2	481.5	496.9	5,838.3
People gains	106.0	95.8	106.7	102.4	106.2	102.5	106.1	106.2	102.8	106.2	103.1	106.4	1,249.8

Operating Costs ($)

End Use	Jan	Feb	Mar	Apr	May	Jun	Jul	Aug	Sep	Oct	Nov	Dec	Year
Cooling or heat pump 1	0	0	99	407	1,886	3,565	4,269	4,215	2,710	876	84	0	17,338
Cooling or heat pump 2	0	0	0	0	0	0	0	0	0	0	0	0	0
Heating system 1	5,367	4,899	3,997	338	76	0	0	0	21	113	2,684	5,149	22,641
Heating system 2	0	0	0	0	0	0	0	0	0	0	0	0	0
Fans and pumps	1,721	1,606	1,726	1,350	2,249	2,989	3,263	3,236	2,658	1,674	1,590	1,793	25,703

Hot water system 1	298	270	300	287	299	286	298	299	289	299	290	299	3,516
Hot water system 2	0	0	0	0	0	0	0	0	0	0	0	0	0
Electrical equipment	6,718	6,205	6,731	6,550	7,420	7,288	7,438	7,445	7,307	6,725	6,569	6,729	83,577
Process equipment	0	0	0	0	0	0	0	0	0	0	0	0	0
Lights	7,003	6,479	7,026	6,832	7,754	7,604	7,764	7,768	7,634	7,016	6,861	7,024	87,239
Total operating cost	21,107	19,459	19,878	15,764	19,685	21,732	23,031	22,963	20,620	16,702	18,078	20,993	240,014

Energy Consumption

End Use	Unit	Jan	Feb	Mar	Apr	May	Jun	Jul	Aug	Sep	Oct	Nov	Dec	Year
Cooling or heat pump 1	kWh	0	0	2,093	8,498	26,096	67,031	81,403	80,376	51,015	18,497	1,763	0	346,751
Cooling or heat pump 2	kWh	0	0	0	0	0	0	0	0	0	0	0	0	0
Heating system 1	Therm	11,146	10,175	8,301	702	159	0	0	0	45	233	5,575	10,694	47,028
Heating system 2	kWh	0	0	0	0	0	0	0	0	0	0	0	0	0
Fans and pumps	kWh	36,318	33,167	36,532	28,207	43,042	56,193	62,217	61,701	50,035	35,360	33,353	37,893	514,038
Hot water system 1	Therm	618	560	622	597	621	598	619	620	602	620	603	621	7,301
Hot water system 2	kWh	0	0	0	0	0	0	0	0	0	0	0	0	0
Electrical equipment	kWh	141,751	128,149	142,448	136,806	142,012	137,041	141,829	141,954	137,554	142,036	137,848	142,267	1,671,607
Process equipment	Therm	0	0	0	0	0	0	0	0	0	0	0	0	0
Lights	kWh	147,766	133,804	148,890	142,688	148,407	142,987	148,045	148,110	143,714	148,197	143,972	148,491	1,744,862
Total														
Electric	kWh	325,835	295,103	329,780	316,209	369,557	403,252	433,494	432,127	382,319	344,112	316,935	328,672	4,277,437
Gas	Therm	11,765	10,735	8,923	1,299	781	596	619	620	646	853	6,178	11,315	54,329

Peak Electric Demand (kW)

End Use	Jan	Feb	Mar	Apr	May	Jun	Jul	Aug	Sep	Oct	Nov	Dec
Cooling or heat pump 1	0	0	0	0	339	478	792	492	415	0	0	0
Cooling or heat pump 2	0	0	0	0	0	0	0	0	0	0	0	0
Heating system 2	0	0	0	0	0	0	0	0	0	0	0	0
Fans and pumps	67	67	67	57	235	296	310	306	281	88	66	67
Hot water system 2	0	0	0	0	0	0	0	0	0	0	0	0
Electrical equipment	357	357	357	357	357	357	357	357	357	357	357	357
Lights	396	396	396	396	396	396	396	396	396	396	396	396

Table F-2 Storage System

HVAC Systems Performance (Million Btu or Million cfm·hr)

System #1	Jan	Feb	Mar	Apr	May	Jun	Jul	Aug	Sep	Oct	Nov	Dec	Year
Cooling	0.0	0.0	30.2	69.1	211.2	316.5	389.5	366.4	259.2	161.7	38.3	0.3	1,862.1
Heating	1,180.0	1,077.6	881.6	78.5	17.8	0.0	0.0	0.0	5.0	26.0	593.8	1,137.2	4,997.4
Heat pump	0.0	0.0	0.0	0.0	0.0	0.0	0.0	0.0	0.0	0.0	0.0	0.0	0.0
Fans and pumps	38.6	35.7	48.3	39.5	79.1	115.6	121.1	121.3	99.7	69.0	43.9	39.3	850.7
Hot water	61.8	56.0	62.2	59.7	62.1	59.8	62.0	62.0	60.1	62.0	60.2	62.1	730.0
Supply fan energy	49.1	44.6	50.5	47.3	48.4	49.9	55.8	55.1	46.2	44.2	49.7	51.8	592.7
Supply airflow	40.5	36.7	39.5	34.8	32.7	30.9	33.1	32.9	30.6	31.9	37.1	41.4	421.8
Outstanding airflow	31.4	28.2	32.9	32.4	26.9	18.6	16.7	16.5	22.2	30.3	33.8	32.7	322.7
Return fan energy	8.4	7.7	9.2	9.4	10.4	11.3	12.9	12.7	10.0	9.0	9.5	9.2	119.4
Sensible cooling load	0.0	0.0	49.2	186.1	503.8	833.8	1,002.79	91.4	652.0	318.4	56.8	0.0	4,594.6
Sensible heating load	644.9	581.6	476.6	220.4	67.9	0.6	0.1	0.1	24.6	87.6	336.1	602.5	3,043.1
Latent cooling load	0.0	0.0	22.6	84.6	251.7	411.5	480.5	470.5	332.6	176.4	22.9	0.0	2,253.4
Humidification load	1.5	1.8	0.0	0.0	0.0	0.0	0.0	0.0	0.0	0.0	0.0	0.2	3.2

Building Loads (Million Btu)

Item	Jan	Feb	Mar	Apr	May	Jun	Jul	Aug	Sep	Oct	Nov	Dec	Year
Transmission	-501.9	-449.5	-393.3	-254.0	-198.0	-136.6	-108.7	-111.4	-178.7	-246.3	-299.9	-448.1	-3,326.5
Infiltration	-56.0	-49.9	-58.3	-64.7	-63.9	-41.2	-26.3	-32.8	-51.3	-75.6	-60.6	-50.2	-631.5
Solar loads	170.8	154.9	188.4	202.3	228.6	235.7	247.4	246.7	231.0	222.7	192.2	178.1	2,499.3
Equipment gain	314.2	283.8	315.5	303.2	314.2	303.6	314.2	314.6	304.5	314.7	305.2	315.2	3,702.0
Lighting gains	495.2	447.5	497.5	477.9	495.9	478.7	495.4	495.8	480.3	496.2	481.4	496.8	5,838.5
Peole gains	106.0	95.8	106.5	102.4	106.2	102.5	106.1	106.1	102.8	106.3	103.0	108.3	1,250.3

Operating Costs ($)

End Use	Jan	Feb	Mar	Apr	May	Jun	Jul	Aug	Sep	Oct	Nov	Dec	Year
Cooling or heat pump 1	0	0	416	1,225	3,042	4,446	5,365	5,063	3,700	2,350	530	5	26,115
Cooling or heat pump 2	0	0	0	0	0	0	0	0	0	0	0	0	0
Heating system 1	5,681	5,189	4,245	378	85	0	0	0	24	125	2,859	5,476	24,064
Heating system 2	0	0	0	0	0	0	0	0	0	0	0	0	0
Fans and pumps	1,339	1,254	1,489	1,323	1,985	2,484	2,615	2,613	2,225	1,777	1,426	1,394	21,932
Hot water system 1	298	269	300	288	299	288	298	298	290	298	290	299	3,512

Hot water system 2	0	0	0	0	0	0	0	0	0	0	0	0	0
Electrical equipment	6,744	6,231	6,704	6,423	6,984	6,573	6,670	6,696	6,705	7,048	6,504	6,758	80,047
Process equipment	0	0	0	0	0	0	0	0	0	0	0	0	0
Lights	7,031	6,506	6,999	6,999	7,299	6,858	6,962	6,986	7,005	7,353	6,793	7,054	83,558
Total operating cost	21,093	19,448	20,152	16,336	19,695	20,649	21,911	21,658	19,949	18,950	18,402	20,985	239,227

Energy Consumption

End Use	Unit	Jan	Feb	Mar	Apr	May	Jun	Jul	Aug	Sep	Oct	Nov	Dec	Year
Cooling or heat pump 1	kWh	0	0	8,833	26,097	61,860	92,705	114,076	107,328	75,918	47,353	11,218	102	545,429
Cooling or heat pump 2	kWh	0	0	0	0	0	0	0	0	0	0	0	0	0
Heating system 1	Therm	11,800	10,776	8,816	785	178	0	0	0	50	260	5,936	11,372	49,974
Heating system 2	kWh	0	0	0	0	0	0	0	0	0	0	0	0	0
Fans and pumps	kWh	28,136	25,781	31,640	28,176	40,364	51,789	55,602	55,389	45,646	35,804	30,230	29,347	457,967
Hot water system 1	Therm	618	560	622	597	621	598	620	620	601	620	602	621	7,300
Hot water system 2	kWh	0	0	0	0	0	0	0	0	0	0	0	0	0
Electrical equipment	kWh	141,740	128,146	142,446	136,813	142,015	137,034	141,833	141,950	137,557	142,034	137,857	142,269	1,671,645
Process equipment	Therm	0	0	0	0	0	0	0	0	0	0	0	0	0
Lights	kWh	147,770	133,802	148,699	142,684	148,415	142,983	146,043	148,101	143,712	148,198	143,969	148,499	1,744,965
Total														
Electric	kWh	317,646	287,729	331,618	333,770	392,641	424,510	459,539	452,767	402,834	373,400	323,275	320,217	4,420,006
Gas	Therm	12,418	11,336	9,438	1,381	799	598	620	620	651	880	6,540	11,994	57,273

Peak Electric Demand (kW)

End Use	Jan	Feb	Mar	Apr	May	Jun	Jul	Aug	Sep	Oct	Nov	Dec
Cooling or heat pump 1	0	0	0	0	253	253	253	253	253	253	0	0
Cooling or heat pump 2	0	0	0	0	0	0	0	0	0	0	0	0
Heating system 2	0	0	0	0	0	0	0	0	0	0	0	0
Fans and pumps	57	57	59	39	138	183	199	194	166	112	50	58
Hot water system 2	0	0	0	0	0	0	0	0	0	0	0	0
Electrical equipment	357	357	357	357	357	357	357	357	357	357	357	357
Lights	396	396	396	396	396	396	396	396	396	396	396	396

Thermal energy storage (TES) cooling systems use electric chillers to cool water or salts or to make ice, during times of the day when electric rates are lower (off-peak). The cooled mass is then used for air conditioning when the demand (and the price) for electricity is higher (on-peak). Figures F-2 and F-3 show a comparison of electricity use for a conventional cooling system and the recommended TES option. Note that the TES system shifts much of the electric use for air conditioning from on-peak to off-peak periods.

F-5 MONTHLY OFF-PEAK COOLING ANALYSIS

The on-peak period was assumed to be 7 A.M. to 9 P.M., Monday through Friday. All other times were assumed to be off-peak. Tables F-3 and F-4 compare the monthly breakdown of energy use during time-of-day periods for each HVAC system.

Table F-3 Monthly Electric Use for Air-Conditioning — Conventional

	Peak Demand		Energy Consumption, kWh			Total
Month	Demand, kW	Hour	Off-Peak	Semi-Peak	On-Peak	Energy, kWh
January	67	9	11,472	0	24,855	36,327
February	67	9	10,480	0	22,689	33,169
March	67	9	12,499	0	26,127	38,626
April	57	10	3,541	0	33,164	36,705
May	574	14	0	0	79,144	79,144
June	774	14	0	0	123,229	123,229
July	802	14	0	0	143,629	143,629
August	798	14	0	0	142,074	142,074
September	696	14	0	0	101,047	101,047
October	88	14	7,707	0	46,148	53,855
November	66	10	11,202	0	23,906	35,108
December	67	9	12,020	0	25,878	37,897
All year	802		68,909	0	791,865	860,773

Table F-4 Monthly Electric Use for Air-Conditioning — Ice Storage

Month	Peak Demand		Energy Consumption, kWh			Total Energy, kWh
	Demand, kW	Hour	Off-Peak	Semi-Peak	On-Peak	
January	57	9	8,886	0	19,257	28,143
February	57	9	8,141	0	17,643	25,784
March	59	9	13,623	0	26,844	40,467
April	39	10	10,873	0	43,398	54,271
May	391	14	14,873	0	87,353	102,227
June	436	10	46,724	0	97,769	144,493
July	452	10	60,501	0	109,180	169,682
August	447	10	57,746	0	104,974	162,721
September	421	14	31,641	0	89,926	121,567
October	365	14	8,883	0	74,270	83,153
November	50	10	14,266	0	27,185	41,451
December	58	9	9,296	0	20,155	29,451
All year	452		285,435	0	718,012	1,003,446

Appendix G CONTROLS

This appendix describes the sequence of control operations for a number of HVAC systems. The symbols for all figures and the control philosophy for these systems are given in Chapter 10.

G-1 VAV SYSTEM, FAN SPEED CONTROL, RADIATION HEAT, NO RETURN FAN — SEQUENCE OF OPERATIONS (FIGURES G-1 AND G-2)

A. Start-up

1. The supply fan may be started and stopped in one of two ways:

Manually. The "HAND, OFF, AUTO" (H-O-A) switch on the fan motor starter can be set in the "HAND" (or "ON") position to start the fan and in the "OFF" position to stop the fan.

Automatically. With the fan motor starter switch in the "AUTO" position, the VAV heating and air-conditioning system has three modes of operation—off, normal, and warm-up.

2. Normal operation is initiated (usually at the beginning of the working day) by a contact closure from local time clock switch TC1 or, in some applications, a remote energy monitoring and control system (EMCS) may be used. This contact closure provides power to the fan motor starter. Power to the hot water temperature control panel, VAV temperature control panel, and static pressure control panel is supplied through auxiliary contacts M1-1, M1-2, and M1-3 on the supply fan motor starter.

3. At the end of the occupied period, time clock switch TC1, or the remote EMCS, deenergizes the fan starter relay, consequently interrupting power to all the control panels.

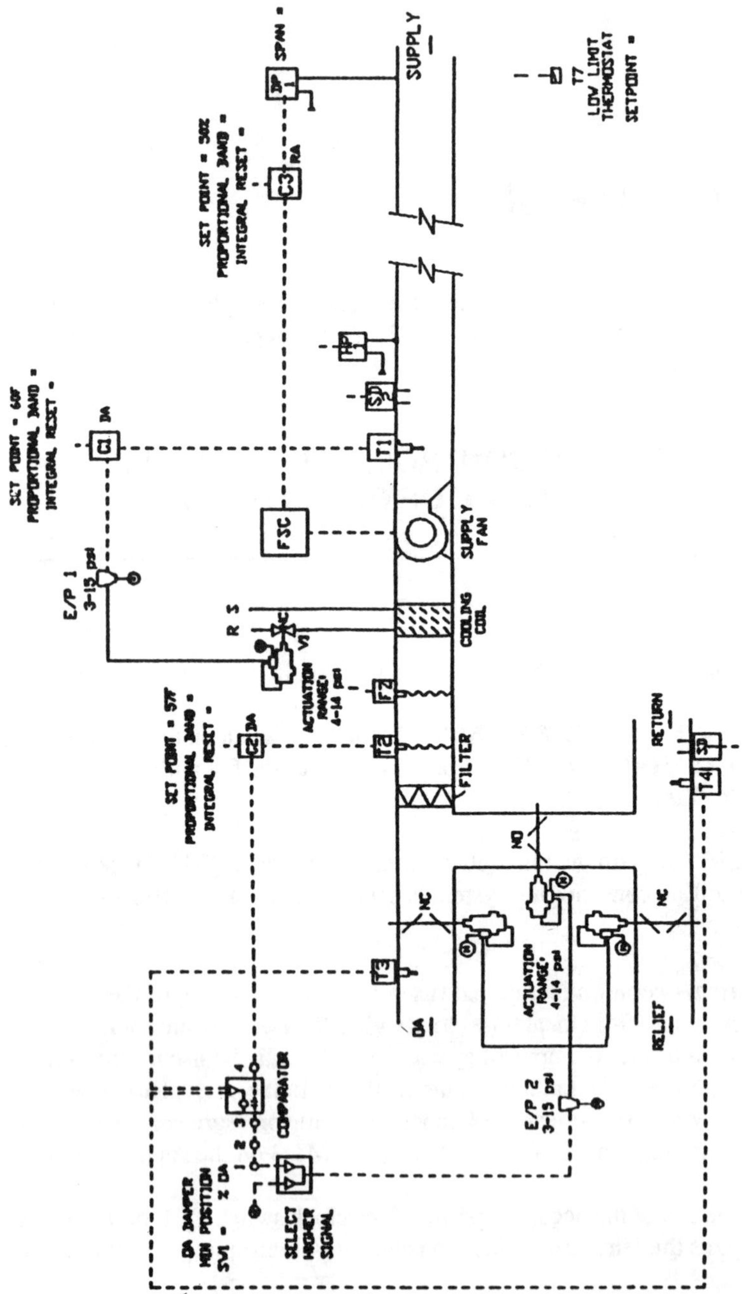

Fig. G-1 VAV Fan Speed Control (FSC) System

4. To guard against freezing in the building, two-position low-limit electric freeze protection thermostats (T7) are located as shown in the drawings. If the temperature drops below the set point of any of the freeze protection thermostat(s), relay R1 is energized. Through auxiliary contact R1-1, power is supplied to the hot water temperature control panel and, through an electric- to-pneumatic switch (not shown), pressure is supplied to the room thermostats. Auxiliary contact R1-2 (normally closed) opens, and the VAV temperature control panel remains deenergized. Through auxiliary contact R1-3, the fan motor starter is energized, and power is supplied to the static pressure control panel (through auxiliary contact M1-3). With no power to the VAV temperature control panel, the outdoor air dampers and the cooling coil valve remain closed. Heat is delivered through a baseboard heater in the zone controlled by the room thermostat. With the fan on, heat is distributed evenly

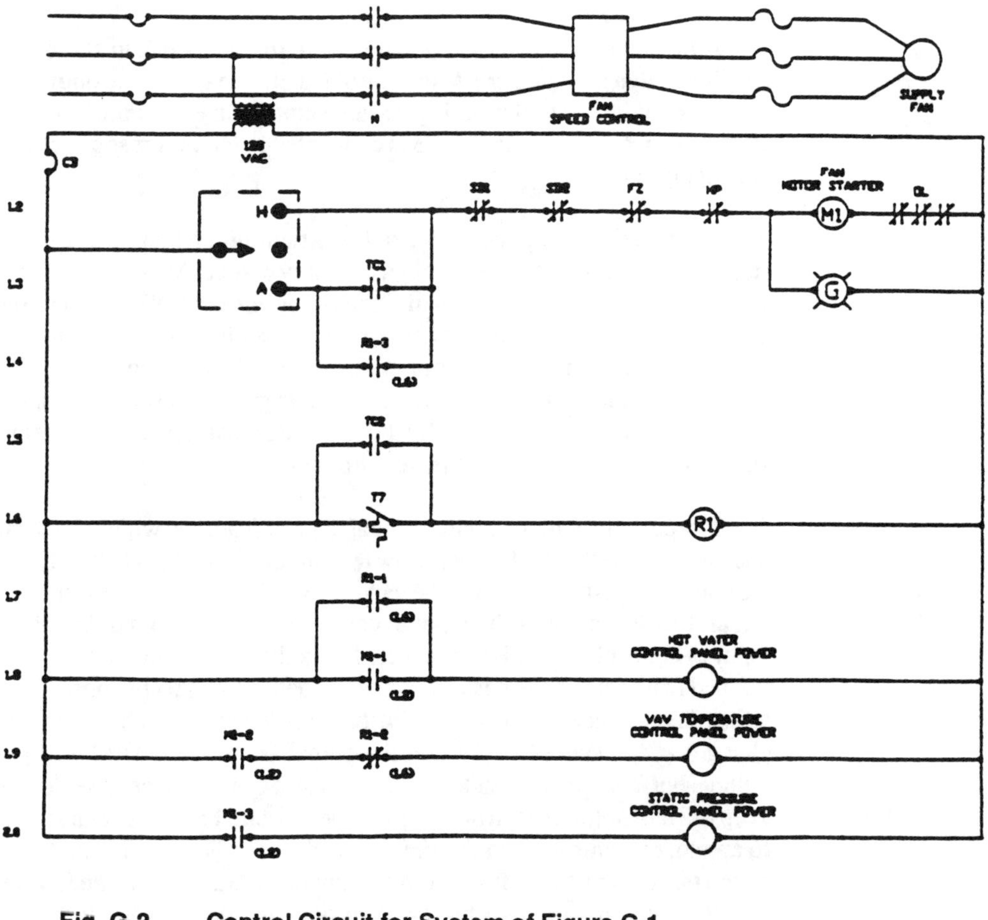

Fig. G-2 Control Circuit for System of Figure G-1

throughout the building. On a rise in room temperature, the contacts on the freeze protection thermostats open, returning the system to the off state.

5. To provide a preoccupancy warm-up cycle, an auxiliary contact on time clock switch TC2 or, in some applications, remotely energized contacts from an EMCS system, are energized for a period before occupancy. Closure of these contacts has exactly the same effect as the closure of contacts on one of the freeze protection thermostats. At the end of the warm-up period, the system begins normal operation.

B. Static Pressure Control

1. The static pressure at a representative point in the ductwork is held constant by the static pressure control panel.

2. A tube connects a static pressure tap in the ductwork to static pressure sensor DP1 located in the static pressure control panel. The electronic output of the pressure sensor is compared to the static pressure setpoint in electronic PI controller C3. the electronic output of controller C3 is connected to fan motor speed controller FSC on the supply fan.

3. The panel is equipped with a soft start circuit. When power is supplied to the static pressure control panel through auxiliary contact M1-3 on the fan motor starter, a delay of about 15 s occurs while the output of controller C3 is ramped to zero. During this delay, the output of controller C3 is disconnected from controller FSC. After the delay, the output from controller C3 is reconnected to controller FSC, allowing the supply fan to run. The setpoint is gradually ramped from zero up to the desired setpoint (as determined by the setpoint adjustment knob), after which the soft start circuit no longer affects the control system.

4. The panel is equipped with manual control features, which, by turning the timer and pressing the "SET" button, allow the output of the panel to be adjusted by turning the manual adjust knob. When the panel is switched to the manual mode, the soft start circuit first disconnects the input to controller FSC. After a delay of about 15 s, the input to controller FSC is reconnected, and the voltage output to controller FSC is ramped from zero to the desired voltage as determined by the position of the manual adjust knob. Once the manual adjustment voltage is reached, any manual output changes (*i.e.*, manually changing the output from 0 to 100%) will pass through a voltage buffer to prevent sudden changes from causing excessive duct pressures and keep the fan motor drive from tripping their circuit breakers. The system is returned to automatic control when the timer runs down or when the reset button is pushed. When this occurs, the soft start circuit functions exactly as if the fan were just being started. The timer or reset button shall only enable the soft start circuit when the

system is being switched from manual to automatic control. This shall prevent the timer from cycling the system after it has been placed in the automatic mode via the reset button.

C. Supply and Mixed-Air Temperature Control

1. The supply air temperature is controlled by electronic PI controller C1, located in the VAV temperature control panel.

2. The mixed-air temperature is controlled by electronic PI controller C2, located in the VAV temperature control panel.

3. The cooling coil chilled-water valve V1 is modulated by controller C1 using the sensed supply air temperature from temperature sensor T1. The output from controller C1 is connected to electronic-to-pneumatic transducer E/P-1, which produces a pneumatic signal to position valve V1, maintaining a constant supply air temperature. The setpoint of controller C1 is 60°F.

4. The outside, relief, and return air dampers are modulated by controller C2 using the sensed mixed-air temperature from temperature sensor T2. The output from controller C2 is connected to the comparator relay, which will pass the controller signal only if the outdoor air temperature is less than the return air temperature. The electronic high signal selector compares the signal from the comparator with the voltage produced by the minimum positioning adjustment knob SW1. If the output from the comparator is less than the output from SW1, the minimum positioning signal will be passed to electronic-to-pneumatic transducer E/P-2, which produces a pneumatic signal to hold the outside and relief air dampers at their minimum position. When more than minimum outdoor air is economical, the signal from controller C2 will be higher than the signal from SW1, its value will be passed to E/P-2, which will produce a pneumatic signal to position the outside, relief, and return air dampers accordingly. The setpoint of controller C2 is 60°F minus the temperature rise across the fan (under full airflow conditions).

5. Remote EMCS economizer control is possible through the use of economizer contacts 1, 2, 3, and 4. With contacts 3 and 4 open and 1 and 2 shorted, the economizer works as described above. With all contacts shorted, the comparator is bypassed, and the control signal is sent directly to the high signal selector. With 1 and 2 open, the minimum position signal holds the dampers in their minimum position.

6. Hysteresis is required in the comparator relay circuit. If the comparator relay is open, the outdoor air temperature must fall to approximately 2.0°F below the return

air temperature before the relay closes; if the relay is closed, the outdoor air temperature must rise to equal the return air temperature before the relay opens.

7. When the system is off (no power to the control panels), the outdoor air dampers return to their normally closed position.

D. Room Air Temperature Control

1. Room air temperature control is achieved by pneumatic room thermostats modulating individual VAV boxes. When heating is required, the room thermostat also modulates baseboard heater valve V2.

2. Thermostat calibration and selection of actuator spring ranges shall be coordinated to provide the control action shown in Figures 10-7 and 10-8, Chapter 10.

E. Hot Water Temperature Control

The temperature of hot water supplied to the baseboard heaters is controlled by the hot water temperature control panel as described on the sheet labeled Hot Water Control System.

F. Interlocks

Smoke detectors (SD1 and SD2), the low-temperature safety switch (FZ), the high-pressure limit switch (HP1) are wired in series with fan motor starter relay M1 to stop the fan in the event of smoke, extremely low temperatures, or damagingly high pressures. The ladder schematic on the drawings shows how equipment is to be interlocked.

G-2 VAV SYSTEM, FAN SPEED CONTROL, REHEAT, NO RETURN FAN SEQUENCE OF OPERATIONS (FIGURES G-1 AND G-2)

The sequence of operations is identical to that described in Section G-1, with the following exceptions:

1. The last three sentences under Section A.4 are replaced with the following:

"Heat is provided through a reheat coil in the zone's duct controlled by the room thermostat. On a rise in room temperature, the contacts on the freeze protection thermostats open, returning the system to the off state."

2. In Section D, line 3, the words "baseboard heater" are replaced with the words "reheat coil."

3. In Section E, line 1, the words "baseboard heaters" are replaced with the words "reheat coils."

G-3 VAV SYSTEM, FAN SPEED CONTROL, RADIATION HEAT, RETURN FAN — SEQUENCE OF OPERATIONS (FIGURES G-3 AND G-4)

A. Start-up

1. The supply fan and return fan may be started and stopped in one of two ways:

Manually. The "HAND, OFF, AUTO" (H-O-A) switch on the fan motor starter can be set in the "HAND" (or "ON") position to start the fan and in the "OFF" position to stop the fan.

Automatically. With the fan motor starter switch in the "AUTO" position, the VAV heating and air-conditioning system has three modes of operation—off, normal, and warm-up.

2. Normal operation is initiated (usually at the beginning of the working day) by a contact closure from local time clock switch TC1 or, in some applications, a remote energy monitoring and control system (EMCS) may be used. This contact closure provides power to the fan motor starters, M1 and M2, and relay R2. Power to the hot water temperature control panel, the VAV temperature control panel, and the static pressure control panel is supplied through auxiliary contacts R2-1, R2-2, and R2-3.

3. At the end of the occupied period, time clock switch TC1, or the remote EMCS, deenergizes the fan starter relays, consequently interrupting power to all the control panels.

4. To guard against freezing in the building, two-position low-limit electric freeze protection thermostat(s) (T7) are located as shown in Figures G-3 and G-4. If the temperature drops below the setpoint of any of the freeze protection thermostat(s), relay R1 is energized. Through auxiliary contact R1-1, power is supplied to the hot

water temperature control panel and, through an electric-to-pneumatic switch (not shown), pressure is supplied to the room thermostats. Auxiliary contact R1-2 (normally closed) opens, and the VAV temperature control panel remains deenergized. Through auxiliary contacts R1-3 and R1-4, the fan motor starters are energized, and power is supplied to the static pressure control panel (through auxiliary contact R2-3). With no power to the VAV temperature control panel, the outdoor air dampers and the cooling coil valve remain closed. Heat is delivered through a baseboard heater in the zone controlled by the room thermostat. With the fan on, heat is distributed evenly throughout the building. On a rise in room temperature, the contacts on the freeze protection thermostats open, returning the system to the off state.

5. To provide a preoccupancy warm-up cycle, an auxiliary contact on time clock switch TC2 or, in some applications, remotely energized contacts from an EMCS system are energized for a period before occupancy. Closure of these contacts has exactly the same effect as the closure of contacts on one of the freeze protection thermostats. At the end of the warm-up period, the system begins normal operation.

B. Static Pressure Control

1. The static pressure at a representative point in the ductwork is held constant by the static pressure control panel.

2. A tube connects a static pressure tap in the ductwork to static pressure sensor DP1 located in the static pressure control panel. The electronic output of the pressure sensor is compared to the static pressure setpoint in electronic PI controller C3. the electronic output of controller C3 is connected to fan motor speed controller FSC on the supply fan.

3. The panel is equipped with a soft start circuit. When power is supplied to the static pressure control panel through auxiliary contact M1-3 on the fan motor starter, a delay of about 15 s occurs while the output of controller C3 is ramped to zero. During this delay, the output of controller C3 is disconnected from controller FSC. After the delay, the output from controller C3 is reconnected to controller FSC, allowing the supply fan to run. The setpoint is gradually ramped from zero up to the desired setpoint (as determined by the setpoint adjustment knob), after which the soft start circuit no longer affects the control system.

4. The panel is equipped with manual control features, which, by turning the timer and pressing the "SET" button, allow the output of the panel to be adjusted by turning the manual adjust knob. When the panel is switched to the manual mode, the soft start circuit first disconnects the input to controller FSC. After a delay of about 15 s, the

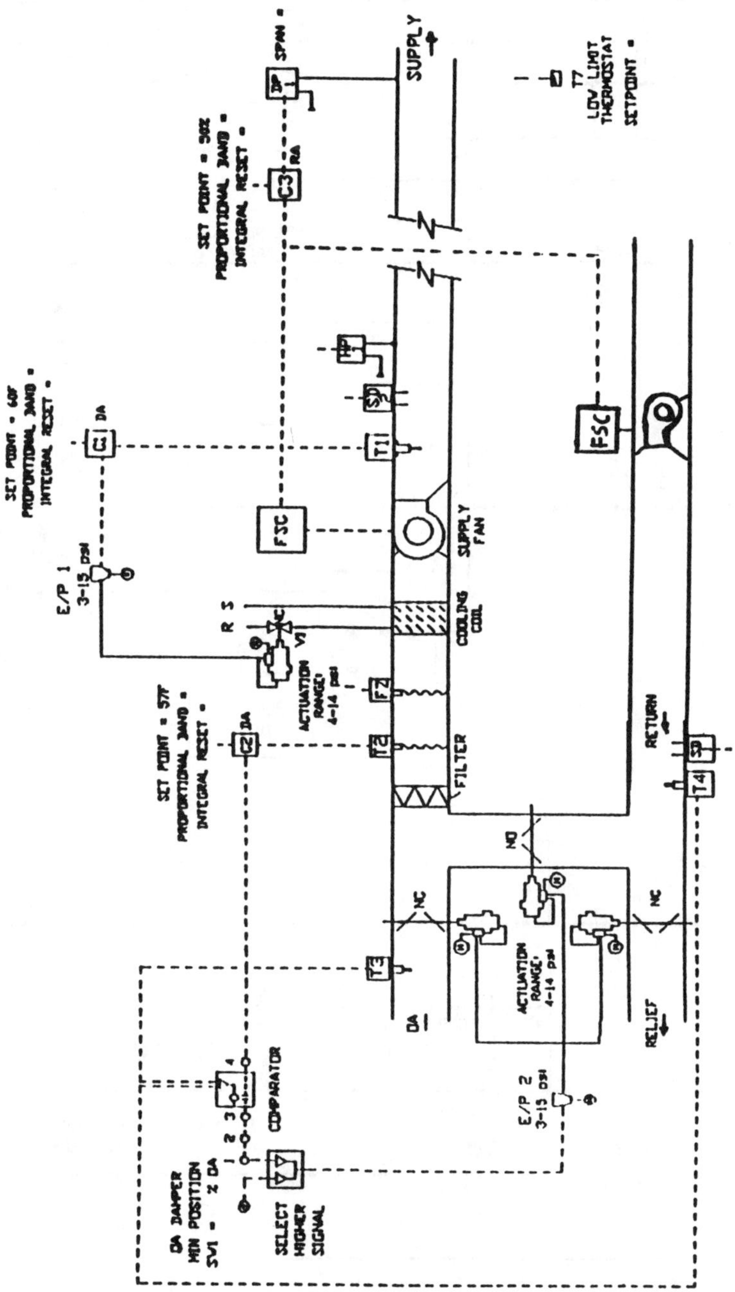

Fig. G-3 VAV Fan Speed Control (FSC) System with Return Fan

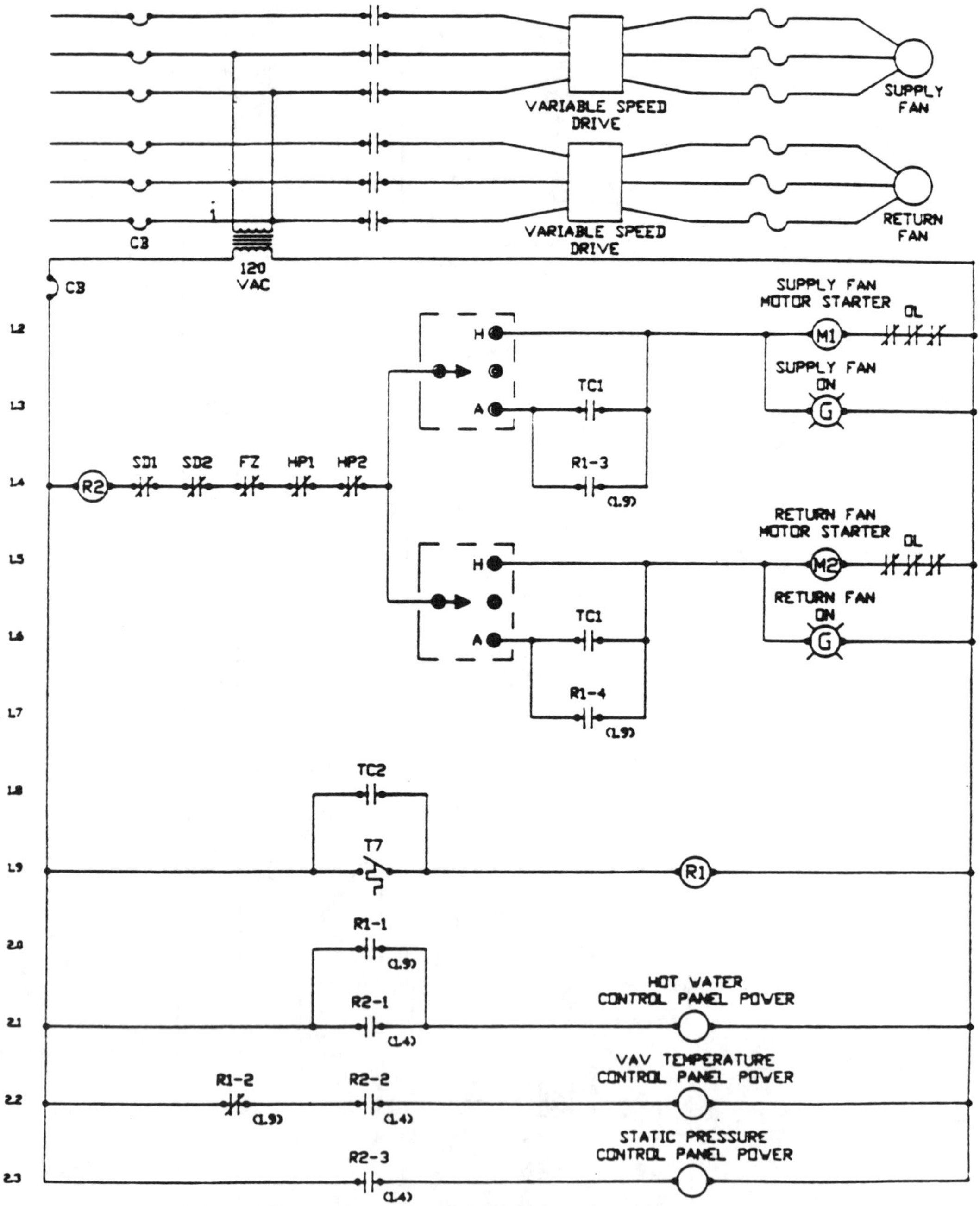

Fig. G-4 Control Circuit for System of Figure G-3

input to controller FSC is reconnected, and the voltage output to controller FSC is ramped from zero to the desired voltage as determined by the position of the manual adjust knob. Once the manual adjustment voltage is reached, any manual output changes (*i.e.*, manually changing the output from 0 to 100%) will pass through a voltage buffer to prevent sudden changes from causing excessive duct pressures and keep the fan motor drive from tripping their circuit breakers. The system is returned to automatic control when the timer runs down or when the reset button is pushed. When this occurs, the soft start circuit functions exactly as if the fan were just being started. The timer or reset button shall only enable the soft start circuit when the system is being switched from manual to automatic control. This shall prevent the timer from cycling the system after it has been placed in the automatic mode via the reset button.

C. Supply and Mixed-Air Temperature Control

1. The supply air temperature is controlled by electronic PI controller C1, located in the VAV temperature control panel.

2. The mixed-air temperature is controlled by electronic PI controller C2 located in the VAV temperature control panel.

3. The cooling coil chilled-water valve V1 is modulated by controller C1 using the sensed supply air temperature from temperature sensor T1. The output from controller C1 is connected to electronic-to-pneumatic transducer E/P-1, which produces a pneumatic signal to position valve V1, maintaining a constant supply air temperature. The setpoint of controller C1 is 60°F.

4. The outside, relief, and return air dampers are modulated by controller C2 using the sensed mixed-air temperature from temperature sensor T2. The output from controller C2 is connected to the comparator relay, which will pass the controller signal only if the outdoor air temperature is less than the return air temperature. The electronic high signal selector compares the signal from the comparator with the voltage produced by the minimum positioning adjustment knob SW1. If the output from the comparator is less than the output from SW1, the minimum positioning signal will be passed to electronic-to-pneumatic transducer E/P-2, which produces a pneumatic signal to hold the outside and relief air dampers at their minimum position. When more than minimum outdoor air is economical, the signal from controller C2 will be higher than the signal from SW1, its value will be passed to E/P-2, which will produce a pneumatic signal to position the outside, relief, and return air dampers accordingly. The setpoint of controller C2 is 60°F minus the temperature rise across the fan (under full airflow conditions).

5. Remote EMCS economizer control is possible through the use of economizer contacts 1, 2, 3, and 4. With contacts 3 and 4 open and 1 and 2 shorted, the economizer works as described above. With all contacts shorted, the comparator is bypassed, and the control signal is sent directly to the high signal selector. With 1 and 2 open, the minimum position signal holds the dampers in their minimum position.

6. Hysteresis is required in the comparator relay circuit. If the comparator relay is open, the outdoor air temperature must fall to approximately 2.0°F below the return air temperature before the relay closes; if the relay is closed, the outdoor air temperature must rise to equal the return air temperature before the relay opens.

7. When the system is off (no power to the control panels), the outdoor air dampers return to their normally closed position.

D. Room Air Temperature Control

1. Room air temperature control is achieved by pneumatic room thermostats modulating individual VAV boxes. When heating is required, the room thermostat also modulates baseboard heater valve V2.

2. Thermostat calibration and selection of actuator spring ranges shall be coordinated to provide the control action shown in Figures 10-7 and 10-8, Chapter 10.

E. Hot Water Temperature Control

The temperature of hot water supplied to the baseboard heaters is controlled by the hot water temperature control panel.

F. Interlocks

Smoke detectors (SD1 and SD2), the low-temperature safety switch (FZ), and the high-pressure limit switches (HP1 and HP2) are wired in series with the fan motor starter relay M1 to stop the fan in the event of smoke, extremely low temperatures, or damagingly high pressures. The ladder schematic on Figures G-3 and G-4 shows how equipment is to be interlocked.

G-4 VAV SYSTEM, FAN SPEED CONTROL, REHEAT, RETURN FAN SEQUENCE OF OPERATIONS (FIGURES G-3 AND G-4)

The sequence of operations is identical to that described in Section G-1, with the following exceptions:

1. The last three sentences under Section A.4 are replaced with the following:

> "Heat is provided through a reheat coil in the zone's duct controlled by the room thermostat. On a rise in room temperature, the contacts on the freeze protection thermostats open, returning the system to the off state."

2. In Section D, line 3, the words "baseboard heater" are replaced with the words "reheat coil."

3. In Section E, line 1, the words "baseboard heaters" are replaced with the words "reheat coils."

G-5 VAV SYSTEM, FAN INLET GUIDE VANE CONTROL — SEQUENCE OF OPERATIONS (FIGURES G-5 AND G-6)

The sequence of operations is identical to that described in Sections G-1 through G-4, except when fan speed control is replaced by control of the inlet guide vanes (see Figures G-5 and G-6). Replace the words "controller FSC" with the words "E/P-3" in Sections G-1 through G-4.

G-6 SMALL SINGLE-ZONE SYSTEM, ONE CONTROLLER — SEQUENCE OF OPERATIONS (FIGURE G-7)

A. Start-up

1. The supply fan and return fan may be started and stopped in one of two ways:

Manually. The "HAND, OFF, AUTO" (H-O-A) switch on the fan motor starter can be set in the "HAND" (or "ON") position to start the fan and in the "OFF" position to stop the fan.

Automatically. With the fan motor starter switch in the "AUTO" position, the VAV heating and air-conditioning system has three modes of operation—off, normal, and warm-up.

2. Normal operation is initiated (usually at the beginning of the working day) by a contact closure from local time clock switch TC1 or, in some applications, a remote energy monitoring and control system (EMCS) may be used. This contact closure provides power to the fan motor starter. Power to the single-zone temperature control panel and the hot water temperature control panel is supplied through auxiliary contacts M1-1 and M1-2 on the supply fan motor starter.

3. At the end of the occupied period, time clock switch TC1, or the remote EMCS, deenergizes the fan starter relay, consequently interrupting power to all the control panels.

4. To guard against freezing in the building, two-position low-limit electric freeze protection thermostat(s) (T3) are located as shown in the drawings. If the temperature drops below the setpoint of any of the freeze protection thermostat(s), relay R1 is energized. Through auxiliary contact R1-1, power is supplied to the fan motor starter, and the single-zone temperature control panel and the hot water temperature control panel are energized (through auxiliary contacts M1-1 and M1-2). Auxiliary contact R1-2 (normally closed) opens and deenergizes electrically actuated pneumatic switch EPS1, returning it to its normal position. Thus EPS1 disconnects the damper actuators from their pneumatic control signal and allows the dampers to return to their normal states, preventing outside air from entering the system. Heat is provided through heating coil valve V1, which is regulated by controller C1. On a rise in room temperature, the contacts on the freeze protection thermostats open, returning the system to the off state.

5. To provide a preoccupancy warm-up cycle, an auxiliary contact on time clock switch TC2 or, in some applications, remotely energized contacts from an EMCS system are energized for a period before occupancy. Closure of these contacts has exactly the same effect as the closure of contacts on one of the freeze protection thermostats. At the end of the warm-up period, the system begins normal operation.

B. Supply and Mixed-Air Temperature Control

1. The mixed-air temperature and the supply air temperature are controlled by electronic proportional controller C1, located in the single-zone temperature control panel.

2. The heating coil hot water valve V1 and the cooling coil chilled-water valve V2 are modulated by controller C1 using the sensed air temperature from temperature sensor T1. The output from controller C1 is connected to electronic-to-pneumatic transducer E/P-1, which produces a pneumatic signal to position valves V1 and V2

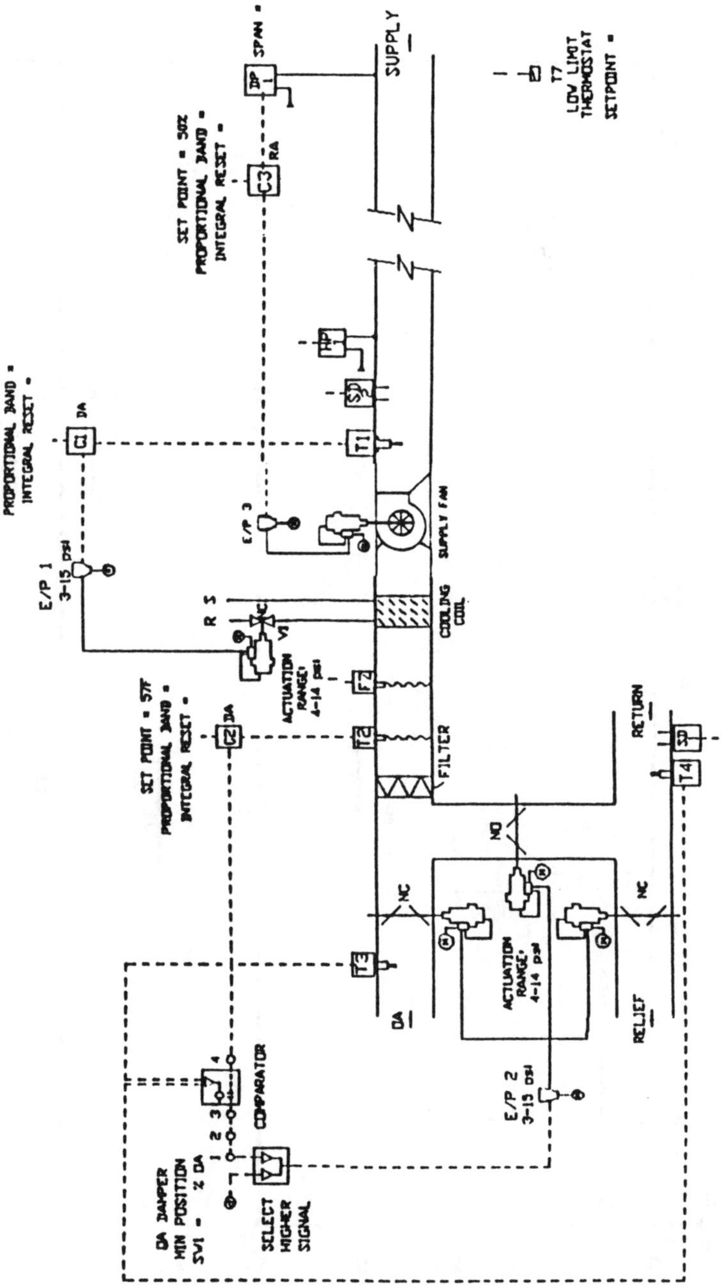

Fig. G-5 VAV Inlet Guide Vane (IGV) Control System

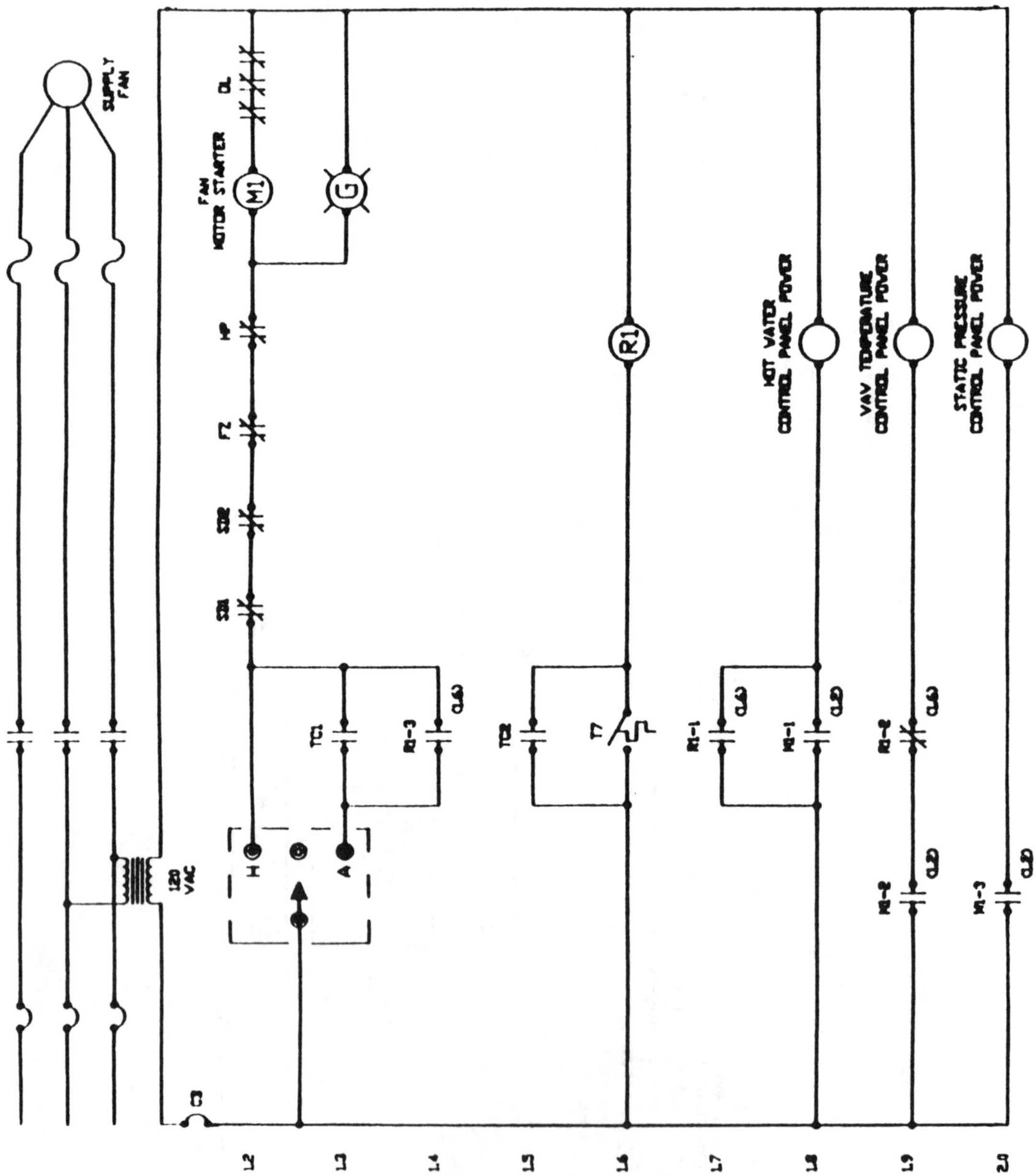

Fig. G-6 Control Circuit for System of Figure G-5

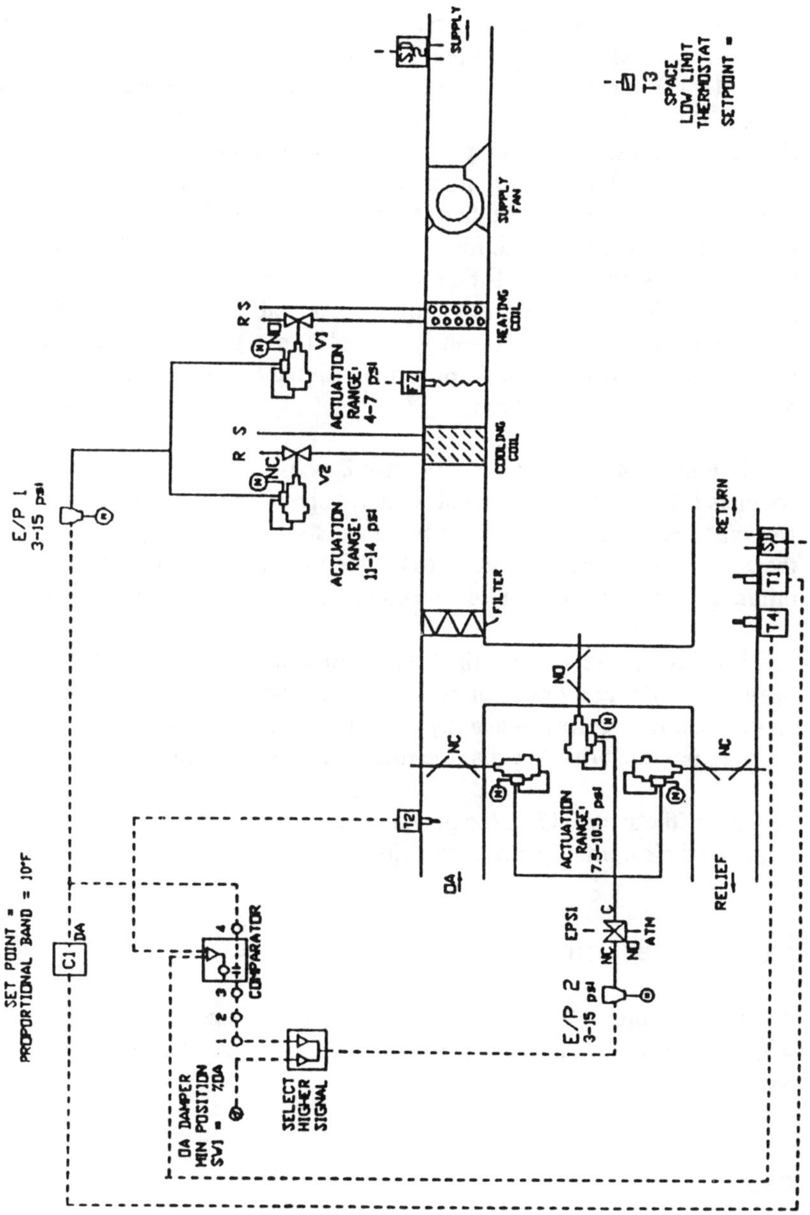

Fig. G-7 Single-Zone System with Simple Control

accordingly, maintaining a constant supply air temperature. The setpoint of controller C1 is the desired room temperature (72 to 74°F).

3. The outside, relief, and return air dampers are also modulated by controller C1 using the sensed air temperature from temperature sensor T1. The output from controller C1 is connected to the comparator relay, which will pass the controller signal only if the outdoor air temperature is less than the return air temperature. The electronic high signal selector compares the signal from the comparator with the voltage produced by the minimum positioning adjustment knob SW1. If the output from the comparator is less than the output from SW1, the minimum positioning signal will be passed to electronic-to-pneumatic transducer E/P-2, which produces a pneumatic signal to hold the outside and relief air dampers at their minimum position. When more than minimum outdoor air is economical, the signal from controller C1 will be higher than the signal from SW1, and its value will be passed to E/P-2, which will produce a pneumatic signal to position the outside, relief, and return air dampers accordingly.

4. Remote EMCS economizer control is possible through the use of economizer contacts 1, 2, 3, and 4. With contacts 3 and 4 open and 1 and 2 shorted, the economizer works as described earlier. With all contacts shorted, the comparator is bypassed, and the control signal is sent directly to the high signal selector. With 1 and 2 open, the minimum position signal holds the dampers in their minimum position.

5. Hysteresis is required in the comparator relay circuit. If the comparator relay is open, the outdoor air temperature must fall to approximately 2.0°F below the return air temperature before the relay closes; if the relay is closed, the outdoor air temperature must rise to equal the return air temperature before the relay opens.

6. When the system is off (no power to the control panels), the outdoor air dampers return to their normally closed position.

C. Hot Water Temperature Control

The temperature of hot water supplied to the heating coil is controlled by the hot water temperature control panel as described on the sheet labeled Hot Water Control System.

D. Interlocks

Smoke detectors (SD1 and SD2) and the low-temperature safety switch (FZ) are wired in series with fan motor starter relay M1 to stop the fan in the event of smoke or extremely low temperatures. The ladder schematic on the drawings shows how equipment is to be interlocked.

G-7 SINGLE-ZONE SYSTEM, SEPARATE HEATING AND COOLING CONTROLLERS — SEQUENCE OF OPERATIONS (FIGURE G-8)

A. Start-up

The start-up procedure is identical to the one in Section G-6.

B. Mixed-Air Temperature Control

1. The mixed-air temperature is controlled by electronic PI controller C3, located in the single-zone temperature control panel.

2. The outside, relief, and return air dampers are modulated by controller C3 using the sensed mixed-air temperature from temperature sensor T3. The output from controller C3 is connected to the comparator relay, which will pass the controller signal only is the outdoor air temperature is less than the return air temperature. The electronic high signal selector compares the signal from the comparator with the voltage produced by the minimum positioning adjustment knob SW1. If the output from the comparator is less than the output from SW1, the minimum positioning signal will be passed to electronic-to-pneumatic transducer E/P-3, which produces a pneumatic signal to hold the outside and relief air dampers at their minimum position. When more than minimum outdoor air is economical, the signal from controller C3 will be higher than the signal from SW1, its value will be passed to E/P-3, which will produce a pneumatic signal to position the outside, relief, and return air dampers accordingly. The setpoint of controller C3 is determined by the output of electronic proportional controller C4 using the sensed return air temperature from temperature sensor T4—the mixed-air temperature controller is reset on the basis of return air temperature.

3. Remote EMCS economizer control is possible through the use of economizer contacts 1, 2, 3, and 4. With contacts 3 and 4 open and 1 and 2 shorted, the economizer works as described above. With all contacts shorted, the comparator is bypassed, and the control signal is sent directly to the high signal selector. With 1 and 2 open, the minimum position signal holds the dampers in their minimum position.

4. Hysteresis is required in the comparator relay circuit. If the comparator relay is open, the outdoor air temperature must fall to approximately 2.0°F below the return

air temperature before the relay closes; if the relay is closed, the outdoor air temperature must rise to equal the return air temperature before the relay opens.

5. When the system is off (no power to the control panels), the outdoor air dampers return to their normally closed position.

C. Supply Air Temperature Control

1. The supply air temperature is controlled by the single-zone temperature control panel.

2. The heating coil hot water valve V2 is modulated by electronic PI controller C2 using the sensed air temperature from temperature sensor T2. The output from controller C2 is connected to electronic-to-pneumatic transducer E/P-2, which produces a temperature leaving the heating coil. The setpoint of controller C2 is determined by the output of electronic proportional controller C4 using the sensed return air temperature from temperature sensor T4—the heating coil controller is reset on the basis of return air temperature.

3. The cooling coil chilled-water valve V1 is modulated by electronic PI controller C1 using the sensed supply air temperature from temperature sensor T1. The output from controller C1 is connected to electronic-to-pneumatic transducer E/P-1, which produces a pneumatic signal to position valve V1, maintaining a constant supply air temperature. The setpoint of controller C1 is determined by the output of electronic proportional controller C4 using the sensed return air temperature from temperature sensor T4—the cooling coil controller is reset on the basis of return air temperature.

D. Hot Water Temperature Control

The temperature of hot water supplied to the heating coil is controlled by the hot water temperature control panel.

E. Interlocks

Smoke detectors (SD1 and SD2) and the low-temperature safety switch (FZ) are wired in series with fan motor starter relay M1 to stop the fan in the event of smoke or extremely low temperatures. The ladder schematic on Figure G-8 shows how equipment is to be interlocked.

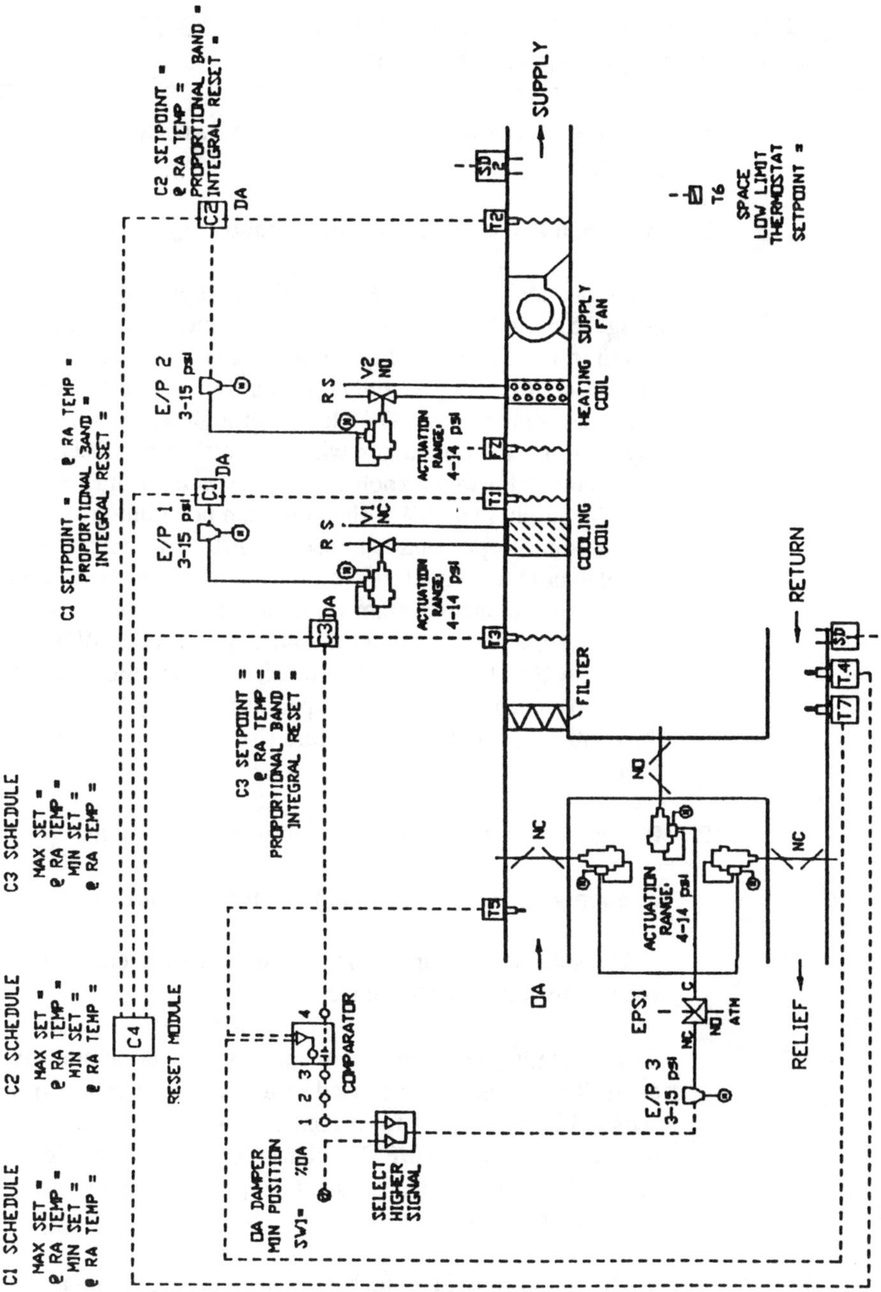

Fig. G-8 Single-Zone System with Separate Controllers for Heating and Cooling (Cascade Control)

G-8 SINGLE-ZONE SYSTEM WITH HUMIDITY CONTROL — SEQUENCE OF OPERATIONS (FIGURE G-9)

The sequence of control operations is the same as in Section G-7, with the following exceptions:

1. Paragraph B-2 is replaced with the following paragraph:

   2. The outside, relief, and return air dampers are modulated by controller C1 using the sensed mixed-air temperature from air temperature sensor T1. The output from controller C1 is connected to the comparator relay, which will pass the controller signal only if the outdoor air temperature is less than the return air temperature. The electronic high signal selector compares the signal from the comparator with the voltage produced by the minimum positioning adjustment knob SW1. If the output from the comparator is less than the output from SW1, the minimum positioning signal will be passed to electronic-to-pneumatic transducer E/P-1, which produces a pneumatic signal to hold the outside and relief air dampers at their minimum position. When more than minimum outdoor air is economical, the signal from controller C1 will be higher than the signal from SW1; its value will be passed to E/P-1, which will produce a pneumatic signal to position the outside, relief, and return air dampers accordingly. The setpoint of controller C1 is 60°F minus the temperature rise across the fan (under full airflow conditions).

2. Paragraph C is replaced by the following paragraphs:

   C. Supply Air Temperature and Humidity Control

   1. The supply air temperature and humidity are controlled by the single-zone temperature control panel.

   2. The heating coil hot water valve V2 is modulated by electronic PI controller C2 using the sensed return air temperature from temperature sensor T2. The output from controller C2 is connected to electronic-to-pneumatic transducer E/P-2, which produces a pneumatic signal to position valve V2, maintaining a constant supply air temperature. The setpoint of controller C2 is 68°F.

   3. The cooling coil face and bypass dampers are modulated by electronic PI controller C3 using the sensed air temperature from temperature sensor T3. The output from controller C3 is connected to electronic-to-pneumatic transducer E/P-3, which produces a pneumatic signal to position face and

bypass dampers accordingly, maintaining a constant supply air temperature. The setpoint of controller C3 is 60°F.

4. The humidifier valve V1 is modulated by electronic proportional controller C4 using the signal from supply air humidity sensor H2. The output of controller C4 goes to electronic-to-pneumatic transducer E/P-4, which produces a pneumatic signal to position valve V1. The setpoint of controller C4 is reset from electronic PI controller C5, which uses the signal from return air humidity sensor H1.

G-9 MULTIZONE SYSTEM — SEQUENCE OF OPERATIONS (FIGURES G-10 AND G-11)

A. Start-up

1. The supply fan and return fan may be started and stopped in one of two ways:

Manually. The "HAND, OFF, AUTO" (H-O-A) switch on the fan motor starter can be set in the "HAND" (or "ON") position to start the fan and in the "OFF" position to stop the fan.

Automatically. With the fan motor starter switch in the "AUTO" position, the VAV heating and air-conditioning system has three modes of operation—off, normal, and warm-up.

2. Normal operation is initiated (usually at the beginning of the working day) by a contact closure from local time clock switch TC1 or, in some applications, a remote energy monitoring and control system (EMCS) may be used. This contact closure provides power to the fan motor starter. Power to the multizone temperature control panel and the hot water temperature control panel is supplied through auxiliary contacts M1-1 and M1-2 on the supply fan motor starter.

3. At the end of the occupied period, time clock switch TC1, or the remote EMCS, deenergizes the fan starter relay, consequently interrupting power to all the control panels.

4. To guard against freezing in the building, two-position low-limit electric freeze protection thermostat(s) (T7) are located as shown in the drawings. If the temperature drops below the setpoint of any of the freeze protection thermostat(s), relay R1 is energized. Through auxiliary contact R1-1, power is supplied to the fan motor starter. Auxiliary contact R1-2 (normally closed) opens, and the multizone temperature control panel remains deenergized. Through auxiliary contact M1-2, the hot

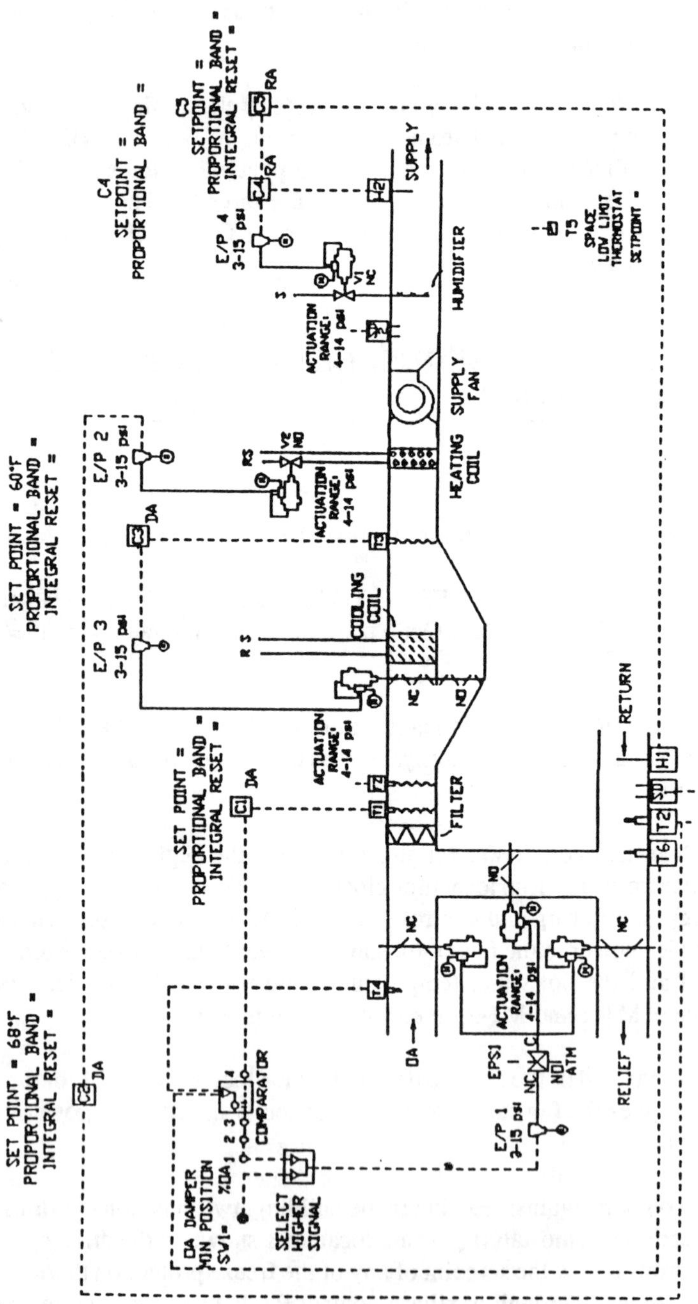

Fig. G-9 Single-Zone System with Humidity Control

water temperature control panel is energized and, through an electric-to-pneumatic switch (not shown), pressure is supplied to the room thermostats. With no power to the multizone temperature control panel, the outdoor air dampers and the cooling coil valve remain closed. The heating coil valve remains in its normally open state, and the fan delivers warm air to the zones. On a rise in room temperature, the contacts on the freeze protection thermostats open, returning the system to the off state.

5. To provide a preoccupancy warm-up cycle, an auxiliary contact on time clock switch TC2 or, in some applications, remotely energized contacts from an EMCS system are energized for a period before occupancy. Closure of these contacts has exactly the same effect as the closure of contacts on one of the freeze protection thermostats. At the end of the warm-up period, the system begins normal operation.

B. Mixed-Air Temperature Control

1. The mixed-air temperature is controlled by electronic PI controller C3, located in the multizone temperature control panel.

2. The outside, relief, and return air dampers are modulated by controller C3 using the sensed mixed-air temperature from temperature sensor T3. The output from controller C3 passes through the "off temperature" relay contact, which disables economizer operation during cold periods. The "off period" relay is connected to the comparator relay, which will pass the controller signal only if the outdoor air temperature is less than the return air temperature. The electronic high signal selector compares the signal from the comparator with the voltage produced by the minimum positioning adjustment knob SW1. If the output from the comparator is less than the output from SW1, the minimum positioning signal will be passed to electronic-to-pneumatic transducer E/P-3, which produces a pneumatic signal to hold the outside and relief air dampers at their minimum position. When more than minimum outdoor air is economical, the signal from controller C3 will be higher than the signal from SW1; its value will be passed to E/P-3, which will produce a pneumatic signal to position the outside, relief, and return air dampers accordingly. The setpoint of controller C3 is 60°F.

3. Remote EMCS economizer control is possible through the use of economizer contacts 1, 2, 3, and 4. With contacts 3 and 4 open and 1 and 2 shorted, the economizer works as described earlier. With all contacts shorted, the comparator is bypassed, and the control signal is sent directly to the high signal selector. With 1 and 2 open, the minimum position signal holds the dampers in their minimum position.

4. Hysteresis is required in the comparator relay circuit. If the comparator relay is open, the outdoor air temperature must fall to approximately 2.0°F below the return

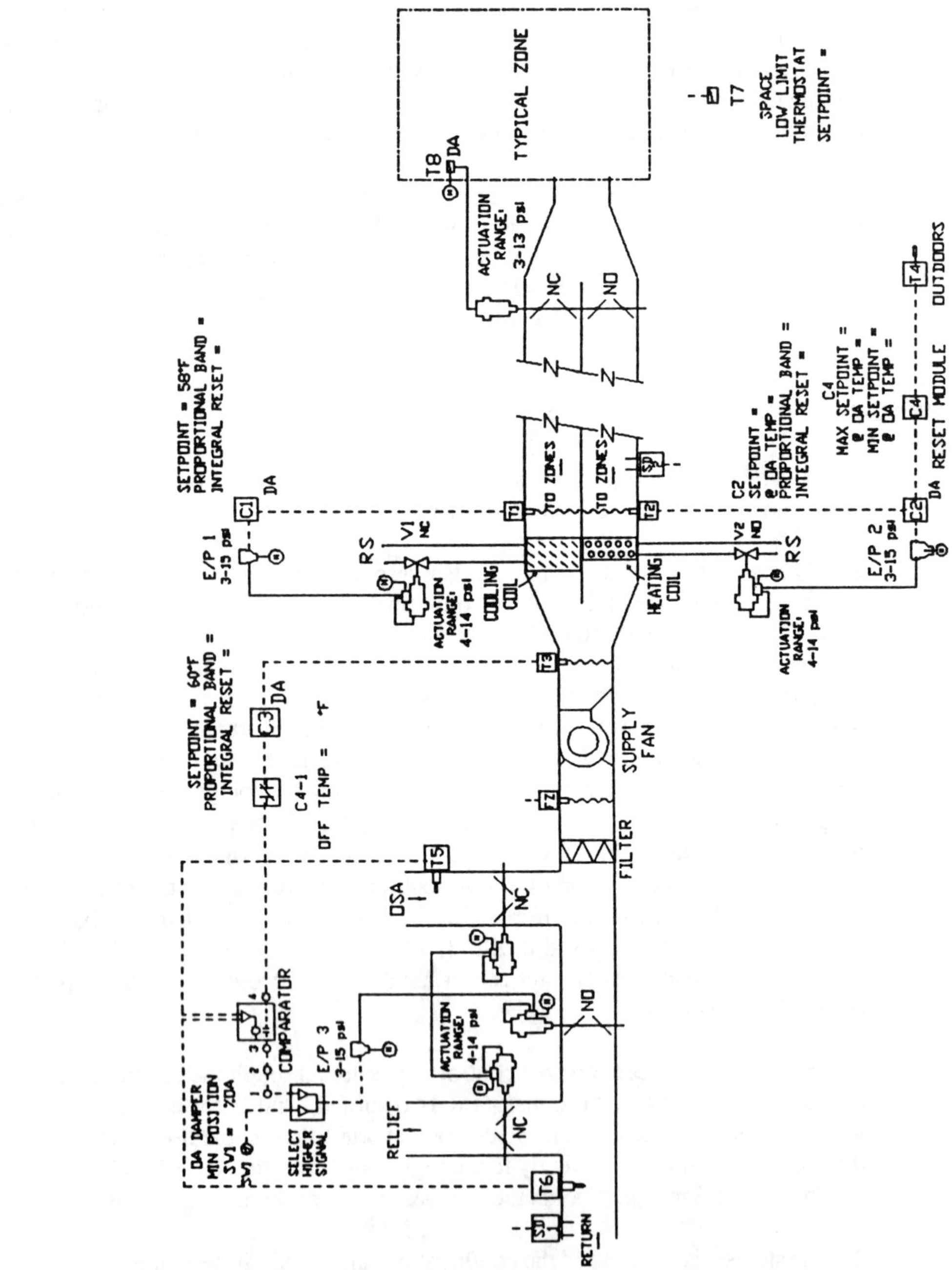

Fig. G-10 Multizone System Control

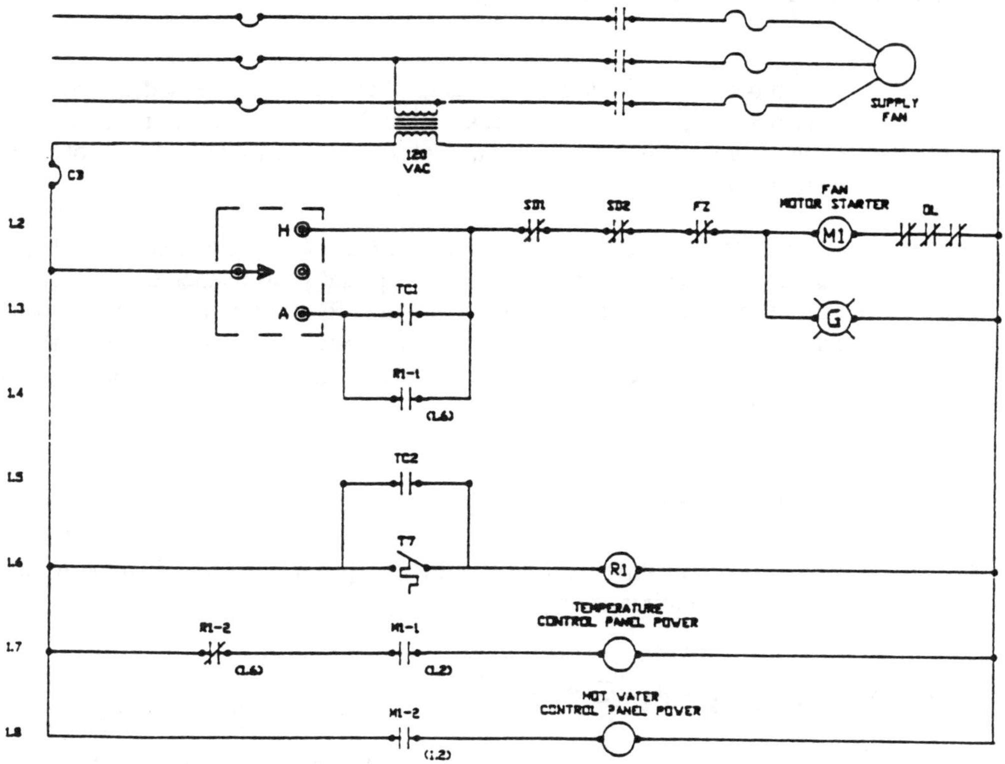

Fig. G-11 Control Circuit for System of Figure G-10

air temperature before the relay closes; if the relay is closed, the outdoor air temperature must rise to equal the return air temperature before the relay opens.

5. When the system is off (no power to the control panels), the outdoor air dampers return to their normally closed position.

C. Supply Air Temperature Control

1. The supply air temperature is controlled by the multizone temperature control panel.

2. The heating coil hot water valve V2 is modulated by electronic PI controller C2 using the sensed hot deck air temperature from temperature sensor T2. The output from controller C2 is connected to electronic-to-pneumatic transducer E/P-2, which produces a pneumatic signal to position valve V2, maintaining a constant hot deck

air temperature. The setpoint of controller C2 is determined by the output of electronic proportional controller C4 using the sensed outdoor air temperature from temperature sensor T4—the heating coil controller is reset on the basis of outdoor air temperature.

3. The cooling coil chilled-water valve V1 is modulated by electronic PI controller C1 using the sensed cold deck air temperature from temperature sensor T1. The output from controller C1 is connected to electronic-to-pneumatic transducer E/P-1, which produces a pneumatic signal to position valve V1, maintaining a constant cold deck air temperature. The setpoint of controller C1 is 58°F.

D. Room Air Temperature Control

Room air temperature control is achieved by pneumatic room thermostats modulating individual mixing dampers.

E. Hot Water Temperature Control

The temperature of hot water supplied to the heating coil is controlled by the hot water temperature control panel.

F. Interlocks

Smoke detectors (SD1 and SD2) and the low-temperature safety switch (FZ) are wired in series with fan motor starter relay M1 to stop the fan in the event of smoke or extremely low temperatures. The ladder schematic on the drawings shows how equipment is to be interlocked.